Third Edition

LINEAR ALGEBRA

An Introductory Approach

CHARLES W. CURTIS
University of Oregon

Allyn and Bacon, Inc. Boston

LIBRARY OF CONGRESS CATALOG CARD NO: 73–82269

Second printing . . . November, 1974

PRINTED IN THE UNITED STATES OF AMERICA

To Timothy, Daniel, and Robert

Contents

Preface

Linear algebra is the branch of mathematics that has grown out of a theoretical study of the problem of solving systems of linear equations. The basic concepts of linear algebra are vector spaces, linear transformations, and matrices. The concepts are used increasingly as a language for stating ideas and solving problems in geometry, algebra, and analysis, as well as in applications of mathematics to the natural sciences and the social sciences.

The main purpose of a course on linear algebra, based on this text, will be to provide the student with an easy familiarity with the basic ideas of the subject. Enough examples and applications are included to give an appreciation of the wide variety of applications of these ideas.

A second purpose is to give a careful account of the theoretical development of the subject so that the student will be able to understand, in detail, places where linear algebra is used in other subjects, and to tackle problems of some depth, involving more than a superficial acquaintance with the basic facts.

Finally, since for many students this may be the first course in which the subject is developed in an uncompromising way, without passing over the difficulties, there has been a real attempt to present linear algebra as a coherent part of mathematics, with an integrity and beauty of its own, and not as something subservient to the other disciplines where it is used.

There are two kinds of courses which can be based on this book: a short course of one or two quarters or one semester, or a full course, taking a whole year. For both courses the only prerequisite is a one-year course in calculus and analytic geometry.

For a short course, of a problem-solving rather than a theoretical nature, as much as possible of Chapters 1 to 7 should be included. It will be necessary, however, to omit most of Chapter 1 except for brief glances at parts of Sections 1 and 2 to get some motivation and background, and to omit Sections 10, 14, 17, 19, 20, 21 (except for a brief review of polynomials and complex numbers), and 25. Emphasis should be given to understanding the statements of the theorems, but most of the proofs can be omitted in such a course. The time will be better spent in a careful study of the more than 50 worked examples. The examples appear toward the end of almost every section and are identified as Example A, B, etc. They contain detailed accounts of how the theorems work in particular numerical cases and serve as a basis for solving the exercises.

The exercises include both routine numerical problems and theoretical problems often involving a good understanding of the proofs of the theorems for their solutions. In a short course, in which not many proofs of theorems are included, most of the exercises assigned should be of a computational sort. Answers to the majority of the exercises in the first seven chapters are included at the back of the book. In cases where a simple check is available, no answers are furnished.

A short course of two quarters should still stick to the first seven chapters, and probably omit most of the sections listed above. But there should be time for discussion of the proofs of more theorems, and more of the theoretical problems can be assigned.

A full course, lasting a year, would cover the entire book, and students should become familiar with the proofs of most of the main theorems. The last three chapters contain a large amount of new material, which has been added to make the book's coverage of linear algebra considerably more comprehensive than the earlier editions. In particular, full accounts of the rational and Jordan canonical forms are included, with the proofs of the main results based on the theory of dual vector spaces. An introduction to tensor products is included. Chapter 9 contains a much expanded and now fairly thorough treatment of orthogonal, unitary, and normal transformations, including the principal axis theorem, the spectral theorem, and the polar decomposition theorem. The last chapter contains some of the really substantial applications of linear algebra, to geometry, differential equations, and algebra.

Even in a short course, I believe that the applications (in Chapter 2) of the main theorems on basis and dimension to systems of linear equations should be presented in full. Besides the historical interest, it is a good practice, if one has introduced some really hard ideas, such as linear dependence and the concept of dimension, to give some applications where these ideas are needed in their full generality, as they are in the theory of systems of equations. Nothing is more damaging to one's appreciation of a subject than to see all sorts of general ideas introduced, and then never used in an essential way. In this book all the ideas introduced are put to work!

The instructor thinking about a full course based on this book will first realize that the degree of difficulty of the later chapters means that the course will be aimed at juniors and seniors (who may already have had a computational course, so that not all the introductory material in early chapters need be repeated). Where does this full-year course in linear algebra fit into the students' schedules, if they also have to take a year course in modern algebra, along with analysis, topology, etc.? The answer is, put in the full-year course in linear algebra, in which all the algebraic structures—groups, rings, fields, etc.—appear naturally, and then have a short course on algebraic structures, which will be easy for students to understand after some linear algebra. My experience is that the background in linear algebra of many graduate students in mathematics is weak, because they have never had a course with time enough to develop the material thoroughly. Linear algebra is absolutely essential for graduate work, and there is no time to teach it to graduate students along with everything else they are taking. So it should come in their undergraduate program, and not just as an introduction to vectors and matrices in two and three dimensions!

Something should be said about applications of linear algebra. They are many and fascinating. Some are included in the book, but certainly not enough. The bibliography and notes contain references to excellent accounts of various applications of linear algebra. Many instructors may wish to incorporate some of this material into a course.

Finally, it is a pleasure to acknowledge the encouragement I have received toward this project from students at Wisconsin and Oregon. I have also received inspiration and ideas on how to present different parts of the subject from the books listed in the bibliography, from talks with friends and colleagues, and particularly from thoughtful reviews by instructors who have taught from the earlier editions. And last, I owe another debt of thanks to my family, whose patience has enabled us to survive another year of lost weekends as this edition was written.

Charles W. Curtis

Chapter 1

Introduction to Linear Algebra

Mathematical theories are not invented spontaneously. The theories that have proved to be useful have their beginnings, in most cases, in special problems, which are difficult to understand or to solve without a grasp of the underlying principles. This chapter begins with two such problems, which have a common set of underlying principles. Linear algebra, which is the study of vector spaces, linear transformations, and matrices, is the result of trying to understand the common features of these and other similar problems.

1. SOME PROBLEMS WHICH LEAD TO LINEAR ALGEBRA

This section should be read quickly the first time through, with the objective of getting some motivation, but not a thorough understanding of the details.

Problem A. Everyone is familiar from elementary algebra with the problem of solving two linear equations in two unknowns, for example,

$$\begin{aligned} 3x - 2y &= 1 \\ x + y &= 2. \end{aligned} \tag{1.1}$$

We solve such a system by eliminating one unknown, for example, by multiplying the second equation by 2 and adding it to the first. This yields $5x = 5$, and $x = 1$. Substituting back in the first equation, we see that

$$3 - 2y = 1,$$

or $y = 1$. The solution of the original system of equations is the pair of numbers (1, 1), and there is only one solution.

If we look at the original problem from a geometrical point of view, then it becomes clear what is happening. The equations

$$3x - 2y = 1 \quad \text{and} \quad x + y = 2$$

are equations of straight lines in the (x, y) plane (see Figure 1.1). The

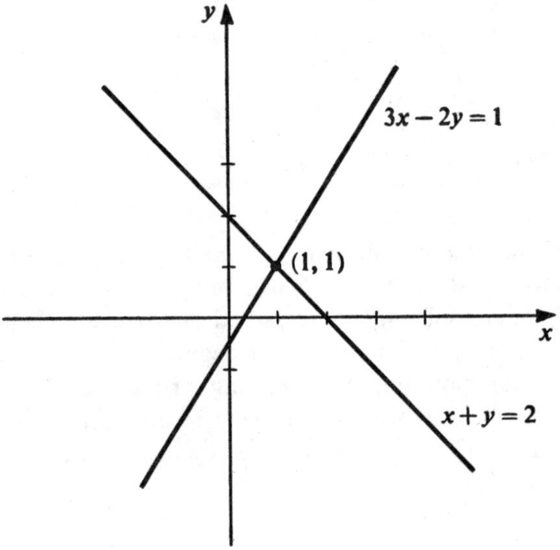

FIGURE 1.1

solution we obtained to the system of equations expresses the fact that the two lines intersect in a unique point, which in this case is (1, 1).

Now let us consider a similar problem, involving two equations in three unknowns.

(1.2)
$$\begin{aligned} x + 2y - z &= 1 \\ 2x + y + z &= 0. \end{aligned}$$

This time it isn't so easy to eliminate the unknowns. Yet we see that if z is given a particular value, say 0, we can solve the resulting system

$$\begin{aligned} x + 2y &= 1 \\ 2x + y &= 0 \end{aligned} \qquad (z = 0)$$

and obtain $x = -\frac{1}{3}$, $y = \frac{2}{3}$; so a solution of the original system (1.2) is $(-\frac{1}{3}, \frac{2}{3}, 0)$. But this time there are many other solutions; for example, letting $z = 1$, we obtain a solution $(-\frac{4}{3}, \frac{5}{3}, 1)$. In fact we can obtain a solution for every value of z. The puzzle of why we get so many solutions is answered again by a geometrical interpretation. The equations in (1.2) are equations of planes in (x, y, z) coordinates (see Figure 1.2).

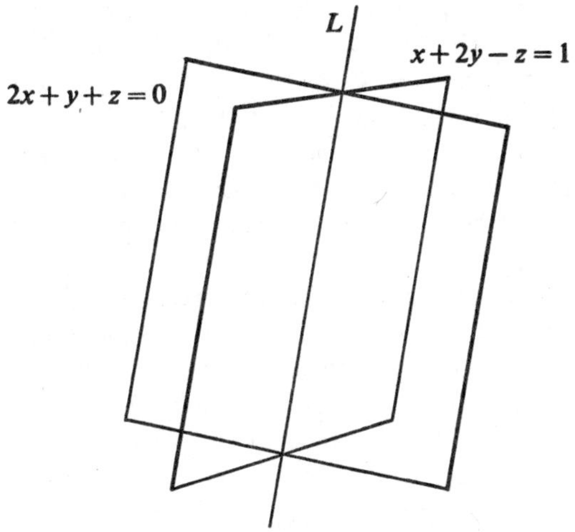

FIGURE 1.2

This time the solutions, viewed as points in (x, y, z) coordinates, correspond to the points on the line of intersection L of the two planes.

The second problem already shows that more work is needed to describe the solutions of the system (1.2) and to give a method for finding all of them. But why stop there? In many types of problems, a solution is needed to systems of equations in more than two or three unknowns; for example,

$$\text{(1.3)} \qquad \begin{aligned} x + 2y - 3z + t &= 1 \\ x + y + z + t &= 0. \end{aligned}$$

In this case and in other more complicated ones, geometric intuition is not available (for most of us, anyway) to give a picture of what the solution should be. At the same time such examples are not farfetched from the applied point of view since, for example, a formula describing the temperature t at a point (x, y, z) involves four variables. One of the tasks of linear algebra is to provide a framework for discussing problems of this nature.

Problem B. In calculus, a familiar type of question is to find a function $y = f(x)$ if its derivative $y' = Df(x)$ is known. For example, if y' is the constant function 1, we know that $y = x + C$, where C is a constant. This type of problem is called a differential equation. Problems in both mechanics and electric circuits lead to differential equations such as

$$y'' + m^2y = 0, \tag{1.4}$$

where m is a positive constant, and y'' is the second derivative of the unknown function y. This time the solution is not so obvious.

Checking through the rules for differentiation of the elementary functions, the functions $y = A \sin mx$ or $y = B \cos mx$ are recognized as solutions, where A and B are arbitrary constants. But here, as in the case of the equations (1.2) or (1.3), there are many other solutions, such as $C \sin (mx + D)$, where C and D are constants. The problem in this case is to find a clear description of all possible solutions of differential equations such as (1.4).

Now let us see what if anything the two types of problems have in common, beyond the fact that both involve solutions of equations. In the case of the equations (1.3), for example, we are interested in ordered sets of four numbers $(\alpha, \beta, \gamma, \delta)$† which satisfy the equations (1.3) when substituted for x, y, z, and t, respectively. We shall call $(\alpha, \beta, \gamma, \delta)$ a *vector* with four entries; later (in Chapter 2) we shall define vectors $(\alpha_1, \ldots, \alpha_n)$ with n entries $\alpha_1, \ldots, \alpha_n$ for $n = 1, 2, 3, \ldots$. A vector $(\alpha, \beta, \gamma, \delta)$ which satisfies the equations (1.3) will be called a *solution vector* of the system (1.3) [or simply a *solution* of (1.3)]. The following statements about solutions of (1.3) can now be made.

(*i*) If $u = (\alpha, \beta, \gamma, \delta)$ and $u' = (\alpha', \beta', \gamma', \delta')$ are both solutions of (1.3), then

$$(\alpha - \alpha',\ \beta - \beta',\ \gamma - \gamma',\ \delta - \delta')$$

is a solution of the system of *homogeneous equations* (obtained by setting the right-hand side equal to zero),

$$\begin{aligned} x + 2y - 3z + t &= 0 \\ x + y + z + t &= 0. \end{aligned} \tag{1.5}$$

(*ii*) If $u = (\alpha, \beta, \gamma, \delta)$ and $u' = (\alpha', \beta', \gamma', \delta')$ are solutions of the homogeneous system (1.5), then so are

$$(\alpha + \alpha', \beta + \beta', \gamma + \gamma', \delta + \delta')$$

and

$$(\lambda\alpha, \lambda\beta, \lambda\gamma, \lambda\delta)$$

for an arbitrary number λ.

† See the list of Greek letters on p. 332.

(*iii*) Suppose $u_0 = (\alpha, \beta, \gamma, \delta)$ is a fixed solution of the nonhomogeneous system (1.3). Then an arbitrary solution v of the nonhomogeneous system has the form

$$v = (\alpha + \xi, \beta + \eta, \gamma + \lambda, \delta + \mu) \tag{1.6}$$

where $(\xi, \eta, \lambda, \mu)$ is an arbitrary solution of the homogeneous system (1.5).

Statements (i) and (ii) are verified by direct substitution in the equations. To check statement (iii), suppose first that $v = (a, b, c, d)$ is an arbitrary solution of the system (1.3). By statement (i), recalling that $u_0 = (\alpha, \beta, \gamma, \delta)$ is our fixed solution of (1.3), we see that

$$(a - \alpha, b - \beta, c - \gamma, d - \delta)$$

is a solution of the homogeneous system, and setting $\xi = a - \alpha$, $\eta = b - \beta$, $\lambda = c - \gamma$, $\mu = d - \delta$, we have

$$v = (a, b, c, d) = (\alpha + \xi, \beta + \eta, \gamma + \lambda, \delta + \mu)$$

as required. A further argument (which we omit) shows that every vector of the form (1.6) is actually a solution of the nonhomogeneous system (1.3).

What have we learned from all this? First, that facts about the solutions of the equations can be expressed in terms of certain operations on vectors:

$$(\alpha, \beta, \gamma, \delta) + (\alpha', \beta', \gamma', \delta') = (\alpha + \alpha', \beta + \beta', \gamma + \gamma', \delta + \delta')$$

and

$$\lambda(\alpha, \beta, \gamma, \delta) = (\lambda\alpha, \lambda\beta, \lambda\gamma, \lambda\delta).$$

Thus we may add two vectors and obtain a new one, and multiply a vector by a number, to give a new vector. In terms of these operations we can then describe the problem of solving the system (1.3) as follows: (1) We find one solution u_0 of the nonhomogeneous system; (2) an arbitrary solution (often called the *general solution*) v of the nonhomogeneous system is given by

$$v = u_0 + u,$$

where u ranges over the set of solutions of the homogeneous system; (3) we find all solutions of the homogeneous system (1.5).

Now let us turn to the differential equation (1.4). We see first that if the functions f_1 and f_2 are solutions of the equation (that is, $f_1'' + m^2 f_1 = 0$, $f_2'' + m^2 f_2 = 0$), then $f_1 + f_2$ and λf_1 are also solutions, where λ is an arbitrary number. But this is the same situation we encountered in statement (ii) above, whereas now we have functions in place of vectors. Let's go a little further. Suppose we take a nonhomogeneous differential equation (by analogy with the linear equations),

$$y'' + m^2 y = F, \tag{1.7}$$

where F is some fixed function. Then we see that all three of our statements about Problem A are satisfied. The difference of two solutions of the nonhomogeneous system (1.7) is a solution of the homogeneous system. The solutions of the homogeneous system satisfy statement (ii), if the operations of adding functions and multiplying functions by constants replace the corresponding operations on vectors. Finally, an arbitrary solution of the nonhomogeneous system is obtained from a particular one f_0 by adding to f_0 a solution of the homogeneous system.

It is reasonable to believe that the analogous behavior of the solutions of Problems A and B is not a pure coincidence. The common framework underlying both problems will be provided by the concept of a vector space, to be introduced in Chapter 2. Before beginning the study of vector spaces, we have one more introductory section in which we review facts about sets and numbers which will be needed.

2. NUMBER SYSTEMS AND MATHEMATICAL INDUCTION

We assume familiarity with the real number system, as discussed in earlier courses in college algebra, trigonometry, or calculus. It will turn out, however, that number systems other than the real numbers (such as complex numbers) also play an important role in linear algebra. At the same time, not all the detailed facts about these number systems (such as absolute value and polar representation of complex numbers) are needed until later in the book.

In order to have a clearly defined starting place, we shall give a set of axioms for a number system, called a *field*, which will provide a firm basis for the theory of vector spaces to be discussed in the next chapter. The real number system is an example of a field. We first have to recall a few of the basic ideas about sets.

We use the word *set* as synonymous with "collection" or "family" of objects of some sort.

Let X be a set of objects. For a given object x, either x belongs to the set X or it does not. If x belongs to X, we write $x \in X$ (read "x is an element of X" or "x is a member of X"); if x does not belong to X, we write $x \notin X$.

A set Y is called a *subset* of a set X if, for all objects, y, $y \in Y$ implies $y \in X$. In other words, every element of Y is also an element of X. If Y is a subset of X, we write $Y \subset X$. If $Y \subset X$ and $X \subset Y$, then we say that the sets X and Y are equal and write $X = Y$. Thus two sets are equal if they contain exactly the same members.

It is convenient to introduce the set containing no elements at all.

We call it the empty set, and denote it by $\varnothing$. Thus, for every object x, $x \notin \varnothing$. For example, the set of all real numbers x for which the inequalities $x < 0$ and $x > 1$ hold simultaneously is the empty set. The reader will check that from our definition of subset it follows logically that the empty set $\varnothing$ is a subset of every set (why?).

There are two important constructions which can be applied to subsets of a set and yield new subsets. Suppose U and V are subsets of a given set X. We define $U \cap V$ to be the set consisting of all elements belonging to both U and V and call $U \cap V$ the *intersection* of U and V. *Question:* What is the intersection of the set of real numbers x such that $x > 0$ with the set of real numbers y such that $y < 5$? If we have many subsets of X, their intersection is defined as the set of all elements which belong to all the given subsets.

The second construction is the *union* $U \cup V$ of U and V; this is the subset of X consisting of all elements which belong either to U or to V. (When we say "either ... or," it is understood that we mean "either ... or ... or both.")

It is frequently useful to illustrate statements about sets by drawing diagrams. Although they have no mathematical significance, they do give us confidence that we are making sense, and sometimes they suggest important steps in an argument. For example, the statement $X \subset Y$ is illustrated by Figure 1.3. In Figure 1.4 the shaded portion indicates $U \cup V$, while the cross-hatched portion denotes $U \cap V$.

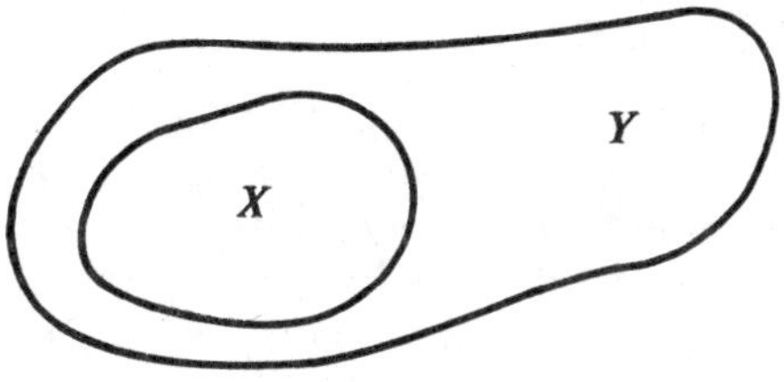

FIGURE 1.3

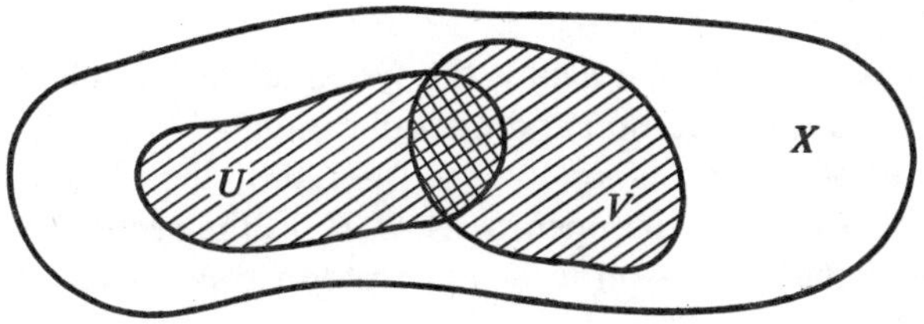

FIGURE 1.4

Examples and Some Notation

We shall often use the notation

$$\{x \in X \mid x \text{ has the property } P\}$$

to denote the set of objects x in a set X which have some given property P. Thus, if R denotes the set of real numbers, $\{x \in R \mid x < 5\}$ denotes the set of all real numbers x such that $x < 5$. Using this notation, the set of solutions of the system of equations (1.1) is described by the statement

$$\{(x, y) \mid 3x - 2y = 1\} \cap \{(x, y) \mid x + y = 2\} = \{(1, 1)\},$$

where $\{(1, 1)\}$ denotes the set containing the single point $(1, 1)$. In general, we shall use the notations $\{a\}$, $\{a, b\}$, $\{a_i\}$, etc., to denote the sets whose members are a, a and b, and a_i, $i = 1, 2, \ldots$, respectively. Examples of sets are plentiful in geometry. Lines and planes are sets of points; the intersection of two lines in a plane is either the empty set (if the lines are parallel) or a single point; the intersection of two planes is either $\varnothing$ or a line.

We now introduce the kind of number system which will be the starting point of our development of vector spaces in Chapter 2.

(2.1) DEFINITION. A *field* is a mathematical system F consisting of a nonempty set F together with two operations, addition and multiplication, which assign to each pair of elements† $\alpha, \beta \in F$ uniquely determined elements $\alpha + \beta$ and $\alpha\beta$ (or $\alpha \cdot \beta$) of F, such that the following conditions are satisfied, for $\alpha, \beta, \gamma \in F$.

1. $\alpha + \beta = \beta + \alpha$, $\alpha\beta = \beta\alpha$ (*commutative laws*).
2. $\alpha + (\beta + \gamma) = (\alpha + \beta) + \gamma$, $(\alpha\beta)\gamma = \alpha(\beta\gamma)$ (*associative laws*).
3. $\alpha(\beta + \gamma) = \alpha\beta + \alpha\gamma$ (*distributive law*).
4. There exists an element 0 in F such that $\alpha + 0 = \alpha$ for all $\alpha \in F$.
5. For each element $\alpha \in F$ there exists an element $-\alpha$ in F such that $\alpha + (-\alpha) = 0$.
6. There exists an element $1 \in F$ such that $1 \neq 0$ and such that $\alpha \cdot 1 = \alpha$ for all $\alpha \in F$.
7. For each nonzero $\alpha \in F$ there exists an element $\alpha^{-1} \in F$ such that $\alpha\alpha^{-1} = 1$.

The first example of a field is the system of real numbers R, with the usual operations of addition and multiplication.

The set of integers $Z = \{\ldots, -2, -1, 0, 1, 2, \ldots\}$ is a subset of R, which is *closed* under the operations of R, in the sense that if m and n belong to Z, then their sum $m + n$ and product mn (which are defined in

† See the list of Greek letters on page 332.

terms of the operations of the larger set R) again belong to Z. But Z is not a field with respect to these operations, since the axiom [2.1(7)] fails to hold. For example, $2 \in Z$ and $2^{-1} \in R$, but $2^{-1} \notin Z$, and in fact there is no element $m \in Z$ such that $2m = 1$. This example suggests the following definition.

(2.2) DEFINITION. A *subfield* F_0 of a field F is a subset F_0 of F, which is closed under the operations defined on F, and which satisfies all the axioms [2.1(1)–(7)] relative to these operations. (Here *closed* means that if $\alpha, \beta \in F_0$, then $\alpha + \beta \in F_0$ and $\alpha\beta \in F_0$.)

Although the integers Z are not a subfield of R, the set of rational numbers Q is a subfield. We recall that the set of rational numbers Q consists of all real numbers of the form

$$mn^{-1}$$

where $m, n \in Z$ and $n \neq 0$. It takes some work to show that the rational numbers actually are a subfield of R. The kinds of things that have to be checked are discussed later in this section.

The fields of both real numbers and rational numbers are *ordered fields* in the sense that each one contains a subset P (called the set of *positive elements*) with the properties

1. $\alpha, \beta \in P$ implies $\alpha + \beta$ and $\alpha\beta \in P$.
2. For each α in the field, one and only one of the following possibilities holds:

$$\alpha \in P, \qquad \alpha = 0, \qquad -\alpha \in P.$$

The field of complex numbers C (to be discussed more fully in Chapter 6) is not an ordered field. It is defined as the set of all pairs of real numbers (α, β), $\alpha, \beta \in R$, where two pairs are defined to be equal, $(\alpha, \beta) = (\alpha', \beta')$, if and only if $\alpha = \alpha'$ and $\beta = \beta'$. The operations in C are defined as follows:

$$(\alpha, \beta) + (\alpha', \beta') = (\alpha + \alpha', \beta + \beta')$$

and

$$(\alpha, \beta)(\alpha', \beta') = (\alpha\alpha' - \beta\beta', \alpha\beta' + \beta\alpha').$$

The proof that the axioms for a field are satisfied is given in Chapter 6.

Consider the set of all complex numbers of the form $\{(\alpha, 0), \alpha \in R\}$. These have the properties that for all $\alpha, \beta \in R$,

(2.3)
$$(\alpha, 0) + (\beta, 0) = (\alpha + \beta, 0),$$
$$(\alpha, 0)(\beta, 0) = (\alpha\beta, 0).$$

It can be checked that these form a subfield R' of C. Moreover, the rule or correspondence $\alpha \leftrightarrow (\alpha, 0)$ which assigns to each $\alpha \in R$ the complex number $(\alpha, 0) \in R'$ has the properties that (a) $(\alpha, 0) = (\alpha', 0)$ if and only if $\alpha = \alpha'$ and that (b) it preserves the operations in the respective number systems, according to the equations (2.3) above. A correspondence between two fields with the properties (a) and (b) is called an *isomorphism*, the word coming from Greek words meaning "to have the same form." The existence of the isomorphism $\alpha \leftrightarrow (\alpha, 0)$ between the fields R and R' means that if we ignore all other properties not given in the definition of a field, then R and R' are indistinguishable. Thus we shall identify the elements of R with the corresponding elements of R', and in this sense we have the following inclusions among the number systems defined so far:

$$Z \subset Q \subset R \subset C.$$

We come now to the principle of mathematical induction, which we shall take as an axiom about the set of positive integers $Z^+ = \{1, 2, \ldots\}$. It is impossible to overemphasize its importance for linear algebra. Almost all the deeper results in this book depend on it in one way or another.

(2.4) PRINCIPLE OF MATHEMATICAL INDUCTION. *Suppose that for each positive integer n there corresponds a statement $E(n)$ which is either true or false. Suppose, further, (A) $E(1)$ is true; and (B), if $E(n)$ is true then $E(n+1)$ is true, for all $n \in Z^+$. Then $E(n)$ is true for all positive integers n.*

Some equivalent forms of the principle of induction are often useful, and we shall list some of them explicitly.

(2.5)A *Let $\{E(n)\}$ be a family of statements defined for each positive integer n. Suppose* (a) *$E(1)$ is true, and* (b) *if $E(r)$ is true for all positive integers $r < n$, then $E(n)$ is true. Then $E(n)$ is true for all positive integers n.*

In discussions where mathematical induction is involved, the statement which plays the role of $E(n)$ will often be called the *induction hypothesis.*

Another equivalent form of (2.4) will be used later, especially in Chapter 6. For a discussion of how this statement can be shown to be equivalent to (2.4), see the book of Courant and Robbins listed in the bibliography.

(2.5)B WELL-ORDERING PRINCIPLE. *Let M be a nonempty set of positive integers. Then M contains a least element, that is, an element m_0 in M exists, which satisfies the condition $m_0 \leq m$, for all $m \in M$.*

Examples of Mathematical Induction

Example A. We shall prove the formula

$$1 + 2 + 3 + \cdots + n = \frac{n(n+1)}{2}, \tag{2.6}$$

for all positive integers n. The induction hypothesis is the statement (2.6) itself. For $n = 1$, the statement reads $1 = 1(2)/2$. Now suppose the statement (2.6) holds for some n. We have to show it is true for $n + 1$. Substituting the right-hand side of (2.6) for $1 + 2 + \cdots + n$, we have

$$\begin{aligned} 1 + 2 + \cdots + n + (n+1) &= \frac{n(n+1)}{2} + (n+1) \\ &= (n+1)\left(\frac{n}{2} + 1\right) = \frac{(n+1)(n+2)}{2}, \end{aligned}$$

which is the statement (2.6) for $n + 1$. By the principle of mathematical induction, we conclude that (2.6) is true for all positive integers n.

Example B. For an arbitrary real number $x \neq 1$,

$$1 + x + x^2 + \cdots + x^{n-1} = \frac{1 - x^n}{1 - x}. \tag{2.7}$$

Again we let (2.7) be the induction hypothesis $E(n)$. For $n = 1$, the statement is

$$1 = \frac{1 - x}{1 - x}.$$

Assume (2.7) for some n. Then using (2.7), we have

$$\begin{aligned} 1 + x + x^2 + \cdots + x^n &= (1 + x + \cdots + x^{n-1}) + x^n \\ &= \frac{1 - x^n}{1 - x} + x^n \\ &= \frac{1 - x^n + x^n(1 - x)}{1 - x} \\ &= \frac{1 - x^{n+1}}{1 - x}, \end{aligned}$$

and the statement (2.7) holds for all n.

We conclude this chapter with some properties of an arbitrary field F. These are all familiar facts about the system of real numbers, but the reader may wish to study some of them closely, to check that they depend only on the axioms for a field. Besides the axioms, the *principle of substitution* is frequently used. This simply means that in a field F, we can, in any formula involving an element α in F, replace α by any other element $\alpha' \in F$ such that $\alpha' = \alpha$.

We shall also use the following familiar properties of the equals relation $\alpha = \beta$:

$$\alpha = \alpha;$$
$$\alpha = \beta \text{ implies } \beta = \alpha;$$
$$\alpha = \beta \text{ and } \beta = \gamma \text{ imply } \alpha = \gamma.$$

Not all of the statements given below are proved in detail. Those appearing with starred numbers [for example, (2.9)*] are left as exercises for the reader.

(2.8) *If $\alpha + \beta = \alpha + \gamma$, then $\beta = \gamma$ (cancellation law for addition).*

Proof. We should like to say "add $-\alpha$ to both sides." We accomplish this by using the substitution principle as follows. From the assumption that $\alpha + \beta = \alpha + \gamma$, we have

$$(-\alpha) + (\alpha + \beta) = (-\alpha) + (\alpha + \gamma).$$

On the one hand, we have by the field axioms (which ones?),

$$(-\alpha) + (\alpha + \beta) = [(-\alpha) + \alpha] + \beta = 0 + \beta = \beta.\dagger$$

Similarly,

$$(-\alpha) + (\alpha + \gamma) = \gamma.$$

By the properties of the equals relation, we have $\beta = \gamma$, and the result is proved.

An argument like this cannot be read like a newspaper article; the reader will find that he must have paper and pencil handy, and write out the steps, checking the references to the axioms, until he sees exactly what has been done.

(2.9) *For arbitrary α and β in F, the equation $\alpha + x = \beta$ has a unique solution.*

Proof. This result is two theorems in one. We are asked to show, first, that there exists at least one element x which satisfies the equation and, second, that there is at most one solution. Both statements are easy, however. First, from the field axioms we have

$$\alpha + [(-\alpha) + \beta] = [\alpha + (-\alpha)] + \beta = 0 + \beta = \beta$$

† A statement of the form $a = b = c = d$ is really shorthand for the separate statements $a = b$, $b = c$, $c = d$. The properties of the equals relation imply that from the separate statements we can conclude that all the objects a, b, c, and d are equal to each other, and this is the meaning of the abbreviated statement $a = b = c = d$.

and we have shown that there is at least one solution, namely, $x = (-\alpha) + \beta$.

Now let γ and γ' be solutions of the equation. Then we have $\alpha + \gamma = \beta$, $\alpha + \gamma' = \beta$, and hence $\alpha + \gamma = \alpha + \gamma'$ (why?). By the cancellation law (2.8) we have $\gamma = \gamma'$, and we have proved that if γ is one solution of the equation, then any other solution is equal to γ.

(2.10) DEFINITION. We write $\beta - \alpha$ to denote the unique solution of the equation $\alpha + x = \beta$.

We have $\beta - \alpha = \beta + (-\alpha)$, and we call $\beta - \alpha$ the result of *subtracting* α from β. (The reader should observe that there is no question of "proving" a statement like the last. A careful inspection of the field axioms shows that no meaning has been attached to the formula $\beta - \alpha$; the matter has to be taken care of by a definition.)

(2.11) $-(-\alpha) = \alpha$.

Proof. The result comes from examining the equation $\alpha + (-\alpha) = 0$ from a different point of view. For we have also $-(-\alpha) + (-\alpha) = 0$; by the cancellation law we obtain $-(-\alpha) = \alpha$.

Now we come to the properties of multiplication, which are exactly parallel to the properties of addition. We remind the reader that he must supply the proofs of the starred statements.

(2.12)* *If* $\alpha \neq 0$ *and* $\alpha\beta = \alpha\gamma$, *then* $\beta = \gamma$.

(2.13)* *The equation* $\alpha x = \beta$, *where* $\alpha \neq 0$, *has a unique solution.*

(2.14) DEFINITION. We denote by $\dfrac{\beta}{\alpha}$ (or β/α) the unique solution of the equation $\alpha x = \beta$, and will speak of β/α or $\dfrac{\beta}{\alpha}$ as the result of *division* of β by α.

Thus $1/\alpha = \alpha^{-1}$, because of Axiom [2.1(6)].

(2.15)* $(\alpha^{-1})^{-1} = \alpha$ *if* $\alpha \neq 0$.

Thus far we haven't used the distributive law [2.1(3)]. In a way, this is the most powerful axiom and most of the more exotic theorems in elementary algebra, such as $(-1)(-1) = 1$, follow from the distributive law.

(2.16) *For all* α *in* F, $\alpha \cdot 0 = 0$.

Proof. We have $0 + 0 = 0$. By the distributive law [2.1(3)] we have $\alpha \cdot 0 + \alpha \cdot 0 = \alpha \cdot 0$. But $\alpha \cdot 0 = \alpha \cdot 0 + 0$ by [2.1(4)], and by the cancellation law we have $\alpha \cdot 0 = 0$.

(2.17) $(-\alpha)\beta = -(\alpha\beta)$, *for all* α *and* β.

Proof. From $\alpha + (-\alpha) = 0$ we have, by the substitution principle, $[\alpha + (-\alpha)]\beta = 0 \cdot \beta$. From the distributive law and (2.16) this implies that

$$\alpha\beta + (-\alpha)\beta = 0.$$

Since $\alpha\beta + [-(\alpha\beta)] = 0$, the cancellation law gives us $(-\alpha)\beta = -(\alpha\beta)$, as we wished to prove.

(2.18) $(-\alpha)(-\beta) = \alpha\beta$, *for all* α, β *in* F.

Proof. We have, by two applications of (2.17) and by the use of the commutative law for multiplication,

$$(-\alpha)(-\beta) = -[\alpha(-\beta)] = -[-(\alpha\beta)].$$

Finally, $-[-(\alpha\beta)] = \alpha\beta$ by (2.11).

As a consequence of (2.18) we have:

(2.19) $(-1)(-1) = 1$.

(It is interesting to construct a direct proof from the original axioms.)

EXERCISES

1. Prove the following statements by mathematical induction:
   a. $1 + 3 + 5 + \cdots + (2k - 1) = k^2$.
   b. $1^2 + 2^2 + \cdots + n^2 = \dfrac{n(n+1)(2n+1)}{6}$.
2. Show, using the definition of α/β as the solution (for $\beta \neq 0$) of the equation $\beta x = \alpha$, that in an arbitrary field F, the following statements hold, for all $\alpha, \beta, \gamma, \delta$, with $\beta, \delta \neq 0$.
   a. $\dfrac{\alpha}{\beta} + \dfrac{\gamma}{\delta} = \dfrac{\alpha\delta + \beta\gamma}{\beta\delta}$.
   b. $\left(\dfrac{\alpha}{\beta}\right)\left(\dfrac{\gamma}{\delta}\right) = \dfrac{\alpha\gamma}{\beta\delta}$.
   c. $\dfrac{\alpha/\beta}{\gamma/\delta} = \dfrac{\alpha\delta}{\beta\gamma}$, if $\dfrac{\gamma}{\delta} \neq 0$.

3. *The binomial coefficients* $\binom{n}{k}$, for $n = 1, 2, \ldots$ and for each n, $k = 0, 1, 2, \ldots, n$, are certain positive integers defined by induction as follows. $\binom{1}{0} = \binom{1}{1} = 1$. Assuming $\binom{n-1}{k}$ has been defined for some n, and for $k = 0, 1, \ldots, n-1$, the binomial coefficient $\binom{n}{k}$ is defined by $\binom{n}{0} = \binom{n}{1} = 1$, and

$$\binom{n}{k} = \binom{n-1}{k-1} + \binom{n-1}{k}, \qquad k = 1, \ldots, n-1.$$

Show, using mathematical induction, that in any field F, the binomial theorem

$$(\alpha + \beta)^n = \binom{n}{0}\alpha^n + \binom{n}{1}\alpha^{n-1}\beta + \cdots + \binom{n}{n}\beta^n$$

holds for all α, β in F, with the understanding that in F, $2\alpha = \alpha + \alpha$, $3\alpha = 2\alpha + \alpha$, etc.

4. Let F be the system consisting of two elements $\{0, 1\}$, with the operations defined by the tables

+	0	1
0	0	1
1	1	0

·	0	1
0	0	0
1	0	1

Show that F is a field, with the property $2\alpha = \alpha + \alpha = 0$ for all $\alpha \in F$.

Chapter 2

Vector Spaces and Systems of Linear Equations

This chapter contains the basic definitions and facts about vector spaces, together with a thorough discussion of the application of the general results on vector spaces to the determination of the solutions of systems of linear equations. The chapter concludes with an optional section on the geometrical interpretation of the theory of systems of linear equations. Some motivation for the definition of a vector space and the theorems to be proved in this chapter was given in Section 1.

3. VECTOR SPACES

Let us begin with the situation discussed in Section 1. There it was convenient to define a vector u to be a quadruple of real numbers $u = (\alpha, \beta, \gamma, \delta)$, which are ordered in the sense that two vectors

$$u = (\alpha, \beta, \gamma, \delta) \qquad \text{and} \qquad u' = (\alpha', \beta', \gamma', \delta')$$

are equal if and only if

$$\alpha = \alpha', \qquad \beta = \beta', \qquad \gamma = \gamma', \qquad \delta = \delta'.$$

The statement about the ordering is needed because in the problem discussed in Section 1, the quadruples (1, 2, 0, 0) and (2, 1, 0, 0) would have quite different meanings. We then defined the sum of two vectors u and u' as above by

$$u + u' = (\alpha + \alpha', \beta + \beta', \gamma + \gamma', \delta + \delta')$$

and also the product of a vector $u = (\alpha, \beta, \gamma, \delta)$ by a real number λ, by

$$\lambda u = (\lambda\alpha, \lambda\beta, \lambda\gamma, \lambda\delta).$$

With these definitions, and from the field properties of the real numbers, we see that vectors satisfy the following laws with respect to the operations $u + u'$ and λu that we have defined.

1. $u + u' = u' + u$ (*commutative law*).
2. $(u + u') + u'' = u + (u' + u'')$ (*associative law*).
3. There exists a vector $\vec{0}$ such that $u + \vec{0} = u$ for all u.
4. For each vector u, there exists a vector $-u$ such that $u + (-u) = \vec{0}$.
5. $\alpha(\beta u) = (\alpha\beta)u$ for all vectors u and real numbers α and β.
6. $(\alpha + \beta)u = \alpha u + \beta u$ } (*distributive laws*).
7. $\alpha(u + u') = \alpha u + \alpha u'$ }
8. $1 \cdot u = u$ for all vectors u.

The proofs of these facts are immediate from the field axioms for the real numbers, and the definition of equality of two vectors. For example we shall give a proof of (2). Let

$$u = (\alpha, \beta, \gamma, \delta), \qquad u' = (\alpha', \beta', \gamma', \delta), \qquad u'' = (\alpha'', \beta'', \gamma'', \delta'').$$

Then

$$(u + u') + u'' \\ = ((\alpha + \alpha') + \alpha'', (\beta + \beta') + \beta'', (\gamma + \gamma') + \gamma'', (\delta + \delta') + \delta'')$$

and

$$u + (u' + u'') \\ = (\alpha + (\alpha' + \alpha''), \beta + (\beta' + \beta''), \gamma + (\gamma' + \gamma''), \delta + (\delta' + \delta'')).$$

Since we have $(\alpha + \alpha') + \alpha'' = \alpha + (\alpha' + \alpha'')$, etc., by the associative law in R, it follows that the vectors $(u + u') + u''$ and $u'' + (u' + u'')$ are equal, and the associative law for vectors is proved. The other statements can be verified in a similar way.

Now let us recall the second problem discussed in Section 1 (Problem B). In that problem we were interested in functions on the real line. Certain algebraic operations on functions were introduced, namely, taking the sum $f + g$ of two functions f and g and multiplying a function by a real number. More precisely, the definition of the sum $f + g$ is $(f + g)(x) = f(x) + g(x)$, $x \in R$ and for $\lambda \in R$, the function λf is defined by $(\lambda f)(x) = \lambda \cdot f(x)$, $x \in R$. Without much effort, it can be checked that the operations $f + g$ and λf satisfy the conditions (1) to (8) which we just verified for vectors $u = (\alpha, \beta, \gamma, \delta)$.

The common properties of these operations in the two examples leads to the following abstract concept of a vector space over a field. The point is, if we are to investigate the problems in Section 1 as fully as possible, we are practically forced to invent the concept of a vector space.

(3.1) DEFINITION. Let F be an arbitrary field. A *vector space V over F* is a nonempty set V of objects $\{v\}$, called *vectors*, together with two operations, one of which assigns to each pair of vectors v and w a vector $v + w$ called the *sum* of v and w, and the other of which assigns to each element $\alpha \in F$ and each vector $v \in V$ a vector αv called the *product* of v by the element $\alpha \in F$. The operations are assumed to satisfy the following axioms, for $\alpha, \beta \in F$ and for $u, v, w \in V$.

1. $u + (v + w) = (u + v) + w$, and $u + v = v + u$.
2. There is a vector 0 such that $u + 0 = u$ for all $u \in V$.†
3. For each vector u there is a vector $-u$ such that $u + (-u) = 0$.
4. $\alpha(u + v) = \alpha u + \alpha v$.
5. $(\alpha + \beta)u = \alpha u + \beta u$.
6. $(\alpha\beta)u = \alpha(\beta u)$.
7. $1u = u$.

In this text we shall generally use Roman letters $\{x, y, u, v, \ldots\}$ to denote vectors and Greek letters (see p. 332) $\{\alpha, \beta, \gamma, \delta, \xi, \eta, \theta, \lambda, \ldots\}$ to denote elements of the field involved. Elements of the field are often called *scalars* (or *numbers*, when the field is R.)

We can now check that our examples do satisfy the axioms. In the first example, of vectors $u = (\alpha, \beta, \gamma, \delta)$, there is of course nothing special about taking quadruples. We could have taken pairs, triples, etc. To cover all these cases, we introduce the idea of an *n-tuple* of real numbers, for some positive integer n.

(3.2) DEFINITION. Let n be a positive integer. An *n-tuple* of real numbers is a rule which assigns to each positive integer i (for $i = 1, 2, \ldots, n$) a unique real number α_i. An n-tuple will be described by the notation $\langle\alpha_1, \alpha_2, \ldots, \alpha_n\rangle$. In other words, an n-tuple is simply a collection of n numbers given in a definite order, with the first number called α_1, the second α_2, etc. Two n-tuples $\langle\alpha_1, \ldots, \alpha_n\rangle$ and $\langle\beta_1, \ldots, \beta_n\rangle$ are said to be equal, and we write

$$\langle\alpha_1, \ldots, \alpha_n\rangle = \langle\beta_1, \ldots, \beta_n\rangle$$

if and only if $\alpha_1 = \beta_1, \alpha_2 = \beta_2, \ldots, \alpha_n = \beta_n$.

† For simplicity we use the same notation for the zero vector and the field element zero. The meaning of "0" will always be clear from the context.

Notice that nothing is said about the real numbers $\{\alpha_i\}$ in $\langle\alpha_1, \ldots, \alpha_n\rangle$ being different from one another. $\langle 0, 0, 0\rangle$ and $\langle 1, 0, 1\rangle$ are perfectly legitimate 3-tuples. We note also that $\langle 1, 0, 1\rangle \neq \langle 0, 1, 1\rangle$, for example.

(3.3) DEFINITION. THE VECTOR SPACE R_n. The *vector space* R_n over the field of real numbers R is the algebraic system consisting of all n-tuples $a = \langle\alpha_1, \ldots, \alpha_n\rangle$ with $\alpha_i \in R$, together with the operations of addition and multiplication of n-tuples by real numbers, to be defined below. The n-tuples $a \in R_n$ are called *vectors*, and the real numbers α_i are called the *components* of the vector $a = \langle\alpha_1, \ldots, \alpha_n\rangle$. Two vectors $a = \langle\alpha_1, \ldots, \alpha_n\rangle$ and $b = \langle\beta_1, \ldots, \beta_n\rangle$ are said to be equal, and we write $a = b$ if and only if $\alpha_i = \beta_i, i = 1, \ldots, n$. The *sum* $a + b$ of the vectors $a = \langle\alpha_1, \ldots, \alpha_n\rangle$ and $b = \langle\beta_1, \ldots, \beta_n\rangle$ is defined by

$$a + b = \langle\alpha_1 + \beta_1, \ldots, \alpha_n + \beta_n\rangle.$$

The product of the vector a by the real number λ is defined by

$$\lambda a = \langle\lambda\alpha_1, \ldots, \lambda\alpha_n\rangle.$$

It is straightforward to verify that R_n is a vector space over R, according to Definition (3.1). In particular, the vector 0 is given by $0 = \langle 0, \ldots, 0\rangle$, and if $a = \langle\alpha_1, \ldots, \alpha_n\rangle$, then $-a = \langle -\alpha_1, \ldots, -\alpha_n\rangle$.

(3.3)′ DEFINITION. THE VECTOR SPACE F_n. Let F be an arbitrary field. An n-tuple $\langle\alpha_1, \ldots, \alpha_n\rangle$, with α_i in F, is defined as in (3.2). The set of all such n-tuples forms a vector space F_n over F, with the operations defined as in (3.3).

(3.4) DEFINITION. THE VECTOR SPACE OF FUNCTIONS ON THE REAL LINE. This time, let $\mathscr{F}(R)$ be the set of all real-valued functions defined on the set of all real numbers R. Then $\mathscr{F}(R)$ becomes a vector space over the field of real numbers R if we define

$$\begin{aligned} (f + g)(x) &= f(x) + g(x), && x \in R, \quad f, g \in \mathscr{F}(R) \\ (\alpha f)(x) &= \alpha f(x), && x \in R, \quad \alpha \in R, \quad f \in \mathscr{F}(R), \end{aligned}$$

Then, as pointed out earlier, the vector space axioms can be verified.

We return now the general concept of a vector space. Our first theorem states that many of the properties of fields derived in Section 2 also hold in vector spaces.

(3.5) THEOREM. *Let V be a vector space over a field F. Then the following statements hold.*

(i) *If* $u + v = u + w$, *then* $v = w$, *for all* $u, v, w \in V$.
(ii) *The equation* $u + x = v$ *has a unique solution (which we shall denote by* $v - u$*). We have* $v - u = v + (-u)$.
(iii) $-(-u) = u$, *for all* $u \in V$.
(iv) $0 \cdot u = 0$,† *for all* $u \in V$.
(v) $-(\alpha u) = (-\alpha)u = \alpha(-u)$, *for* $\alpha \in F$, $u \in V$.
(vi) $(-\alpha)(-u) = \alpha u$, $\alpha \in F$, $u \in V$.
(vii) *If* $\alpha u = \alpha v$, *with* $\alpha \neq 0$ *in* F, *then* $u = v$, *for all* $u, v \in V$.

The proofs of (i) to (vi) are identical (word for word!) with the proofs of the corresponding facts about fields. We give the proof of (vii). We are given that

$$\alpha u = \alpha v \quad \text{and} \quad \alpha \neq 0.$$

Then α^{-1} exists, and the substitution principle (which applies to vector spaces as well as to fields) implies that

$$\alpha^{-1}(\alpha u) = \alpha^{-1}(\alpha v).$$

By [3.1(6)] we obtain

$$(\alpha^{-1}\alpha)u = (\alpha^{-1}\alpha)v,$$

and since $\alpha^{-1}\alpha = 1$ and $1u = u$, $1v = v$ by [3.1(7)], we have $u = v$ as required.

We proceed to derive a few other consequences of the definition which will be needed in the next section. The associative law states that for a_1, a_2, a_3 in V,

$$(a_1 + a_2) + a_3 = a_1 + (a_2 + a_3).$$

If we have four vectors a_1, a_2, a_3, a_4, there are the following possible sums we can form:

$$\begin{aligned} &a_1 + [\, a_2 + (a_3 + a_4) \,], \\ &a_1 + [\, (a_2 + a_3) + a_4 \,], \\ &[\, a_1 + (a_2 + a_3) \,] + a_4, \\ &[\, (a_1 + a_2) + a_3 \,] + a_4. \end{aligned}$$

The reader may check that all these expressions represent the same vector. More generally, it can be proved by mathematical induction that all possible ways of adding n vectors $a_1, \ldots, a_n$ together to form a single sum yield a uniquely determined vector which we shall denote by

$$a_1 + \cdots + a_n = \sum_{i=1}^{n} a_i.$$

† See the footnote on p. 18.

The commutative law and this "generalized associative law" imply by a further application of mathematical induction that

$$(a_1 + \cdots + a_n) + (b_1 + \cdots + b_n) = (a_1 + b_1) + \cdots + (a_n + b_n)$$

or, more briefly,

$$\sum_{i=1}^{n} a_i + \sum_{i=1}^{n} b_i = \sum_{i=1}^{n} (a_i + b_i). \tag{3.6}$$

The other rules can also be generalized to sums of more than two vectors:

$$\left(\sum_{i=1}^{n} \lambda_i\right)a = \sum_{i=1}^{n} \lambda_i a,$$

$$\lambda\left(\sum_{i=1}^{n} a_i\right) = \sum_{i=1}^{n} \lambda a_i.$$

We have also by (3.6):

$$\left(\sum_{i=1}^{n} \lambda_i a_i\right) + \left(\sum_{i=1}^{n} \mu_i a_i\right) = \sum_{i=1}^{n} (\lambda_i + \mu_i)a_i.$$

The main point is that these rules do require proof from the basic rules (3.1), and the reader will find it an interesting exercise in the use of mathematical induction to supply, for example, a proof of (3.6).

We conclude this introductory section on vector spaces with some examples to show that our definition of the vector space R_n is consistent with the interpretation of vectors sometimes used in geometry and physics. The study of these examples can be skipped or postponed without interrupting the continuity of the discussion of vector spaces.

Example A. We shall give a geometric interpretation of vectors, in terms of directed line segments.

Let's begin with the real line, which we identify with the vector space R_1. We shall denote the vectors in R_1 by capital letters A, B, C, etc., and think of them as points on the line, located by their position vectors from the origin $O = \langle 0 \rangle$. (See Figure 2.1.)

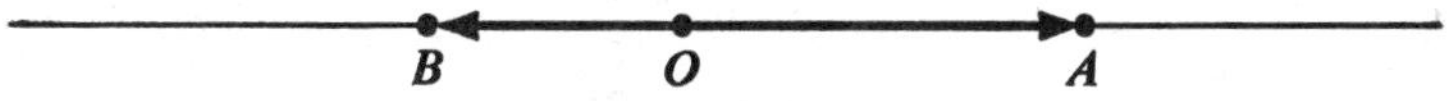

FIGURE 2.1

A directed line segment on R_1 is an ordered pair of points, denoted by $\overrightarrow{AB}$, which is simply the line segment between the points A and B, together with a direction, with A the starting point and B the end point of the segment. Thus $\overrightarrow{BA}$ is the same line segment, with the direction reversed.

We wish to make the concept of a directed line segment involve only length and direction, and not a particular starting point. A case-by-case analysis will show that two directed line segments $\overrightarrow{AB}$ and $\overrightarrow{CD}$ have the same length and direction if and only if $B - A = D - C$. This suggests that a formal definition of the directed line segment $\overrightarrow{AB}$ can be given as

$$\overrightarrow{AB} = B - A$$

(remembering that A and B are vectors in the vector space R_1 so that subtraction is defined). For example, the directed line segments $\overrightarrow{AB}$ and $\overrightarrow{CD}$, with $A = \langle -1 \rangle$, $B = \langle 2 \rangle$, $C = \langle 3 \rangle$, $D = \langle 6 \rangle$, are equal, because $B - A = D - C$ (see Figure 2.2).

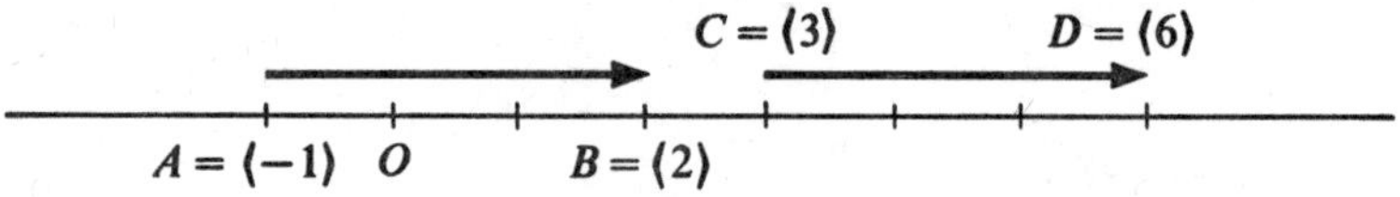

FIGURE 2.2

We shall now define directed line segments in the plane. As in Example A, we start with the vector space R_2, and denote the vectors of R_2 by capital letters $A, B, \ldots,$ thinking of them as points in the plane, given by their position vectors from the origin $O = \langle 0, 0 \rangle$ (see Figure 2.3).

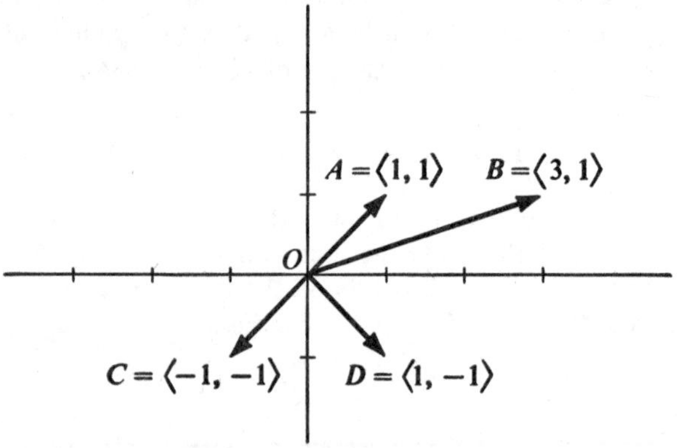

FIGURE 2.3

(3.7) DEFINITION. Let A and B be vectors in R_2. The *directed line segment* $\overrightarrow{AB}$ is defined to be the vector $B - A$.

This definition is suggested by our analysis of directed line segments in R_1. It provides a precise mathematical version of the idea that a "geometric vector" (or directed line segment) should be defined by a length and a direction, without reference to a particular location in the plane. In physics these objects are sometimes called "free vectors."

We represent directed line segments on paper as in Figure 2.4 (where the directed line segments $\overrightarrow{AB}$ and $\overrightarrow{CD}$ are drawn using the same points as in Figure 2.3). Note that the arrows $\overrightarrow{AB}$ and $\overrightarrow{CD}$ have the same

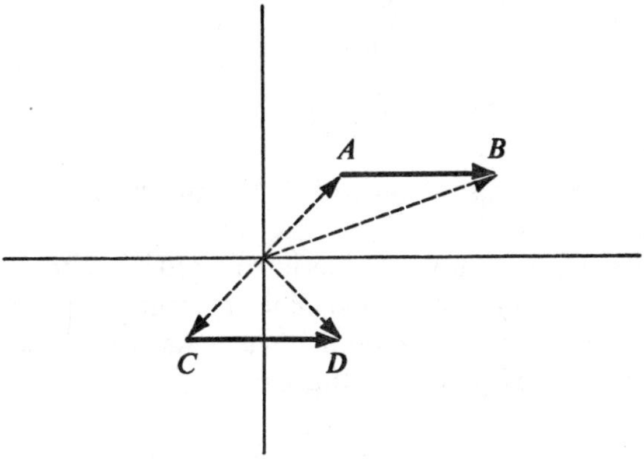

FIGURE 2.4

length and direction, and according to Definition (3.7), they represent the same directed line segment, because $\overrightarrow{AB} = B - A = \langle 2, 0\rangle$ and $\overrightarrow{CD} = D - C = \langle 2, 0\rangle$.

The meaning of Definition (3.7) in general is simply that two directed line segments are equal provided they have the same length and direction.

We can now show that addition of directed line segments is given by the "parallelogram law."

(3.8) THEOREM. *Let A, B, C be points in R_2. Then $\overrightarrow{AB} + \overrightarrow{BC} = \overrightarrow{AC}$. (See Figure 2.5).*

Proof. By Definition (3.7), we have $\overrightarrow{AB} = B - A$, $\overrightarrow{BC} = C - B$, and hence

$$\overrightarrow{AB} + \overrightarrow{BC} = (B - A) + (C - B) = C - A = \overrightarrow{AC},$$

as required.

Theorem (3.8) shows that if $\overrightarrow{AB}$ and $\overrightarrow{BC}$ are directed line segments, arranged so that the end point of the first is the starting point of the second, then their sum is the directed line segment given by the diagonal of the parallelogram with sides $\overline{AB}$ and $\overline{BC}$.

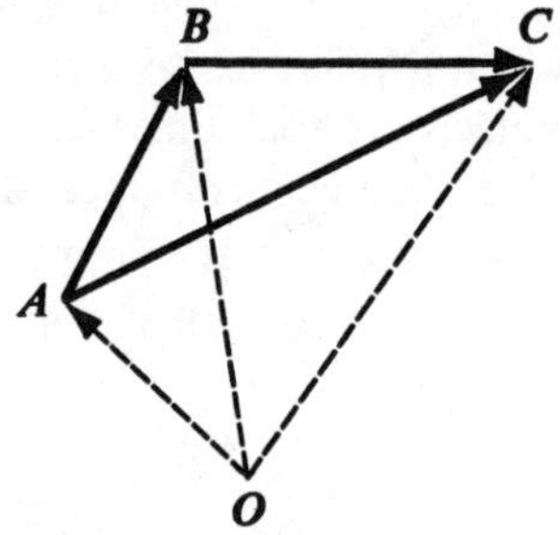

FIGURE 2.5

We conclude this discussion with some application of these ideas to plane geometry.

We shall continue to use the ideas introduced in Example A. We also need the concept of an (undirected) line segment $\overline{AB}$, described by an unordered set of two points. Two line segments are defined to be equal if and only if their end points are the same.

(3.9) DEFINITION. A point X is defined to be the *midpoint of the line segment* $\overline{AB}$ if the directed line segments $\overrightarrow{AX}$ and $\overrightarrow{XB}$ are equal.

(3.10) THEOREM. *The midpoint X of the line segment $\overline{AB}$ is given by*

$$X = \tfrac{1}{2}(A + B).$$

Proof. From Definition (3.9), we have $\overrightarrow{AX} = \overrightarrow{XB}$. Therefore

$$X - A = B - X,$$

and

$$2X = A + B.$$

Multiplying both sides by $\frac{1}{2}$ gives the result.

(3.11) DEFINITION. Four points $\{A, B, C, D\}$ are said to form the vertices of a *parallelogram* if $\overrightarrow{AB} = \overrightarrow{CD}$ (see Figure 2.6). The *sides* of the

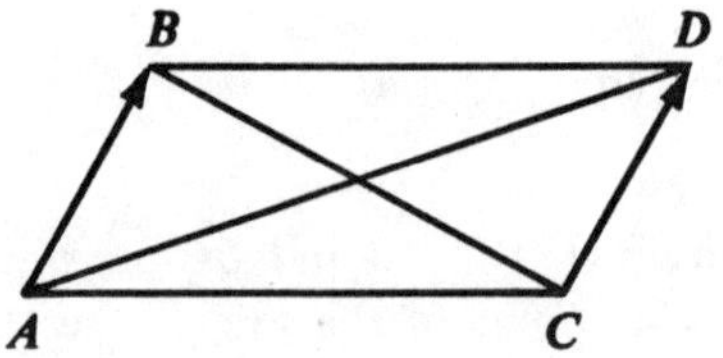

FIGURE 2.6

parallelogram are defined to be the line segments $\overline{AB}$, $\overline{CD}$, $\overline{AC}$, and $\overline{BD}$; the *diagonals* are the line segments $\overline{AD}$ and $\overline{BC}$.

(3.12) THEOREM. *The diagonals of a parallelogram bisect each other.*

Proof. The midpoints of the diagonals of the parallelogram in Figure 2.6 are, by Theorem (3.10),

$$X = \tfrac{1}{2}(A + D), \qquad Y = \tfrac{1}{2}(B + C).$$

We have to show that $X = Y$. Since $\{A, B, C, D\}$ are vertices of a parallelogram, we have $\overrightarrow{AB} = \overrightarrow{CD}$, and hence $B - A = D - C$. Therefore $B + C = A + D$ and $X = Y$ as we wished to prove.

EXERCISES

1. In the vector space R_3, compute (that is, express in the form $\langle\lambda, \mu, \nu\rangle$), the following vectors formed from $a = \langle -1, 2, 1\rangle$, $b = \langle 2, 1, -3\rangle$, $c = \langle 0, 1, 0\rangle$:
   a. $a + b + c$.
   b. $2a - b + c$.
   c. $-a + 2b$.
   d. $\alpha a + \beta b + \gamma c$.
2. In R_3, letting a, b, c be as in Exercise 1, solve the equations below, for $x = \langle\alpha_1, \alpha_2, \alpha_3\rangle$ in R_3.
   a. $x + a = b$.
   b. $2x - 3b = c$.
   c. $b + x = a - 2c$.
3. Check that the axioms for a vector space are satisfied for the vector spaces R_n, F_n, and $\mathscr{F}(R)$.
4. In R_n, let $a = \langle\alpha_1, \ldots, \alpha_n\rangle$, $b = \langle\beta_1, \ldots, \beta_n\rangle$. Show that $b - a = \langle\beta_1 - \alpha_1, \ldots, \beta_n - \alpha_n\rangle$.

The following exercises are based on Example A and (3.9) – (3.12).

5. Let $\overrightarrow{AB}$ be a directed line segment in R_2. Show that $\overrightarrow{AB} = -\overrightarrow{BA}$.
6. Show that the directed line segments $\overrightarrow{AB}$ and $\overrightarrow{CD}$ are equal if and only if there exists a vector X such that $C = A + X$, and $D = B + X$. (The interpretation is that two directed line segments are equal if and only if one is carried onto the other by a translation. A *translation* of R_2 is a rule which sends each vector A to the vector $A + X$, for some fixed vector X.)
7. Find the midpoints of the line segment $\overline{AB}$ in the following cases.
   a. $A = \langle 0, 0\rangle$, $B = \langle 2, 1\rangle$.
   b. $A = \langle 1, -1\rangle$, $B = \langle 3, 2\rangle$.

8. Test the following sets of points to determine which are vertices of parallelograms. The question is whether for some order of the points, the conditions of Definition (3.11) are satisfied.

a. $\langle 0, 0\rangle, \langle 1, 1\rangle, \langle 4, 2\rangle, \langle 3, 1\rangle$.

b. $\langle 0, 0\rangle, \langle 1, 1\rangle, \langle 4, 3\rangle, \langle 2, 1\rangle$.

c. $\langle 1, 1\rangle, \langle 3, 4\rangle, \langle -1, 2\rangle, \langle 1, 5\rangle$.

9. Suppose A, B, C, D are points such that $\overrightarrow{AB} = \overrightarrow{CD}$. Show that $\overrightarrow{AC} = \overrightarrow{BD}$.

10. A *quadrilateral* is a set of four distinct points $\{A, B, C, D\}$; the sides of the quadrilateral are the line segments $\overline{AB}$, $\overline{BC}$, $\overline{CD}$, and $\overline{DA}$. Show that the midpoints of the sides of a quadrilateral are the vertices of a parallelogram.

4. SUBSPACES AND LINEAR DEPENDENCE

This section begins with the concept needed to study the examples given in Section 1.

(4.1) DEFINITION. A *subspace* S of the vector space V is a nonempty set of vectors in V such that:

1. If a and b are in S, then $a + b \in S$.
2. If $a \in S$ and $\lambda \in F$, then $\lambda a \in S$.

The axioms (3.1) for a vector space are clearly satisfied for any subspace S of a vector space V; so we see that a subspace is simply a vector space contained in a larger vector space, in which the operations of addition and multiplication by scalars are the same as those in the larger vector space.

Example A. Consider the system of homogeneous equations discussed in Section 1,

$$\begin{aligned} x + 2y - 3z + t &= 0 \\ x - y + z + t &= 0. \end{aligned}$$

One of the statements derived in that section was that if $u = \langle \alpha, \beta, \gamma, \delta\rangle$ and $u' = \langle \alpha', \beta', \gamma', \delta'\rangle$ are solutions, so are $u + u'$ and λu for $\lambda \in F$. In other words the set of solutions of the system of homogeneous equations is a subspace of R_4, so that the problem of solving the system of equations is to give some clear description of the subspace of solutions.

Example B. Consider the differential equation considered in Problem B, Section 1,

$$y'' + m^2 y = 0. \qquad (4.2)$$

There we observed that if f and g are solutions, so are $f + g$ and λf, for all real numbers λ. This time the solutions of the homogeneous differential equation (4.2) form a subspace of the vector space $\mathscr{F}(R)$ defined in Section 3, and as in Example A, the problem of finding all solutions of the differential equation comes down to finding exactly which functions belong to this subspace of $\mathscr{F}(R)$.

Example C. It may have occurred to the reader by now that the concept of subspace provides the framework for other questions discussed in elementary calculus. For example, some standard results (which ones?) about continuous and differentiable functions show that the following subsets of $\mathscr{F}(R)$ are actually subspaces:

(*i*) The set $C(R)$ of all continuous real functions defined on the real line.

(*ii*) The set $D(R)$ consisting of all differentiable real valued functions on R.

Another familiar example of a subspace of $\mathscr{F}(R)$ is the following one:

(*iii*) The set $P(R)$ of polynomial functions on R, where a *polynomial function* is a function $f \in \mathscr{F}(R)$ such that for some fixed set of real numbers $\alpha_0, \alpha_1, \ldots, \alpha_n$,

$$f(x) = \alpha_0 + \alpha_1 x + \cdots + \alpha_n x^n \qquad \text{for all } x \in R.$$

We now proceed to make a few general remarks about subspaces of a vector space V over a field F. It is convenient to have one more definition.

(4.3) DEFINITION. Let $\{a_1, \ldots, a_m\}$ be a set of vectors in V. A vector $a \in V$ is said to be a *linear combination* of $\{a_1, \ldots, a_m\}$ if it is possible to find elements $\lambda_1, \ldots, \lambda_m$ in F such that

$$a = \lambda_1 a_1 + \cdots + \lambda_m a_m.$$

We can now make the following observation.

(4.4) *If S is a subspace of V containing the vectors $a_1, \ldots, a_m$, then every linear combination of $a_1, \ldots, a_m$ belongs to S.*

Proof. We prove the result by induction on m. If $m = 1$, the result is true by the definition of a subspace. Suppose that any linear combination of $m - 1$ vectors in S belongs to S, and consider a linear combination

$$a = \lambda_1 a_1 + \cdots + \lambda_m a_m$$

of m vectors belonging to S. Letting $a' = \lambda_2 a_2 + \cdots + \lambda_m a_m$, we have

$$a = \lambda_1 a_1 + a'$$

where $\lambda_1 a_1 \in S$, and $a' \in S$ by the induction hypothesis. By part (1) of Definition (4.1), $a \in S$ and (4.4) is proved.

The process of forming linear combinations leads to a method of constructing subspaces, as follows.

(4.5) *Let $\{a_1, \ldots, a_m\}$ be a set of vectors in V, for $m \geq 1$; then the set of all linear combinations of the vectors $a_1, \ldots, a_m$ forms a subspace $S = S(a_1, \ldots, a_m)$. S is the smallest subspace containing $\{a_1, \cdots, a_m\}$, in the sense that if T is any subspace containing $a_1, \ldots, a_m$, then $S \subset T$.*

Proof. Let $a = \sum_{i=1}^{m} \lambda_i a_i$ and $b = \sum_{i=1}^{m} \mu_i a_i$. Then

$$a + b = \sum_{i=1}^{m} (\lambda_i + \mu_i) a_i$$

is again a linear combination of $a_1, \ldots, a_m$. If $\lambda \in F$, then

$$\lambda a = \lambda\left(\sum_{i=1}^{m} \lambda_i a_i\right) = \sum_{i=1}^{m} \lambda(\lambda_i a_i) = \sum_{i=1}^{m} (\lambda\lambda_i) a_i.$$

These computations prove that S is a subspace, and the fact that it is the smallest one containing the given vectors is immediate by (4.4).

(4.6) DEFINITION. The subspace $S = S(a_1, \ldots, a_m)$ defined in (4.5) is called the subspace *generated by* (or, sometimes, *spanned by*) $a_1, \ldots, a_m$, and $a_1, \ldots, a_m$ are called *generators* of S. A subspace S of V is called *finitely generated* if there exist vectors $s_1, \ldots, s_k$ in S such that $S = S(s_1, \ldots, s_k)$.

The problem of solving the system of equations in Example A or the differential equation in Example B now comes into sharper focus. If S is the subspace of solutions in either case, then we would have a satisfactory description of the solutions if we could:

1. Show that S is finitely generated.
2. Find a set of generators of S (in the sense of Definition (4.6).

This idea leads to one more new concept, which will require some close reasoning. Suppose S is a finitely generated subspace of a vector space V over F, with generators $\{a_1, \ldots, a_m\}$. The description of S as the set of linear combinations of $\{a_1, \ldots, a_m\}$ will be especially useful if none of the a_i's are superfluous. What does this mean? It means that no generator a_i can be expressed as a linear combination of the remaining ones. In fact, suppose a_i is a linear combination of $\{a_1, \ldots, a_{i-1}, a_{i+1}, \ldots, a_m\}$. Then $a_i \in S(a_1, \ldots, a_{i-1}, a_i + 1, \ldots, a_m)$ and it follows that $S(a_1, \ldots, a_m) = S(a_1, \ldots, a_{i-1}, a_{i+1}, \ldots, a_m)$; in other words we can replace the set of generators $a_1, \ldots, a_m$ by a smaller set. The statement that a_i is a linear combination of $a_1, \ldots, a_{i-1}, a_{i+1}, \ldots, a_m$ means that there exist $\lambda_1, \ldots, \lambda_{i-1}, \lambda_{i+1}, \ldots, \lambda_m$ in F such that

$$a_i = \lambda_1 a_1 + \cdots + \lambda_{i-1}a_{i-1} + \lambda_{i+1}a_{i+1} + \cdots + \lambda_m a_m.$$

Adding $-a_i = (-1)a_i$ to both sides and using the commutative law, we obtain

$$0 = \lambda_1 a_1 + \cdots + \lambda_{i-1}a_{i-1} + (-1)a_i + \lambda_{i+1}a_{i+1} + \cdots + \lambda_m a_m,$$

and we have shown that there exists elements $\mu_1, \ldots, \mu_m$ in F, not all equal to zero, such that

$$\mu_1 a_1 + \cdots + \mu_m a_m = 0. \tag{4.7}$$

(In this case, $\mu_j = \lambda_j, j \neq i$, and $\mu_i = -1 \neq 0$.) Conversely, suppose a formula such as (4.7) holds, where not all the μ_i are equal to zero. Then we shall prove that some a_j is a linear combination of the remaining vectors $\{a_i, i \neq j\}$. Suppose that $\mu_j \neq 0$. Then using the commutative law and adding $-\mu_j a_j$ to both sides, we obtain

$$-\mu_j a_j = \mu_1 a_1 + \cdots + \mu_{j-1}a_{j-1} + \mu_{j+1}a_{j+1} + \cdots + \mu_m a_m.$$

Multiplying both sides by $-\mu_j^{-1}$ (which exists in F since $\mu_j \neq 0$) we have

$$\begin{aligned}(-\mu_j^{-1})(-\mu_j a_j) = (-\mu_j^{-1})\mu_1 a_1 + \cdots + (-\mu_j^{-1})\mu_{j-1}a_{j-1} \\ + (-\mu_j^{-1})\mu_{j+1}a_{j+1} + \cdots + (-\mu_j^{-1})\mu_m a_m.\end{aligned} \tag{4.8}$$

Using Theorem (3.5), we have, for the left-hand side,

$$(-\mu_j^{-1})(-\mu_j a_j) = \mu_j^{-1}\mu_j a_j = 1a_j = a_j.$$

Substituting this back in (4.8), we have proved that $a_j \in S(a_1, \ldots, a_{j-1}, a_{j+1}, \ldots, a_m)$. To summarize, we have proved the following important result.

(4.9) THEOREM. *Let $\{a_1, \ldots, a_m\}$ be a set of vectors in V, for $m \geq 1$. Some vector a_i can be expressed as a linear combination of the remaining vectors $\{a_1, \ldots, a_{i-1}, a_{i+1}, \ldots, a_m\}$ if and only if there exists elements $\mu_1, \ldots, \mu_m$ in F, not all equal to zero, such that*

$$\mu_1 a_1 + \cdots + \mu_m a_m = 0.$$

The new concept which was mentioned at the beginning of the proof of Theorem (4.9) is the rather subtle but natural condition on the vectors $\{a_1, \ldots, a_m\}$ which emerged in the course of the proof.

(4.10) DEFINITION. *Let* $\{a_1, \ldots, a_m\}$ be an arbitrary finite set of vectors in V. The set of vectors $\{a_1, \ldots, a_m\}$ is said to be *linearly dependent* if there exist elements of F, $\lambda_1, \ldots, \lambda_m$, not all zero, such that

$$\lambda_1 a_1 + \lambda_2 a_2 + \cdots + \lambda_m a_m = 0.$$

Such a formula will be called a *relation of linear dependence.* A set of vectors which is not linearly dependent is said to be *linearly independent.*

Thus, the set $\{a_1, \ldots, a_m\}$ is linearly independent if and only if

$$\lambda_1 a_1 + \cdots + \lambda_m a_m = 0, \qquad \lambda_i \in F,$$

implies that $\lambda_1 = \cdots = \lambda_m = 0$.

Before proceeding, let us look at a few simple cases of linear dependence.

First of all, the set consisting of the zero vector $\{0\}$ in V always forms a linearly dependent set, since for any $\lambda \neq 0$ in F we have

$$\lambda 0 = 0.$$

But if $a \neq 0$ in V, then the set $\{a\}$ alone is a linearly independent set. To see this, we have to prove that if

$$\lambda a = 0, \qquad \lambda \in F,$$

then $\lambda = 0$. This statement is equivalent to proving that if the conclusion ($\lambda = 0$) is false, then the hypothesis ($\lambda a = 0, a \neq 0$) is false also. If $\lambda \neq 0$, then by Axiom 7 of Definition (2.1) there is an element λ^{-1} such that $\lambda^{-1}\lambda = 1$. Since $\lambda a = 0$, we obtain

$$\lambda^{-1}(\lambda a) = (\lambda^{-1}\lambda)a = 0.$$

However, $\lambda^{-1}\lambda = 1$, and hence we have shown that $\lambda a = 0$, where $\lambda \neq 0$, implies $1 \cdot a = a = 0$, contrary to the hypothesis that $a \neq 0$.

Now let $\{a, b\}$ be a set consisting of two vectors in V. We shall prove:

(4.11) *$\{a, b\}$ is a linearly dependent set if and only if $a = \lambda b$ or $b = \lambda' a$ for some λ or λ' in F.*

Proof. First suppose $a = \lambda b$, $\lambda \in F$. Then we have

$$a + [-(\lambda b)] = 0.$$

By Theorem (3.5),

$$-(\lambda b) = (-\lambda)b,$$

and we have

$$1 \cdot a + (-\lambda)b = 0.$$

Since $1 \neq 0$, we have proved that $\{a, b\}$ is a linearly dependent set. Similarly, $b = \lambda' a$ implies that $\{a, b\}$ is a linearly dependent set.

Now suppose that

$$\alpha a + \beta b = 0, \qquad \alpha, \beta \in F \tag{4.12}$$

where either α or $\beta \neq 0$. If $\alpha \neq 0$, then we have

$$\alpha a = (-\beta)b$$

and

$$a = \alpha^{-1}(\alpha a) = \alpha^{-1}(-\beta)b = (-\alpha^{-1}\beta)b,$$

as we wish to prove. If, on the other hand, $\alpha = 0$ and $\beta \neq 0$, the equation (6.3) becomes $\beta b = 0$, and hence $b = 0$. In that case we have $b = \lambda' a$ for $\lambda' = 0$, and (4.11) is completely proved.

Let us have one or two numerical examples.

Example D. Let $a = \langle 1, -1 \rangle$ in R_2. Then $\{a, b\}$ is a linearly independent set if $b = \langle 1, 1 \rangle$ or $\langle 1, 0 \rangle$, and a linearly dependent set if $b = \langle -2, 2 \rangle$ or $\langle 0, 0 \rangle$. [To check these statements, use (4.11).]

Example E. Consider $a = \langle 1, -1 \rangle$, $b = \langle 1, 1 \rangle$, and $c = \langle 2, 1 \rangle$ in R_2. Is $\{a, b, c\}$ linearly dependent or not? Consider a possible relation of linear dependence, $\alpha a + \beta b + \gamma c = 0$, with α, β, γ real numbers. Then

$$\alpha a + \beta b + \gamma c = \langle \alpha + \beta + 2\gamma, -\alpha + \beta + \gamma \rangle.$$

In order for this vector to be zero, we must have

$$\begin{aligned} \alpha + \beta + 2\gamma &= 0 \\ -\alpha + \beta + \gamma &= 0. \end{aligned}$$

Do there exist numbers α, β, γ not all zero, satisfying these equations? Try setting $\gamma = 1$; then the equations become

$$\begin{aligned} \alpha + \beta + 2 &= 0 \\ -\alpha + \beta + 1 &= 0, \end{aligned}$$

and we can solve for α and β, obtaining

$$\alpha = -\tfrac{1}{2}, \qquad \beta = -\tfrac{3}{2}, \qquad \gamma = 1.$$

Thus the vectors are linearly dependent. We shall demonstrate a more systematic way for carrying out this sort of test later in this chapter.

We conclude the section with a numerical example which puts together many of the ideas that have been introduced in this section. The task of checking some of the statements in the discussion of the example is left as an exercise.

Example F. *Describe the set of all solutions of the equation*

$$x_1 + 2x_2 - x_3 = 0.$$

A solution is, first of all, a vector $\langle \alpha_1, \alpha_2, \alpha_3 \rangle$ in R_3, and the set of all solutions is a subspace S of R_3. Setting $x_3 = 0$, we see that

$$u_1 = \langle 1, -\tfrac{1}{2}, 0 \rangle \in S.$$

Similarly, setting $x_1 = 0$, we have

$$u_2 = \langle 0, 1, 2 \rangle \in S.$$

We will show that $S = S(u_1, u_2)$; in other words, that u_1 and u_2 are generators of the subspace consisting of all solutions of the equation. Suppose $u = \langle \alpha, \beta, \gamma \rangle$ is a solution. If $\alpha = \beta = \gamma = 0$, then $u \in S$, and there is nothing further to show. Now suppose $\alpha \neq 0$. Then

$$u - \alpha u_1 = \left\langle 0, \beta + \frac{\alpha}{2}, \gamma \right\rangle = \left\langle 0, \frac{\gamma}{2}, \gamma \right\rangle = \frac{\gamma}{2} u_2,$$

since $\gamma - 2\beta - \alpha = 0$ as a consequence of the assumption that $\langle \alpha, \beta, \gamma \rangle$ satisfies the equation. Thus $u \in S(u_1, u_2)$. Finally, suppose $\alpha = 0$, but $\beta \neq 0$. Then $2\beta - \gamma = 0$, and it is immediate that

$$\langle 0, \beta, \gamma \rangle = \beta u_2.$$

At this point we have proved that the subspace S is finitely generated, and that the vectors u_1 and u_2 are generators. By (4.11), it follows that u_1 and u_2 are linearly independent. Theorem (4.9) then implies that $S = S(u_1, u_2)$, and that neither u_1 nor u_2 can be omitted from the set of generators $\{u_1, u_2\}$. Finally, the statement $S = S(u_1, u_2)$ means that every solution u of the equation

$$x_1 + 2x_2 - x_3 = 0$$

can be expressed as a linear combination $u = \lambda_1 u_1 + \lambda_2 u_2$ of the vectors

$$u_1 = \langle 1, -\tfrac{1}{2}, 0 \rangle \quad \text{and} \quad u_2 = \langle 0, 1, 2 \rangle.$$

EXERCISES

1. Determine which of the following subsets of R_n are subspaces.
   - a. All vectors $\langle \alpha_1, \ldots, \alpha_n \rangle$ such that $\alpha_1 = 1$.
   - b. All vectors $\langle \alpha_1, \ldots, \alpha_n \rangle$ such that $\alpha_1 = 0$.
   - c. All vectors $\langle \alpha_1, \ldots, \alpha_n \rangle$ such that $\alpha_1 + 2\alpha_2 = 0$.
   - d. All vectors $\langle \alpha_1, \ldots, \alpha_n \rangle$ such that $\alpha_1 + \alpha_2 + \cdots + \alpha_n = 1$.
   - e. All vectors $\langle \alpha_1, \ldots, \alpha_n \rangle$ such that $A_1\alpha_1 + \cdots + A_n\alpha_n = 0$ for fixed $A_1, \ldots, A_n$ in R.
   - f. All vectors $\langle \alpha_1, \ldots, \alpha_n \rangle$ such that $A_1\alpha_1 + \cdots + A_n\alpha_n = B$ for fixed $A_1, \ldots, A_n, B$.
   - g. All vectors $\langle \alpha_1, \ldots, \alpha_n \rangle$ such that $\alpha_1^2 = \alpha_2$.

2. Verify that the subsets $C(R)$, $D(R)$, and $P(R)$ of $\mathscr{F}(R)$ defined in Example C, are subspaces of $\mathscr{F}(R)$.
3. Determine which of the following subsets of $C(R)$ are subspaces of $C(R)$.
   a. The set of polynomial functions in $C(R)$.
   b. The set of all $f \in C(R)$ such that $f(\frac{1}{2})$ is a rational number.
   c. The set of all $f \in C(R)$ such that $f(\frac{1}{2}) = 0$.
   d. The set of all $f \in C(R)$ such that $\int_0^1 f(t)\,dt = 1$.
   e. The set of all $f \in C(R)$ such that $\int_0^1 f(t)\,dt = 0$.
   f. The set of all $f \in C(R)$ such that $df/dt = 0$.
   g. The set of all $f \in C(R)$ such that for some $\alpha, \beta, \gamma \in R$

$$\alpha \frac{d^2f}{dt^2} + \beta \frac{df}{dt} + \gamma f = 0.$$

   h. The set of all $f \in C(R)$ such that

$$\alpha \frac{d^2f}{dt^2} + \beta \frac{df}{dt} + \gamma f = g$$

   for a fixed function $g \in C(R)$.
4. Test the following sets of vectors in R_2 and R_3 to determine whether or not they are linearly independent.
   a. $\langle 1, 1\rangle, \langle 2, 1\rangle$.
   b. $\langle 1, 1\rangle, \langle 2, 1\rangle, \langle 1, 2\rangle$.
   c. $\langle 0, 1\rangle, \langle 1, 0\rangle$.
   d. $\langle 0, 1\rangle, \langle 1, 0\rangle, \langle \alpha, \beta\rangle$.
   e. $\langle 1, 1, 2\rangle, \langle 3, 1, 2\rangle, \langle -1, 0, 0\rangle$.
   f. $\langle 3, -1, 1\rangle, \langle 4, 1, 0\rangle, \langle -2, -2, -2\rangle$.
   g. $\langle 1, 1, 0\rangle, \langle 0, 1, 1\rangle, \langle 1, 0, 1\rangle, \langle 1, 1, 1\rangle$.
5. Find a set of linearly independent generators of the subspace of R_3 consisting of all solutions of the equation

$$x_1 - x_2 + x_3 = 0.$$

6. Show that the set of polynomial functions $P(R)$ (see Example C) is not a finitely generated vector space.
7. Is the intersection of two subspaces always a subspace? Prove your answer.
8. Is the union of two subspaces always a subspace? Explain.
9. Let a in R_n be a linear combination of vectors $b_1, \ldots, b_r$ in R_n, and let each vector b_i, $1 \le i \le r$, be a linear combination of vectors $c_1, \ldots, c_s$. Prove that a is a linear combination of $c_1, \ldots, c_s$.
10. Show that a set of vectors, which contains a set of linearly dependent vectors, is linearly dependent. What is the analogous statement about linearly independent vectors?

5. THE CONCEPTS OF BASIS AND DIMENSION

In Example F of Section 4, we showed that the subspace of solutions of the equation

$$x_1 + 2x_2 - x_3 = 0$$

is generated by two vectors, $u_1 = \langle 1, -\frac{1}{2}, 0\rangle$ and $u_2 = \langle 0, 1, 2\rangle$. We have also shown that neither u_1 nor u_2 can be omitted from the set of generators $\{u_1, u_2\}$. But does this show that no set of generators of S contains fewer than two vectors? Not necessarily. It might happen that, by some stroke of luck, somebody would find a single generator for S, which would make our work to find u_1 and u_2 a wasted effort. The purpose of this section is to show that this is not possible. In fact, we shall prove that if S is a subspace of a vector space V having m linearly independent generators, then every other set of generators contains at least m vectors.

It is easy to prove the desired result in the case of our example. We shall use an indirect proof. Suppose S were generated by a single vector u. Then

$$u_1 = \alpha u, \qquad u_2 = \beta u,$$

where α and β are different from zero. Then

$$\beta u_1 - \alpha u_2 = 0,$$

showing that u_1 and u_2 are linearly dependent. But we know that u_1 and u_2 are linearly independent; so we have arrived at a contradiction. Therefore our original assumption, that $S = S(u)$, must have been incorrect, and we conclude that every set of generators of S contains at least two vectors.

Our objective is to use a similar argument to prove the following basic result, which is a general version of what we have just proved in connection with the example. The proof of the theorem is followed by a numerical example, which the reader may prefer to study before tackling the proof.

(5.1) THEOREM. *Let S be a subspace of a vector space V over a field F, such that S is generated by n vectors $\{a_1, \ldots, a_n\}$. Suppose $\{b_1, \ldots, b_m\}$ are vectors in S, with $m > n$. Then the vectors $\{b_1, \ldots, b_m\}$ are linearly dependent.*

Proof. We prove the result by induction on n. For $n = 1$, we have $S = S(a)$, and we are given vectors $\{b_1, \ldots, b_m\}$ in S, with $m > 1$. Then (as in the example discussed before the theorem)

$$b_1 = \lambda_1 a, \qquad b_2 = \lambda_2 a, \ldots, b_m = \lambda_m a$$

for $\lambda_i \in F$. At least one $\lambda_j \neq 0$; otherwise $b_1 = b_2 = \cdots = b_m = 0$, and $b_1 - b_2 = 0$ is a relation of linear dependence. Assume $\lambda_j \neq 0$; then

$$\lambda_j b_1 + 0 \cdot b_2 + \cdots + 0b_{j-1} + (-\lambda_1)b_j + 0b_{j+1} + \cdots + 0b_n$$
$$= \lambda_j\lambda_1 a - \lambda_1\lambda_j a = 0;$$

and since $\lambda_j \neq 0$, we have shown that the vectors $\{b_1, \ldots, b_m\}$ are linearly dependent.

Now suppose, as our induction hypothesis, that the theorem is true for subspaces generated by $n-1$ vectors, and consider m distinct vectors $\{b_1, \ldots, b_m\}$ contained in $S = S(a_1, \ldots, a_n)$, with $m > n$. Then we have

$$\begin{aligned} b_1 &= \lambda_{11}a_1 + \cdots + \lambda_{1n}a_n \\ b_2 &= \lambda_{21}a_1 + \cdots + \lambda_{2n}a_n \\ &\cdots\cdots\cdots\cdots \\ b_m &= \lambda_{m1}a_1 + \cdots + \lambda_{mn}a_n, \end{aligned} \tag{5.2}$$

for some elements $\lambda_{ij} \in F$ (where the index ij of λ_{ij} means that λ_{ij} is the coefficient of a_j in an expression of b_i as a linear combination of the vectors $\{a_1, \ldots, a_n\}$).

We can assume that $\lambda_{i1} \neq 0$ for some i. Otherwise the terms involving a_1 are all missing from the equations (5.2) and $\{b_1, \ldots, b_n\}$ all belong to the subspace $S(a_2, \ldots, a_n)$ with $n-1$ generators. Since $m > n > n-1$, the induction hypothesis implies that $\{b_1, \ldots, b_m\}$ are linearly dependent, and there is nothing further to prove in this case.

Suppose now that some $\lambda_{i1} \neq 0$; by renaming the $\{b_i\}$ (by exchanging b_i for b_1), we may assume that $\lambda_{11} \neq 0$. Then the coefficient of a_1 in $b_2 - \lambda_{21}\lambda_{11}^{-1}b_1$ is $\lambda_{21} - \lambda_{21}\lambda_{11}^{-1} \cdot \lambda_{11} = 0$, so that

$$b_2 - \lambda_{21}\lambda_{11}^{-1}b_1 = \lambda'_{22}a_2 + \cdots + \lambda'_{2n}a_n$$

for some new constants $\{\lambda'_{22}, \ldots, \lambda'_{2n}\}$ in F. Similarly, we have

$$\begin{aligned} b_3 - \lambda_{31}\lambda_{11}^{-1}b_1 &= \lambda'_{32}a_2 + \cdots + \lambda'_{3n}a_n \\ &\cdots\cdots\cdots\cdots \\ b_m - \lambda_m\lambda_{11}^{-1}b_1 &= \lambda'_{m2}a_2 + \cdots + \lambda'_{mn}a_n. \end{aligned}$$

We have shown that the $n-1$ vectors

$$\{b_2 - \lambda_{21}\lambda_{11}^{-1}b_1, \ldots, b_m - \lambda_{m1}\lambda_{11}^{-1}b_1\}$$

all belong to the subspace $S(a_2, \ldots, a_n)$, with $n-1$ generators. Since $m > n, m-1 > n-1$, and it follows from the induction hypothesis that there exist elements $\mu_2, \ldots, \mu_n$ in F, not all zero such that

$$\mu_2(b_2 - \lambda_{21}\lambda_{11}^{-1}b_1) + \cdots + \mu_m(b_m - \lambda_{m1}\lambda_{11}^{-1}b_1) = 0.$$

Rewriting this expression as a linear combination of $\{b_1, \ldots, b_m\}$ we have

$$(-\mu_2\lambda_{21}\lambda_{11}^{-1} + \cdots + (-\mu_m\lambda_{m1}\lambda_{11}^{-1}))b_1 + \mu_2 b_2 + \cdots + \mu_m b_n = 0,$$

where at least one of the coefficients $\{\mu_2, \ldots, \mu_m\}$ is different from zero. This completes the proof that $\{b_1, \ldots, b_m\}$ are linearly dependent, and the theorem is proved.

Example A. Suppose $S = S(a_1, a_2)$ is a subspace of a vector space V over the field of real numbers, and suppose

$$\begin{aligned} b_1 &= 2a_1 + a_2 \\ b_2 &= -a_1 + a_2 \\ b_3 &= a_1 \end{aligned}$$

are three vectors in S. We wish to show that b_1, b_2, b_3 are linearly dependent. The idea is to subtract multiples of b_1 from b_2 and b_3 to make the coefficient of a_1 equal to zero. We have

$$\begin{aligned} b_2 + (\tfrac{1}{2})b_1 &= (\tfrac{3}{2})a_2 \\ b_3 - (\tfrac{1}{2})b_1 &= -(\tfrac{1}{2})a_2 . \end{aligned}$$

We now have two vectors belonging to the space $S(a_2)$ and

$$b_2 + \tfrac{1}{2}b_1 + 3\,[\, b_3 - (\tfrac{1}{2})b_1 \,] = 0.$$

Rewriting as a linear combination of b_1, b_2, b_3, we have

$$[\,\tfrac{1}{2} - (\tfrac{3}{2})\,]b_1 + b_2 + 3b_3 = 0,$$

which is the desired relation of linear dependence among $\{b_1, b_2, b_3\}$.

We conclude this section with another theorem which expresses an important fact about different sets of generators of a finitely generated vector space and permits us to introduce the concepts of *basis* and *dimension* of a finitely generated vector space.

(5.3) THEOREM. *Let S be a subspace of a vector space V, and suppose that $\{a_1, \ldots, a_m\}$ and $\{b_1, \ldots, b_n\}$ are both sets of generators of S, which are linearly independent. Then $n = m$.*

Proof. First, we view $\{b_1, \ldots, b_n\}$ as vectors in $S(a_1, \ldots, a_m)$. Since $b_1, \ldots, b_n$ are assumed to be linearly independent, Theorem (5.1) tells us that $n \le m$. Then viewing $\{a_1, \ldots, a_m\}$ as vectors in $S(b_1, \ldots, b_n)$, we have, by the same reasoning, $m \le n$. It follows that $n = m$, and the theorem is proved.

(5.4) DEFINITION. A finite set of vectors $\{b_1, \ldots, b_k\}$ is said to be a *basis* of a vector space V if $\{b_1, \ldots, b_k\}$ is a linearly independent set of generators of V.

Theorem (5.3) states that if a vector space V has a basis $\{b_1, \ldots, b_k\}$, then every other basis of V must contain exactly k vectors.

(5.5) DEFINITION. Suppose V is a vector space which has a basis $\{b_1, \ldots, b_k\}$. The uniquely determined number of basis vectors, k, is called the *dimension* of the vector space V (notation: $\dim V = k$).

Example B. Let n be a fixed positive integer. *We shall prove that the vector space R_n, defined in Section 2, has dimension n.*

All we have to do is find one basis of R_n, containing n vectors. Let

$$\begin{aligned} e_1 &= \langle 1, 0, 0, \ldots, 0\rangle \\ e_2 &= \langle 0, 1, 0, \ldots, 0\rangle \\ &\cdots\cdots\cdots\cdots\cdots \\ e_n &= \langle 0, 0, \ldots, 0, 1\rangle. \end{aligned}$$

The vectors $\{e_i\}$ will be shown to form a basis of R_n. What has to be checked? First that the vectors are linearly independent. This means that whenever $\{\lambda_1, \ldots, \lambda_n\}$ are elements of R, such that $\sum \lambda_i e_i = 0$, then all $\lambda_i = 0$. Now,

$$\lambda_1 e_1 = \langle \lambda_1, 0, \ldots, 0\rangle, \ldots, \lambda_n e_n = \langle 0, \ldots, 0, \lambda_n\rangle,$$

so that

$$\lambda_1 e_1 + \cdots + \lambda_n e_n = \langle \lambda_1, \ldots, \lambda_n\rangle = 0$$

implies $\lambda_1 = \cdots = \lambda_n = 0$ because of the definition of equality of two vectors in R_n. Next we have to show that $\{e_1, \ldots, e_n\}$ is a set of generators of R_n. The same calculations given above show that if

$$a = \langle \alpha_1, \alpha_2, \ldots, \alpha_n\rangle$$

then

$$a = \alpha_1 e_1 + \alpha_2 e_2 + \cdots + \alpha_n e_n,$$

and the result is established. The vectors $\{e_i\}$ are sometimes called *unit vectors.*

EXERCISES

1. Let $b_1, \ldots, b_r$ be a basis for a vector space V. Prove that $b_i \neq 0$ for $i = 1, \ldots, r$.
2. Prove from the definitions that any set of three vectors in R_2 is a linearly dependent set. (*Hint:* Use Example B to find a set of generators of R_2, and follow the argument of Example A.)
3. Let P_n be the set of all polynomial functions in $P(R)$ of the form $\alpha_0 + \alpha_1 x + \cdots + \alpha_n x^n$, for some fixed positive integer n. Show that P_n is a subspace of $P(R)$, and that $\{1, x, x^2, \ldots, x^n\}$ forms a basis for P_n.
4. Prove that the set of all $f \in C(R)$ such that $df/dt = 0$ is a one-dimensional subspace of $C(R)$. Can you generalize this result? For example, what is the dimension of the subspace consisting of all f such that $d^2f/dt^2 = 0$?
5. Let $a_1, \ldots, a_m$ be linearly independent vectors in V. Prove that $\alpha_1 a_1 + \cdots + \alpha_m a_m = \alpha_1' a_1 + \cdots + \alpha_m' a_m$ if and only if $\alpha_1 = \alpha_1', \ldots, \alpha_m = \alpha_m'$. In other words, the coefficients of a vector expressed as a linear combination of linearly independent vectors are uniquely determined.

6. ROW EQUIVALENCE OF MATRICES

In this section we introduce the main computational technique used to solve practically all problems about linear dependence and systems of linear equations. Let us start with an example.

Example A. *We shall find a basis for the subspace of R_4 generated by the vectors*

$$a = \langle -3, 2, 1, 4\rangle, \qquad b = \langle 4, 1, 0, 2\rangle, \qquad c = \langle -10, 3, 2, 6\rangle.$$

Letting e_1, e_2, e_3, e_4 be the basis of R_4 consisting of the unit vectors

$$e_1 = \langle 1, 0, 0, 0\rangle, \quad \ldots, \quad e_4 = \langle 0, 0, 0, 1\rangle,$$

we have, as in Example B of Section 5,

$$(*) \qquad \begin{aligned} a &= -3e_1 + 2e_2 + e_3 + 4e_4 \\ b &= 4e_1 + e_2 \phantom{{}+ e_3} + 2e_4 \\ c &= -10e_1 + 3e_2 + 2e_3 + 6e_4. \end{aligned}$$

It is not obvious whether a, b, c are linearly independent or not. The best way to find out, and to find a basis for $S(a, b, c)$, is to experiment with a, b, c by adding multiples of the vectors to each other, to arrive at a new set of generators of $S(a, b, c)$ for which the relations $(*)$ are as simple as possible. Some operations that lead to new sets of generators are the following:

(I) *Exchanging one vector for another* [for example, $S(b, a, c) = S(a, b, c)$, since every linear combination of the vectors $\{b, a, c\}$ is certainly a linear combination of the vectors a, b, c and conversely].

(II) *Replacing one vector by the sum of that vector and a multiple of another vector by a scalar.* For example, $S(a - 2b, b, c) = S(a, b, c)$. To check this statement, every linear combination of $\{a - 2b, b, c\}$ is certainly a linear combination of $\{a, b, c\}$ (why?). Therefore $S(a - 2b, b, c) \subset S(a, b, c)$. Conversely, suppose

$$x = \lambda a + \mu b + \nu c$$

is a linear combination of $\{a, b, c\}$. Since

$$a = (a - 2b) + 2b,$$

we have

$$x = \lambda(a - 2b) + (2\lambda + \mu)b + \nu c,$$

and we have shown that $x \in S(a - 2b, b, c)$.

The next thing to observe is that these operations on vectors really involve only the coefficients in the equations $(*)$.

A good way to visualize the situation is to introduce the concept of a matrix.

(6.1) DEFINITION. An ordered set $\{r_1, \ldots, r_m\}$ of m vectors from R_n is called an *m-by-n matrix*. The vectors $\{r_1, \ldots, r_m\}$ are called the *rows* of the matrix.

We shall use the following notation for matrices. In Example A we are interested in the vectors $r_1 = \langle -3, 2, 1, 4\rangle$, $r_2 = \langle 4, 1, 0, 2\rangle$ and $r_3 = \langle -10, 3, 2, 6\rangle$. The 3-by-4 matrix with rows r_1, r_2, r_3 is denoted by

$$\begin{pmatrix} -3 & 2 & 1 & 4 \\ 4 & 1 & 0 & 2 \\ -10 & 3 & 2 & 6 \end{pmatrix}. \tag{6.2}$$

In general, the rows $\{r_1, r_2, r_3\}$ of a 3-by-4 matrix will be given as follows:

$$r_1 = \langle \alpha_{11}, \alpha_{12}, \alpha_{13}, \alpha_{14}\rangle,$$
$$r_2 = \langle \alpha_{21}, \alpha_{22}, \alpha_{23}, \alpha_{24}\rangle,$$
$$r_3 = \langle \alpha_{31}, \alpha_{32}, \alpha_{33}, \alpha_{34}\rangle.$$

The numbers $\{\alpha_{ij}\}$ are called the *entries* of the matrix. The notation is chosen so that the entry α_{ij} is the jth entry in the ith row. For example, if we apply this notation to the matrix (6.2), $\alpha_{11} = -3$, $\alpha_{21} = 4$, $\alpha_{33} = 2$, $\alpha_{14} = 4$, etc.

Returning to our discussion, the operation of exchanging two vectors simply amounts to exchanging two rows of the matrix (6.2); it is indicated as follows:

$$\begin{pmatrix} -3 & 2 & 1 & 4 \\ 4 & 1 & 0 & 2 \\ -10 & 3 & 2 & 6 \end{pmatrix} \overset{\text{I}}{\sim} \begin{pmatrix} 4 & 1 & 0 & 2 \\ -3 & 2 & 1 & 4 \\ -10 & 3 & 2 & 6 \end{pmatrix}.$$

Replacing a by $a - 2b$ is described by the notation

$$\begin{pmatrix} -3 & 2 & 1 & 4 \\ 4 & 1 & 0 & 2 \\ -10 & 3 & 2 & 6 \end{pmatrix} \overset{\text{II}}{\sim} \begin{pmatrix} -11 & 0 & 1 & 0 \\ 4 & 1 & 0 & 2 \\ -10 & 3 & 2 & 6 \end{pmatrix}.$$

We shall call these operations of types I and II *elementary row operations* on the matrix and write

$$\begin{pmatrix} -3 & 2 & 1 & 4 \\ 4 & 1 & 0 & 2 \\ -10 & 3 & 2 & 6 \end{pmatrix} \sim \begin{pmatrix} \alpha_{11} & \alpha_{12} & \alpha_{13} & \alpha_{14} \\ \alpha_{21} & \alpha_{22} & \alpha_{23} & \alpha_{24} \\ \alpha_{31} & \alpha_{32} & \alpha_{33} & \alpha_{34} \end{pmatrix} \tag{6.3}$$

to mean that the matrix

$$\mathbf{A}' = \begin{pmatrix} \alpha_{11} & \alpha_{12} & \alpha_{13} & \alpha_{14} \\ \alpha_{21} & \alpha_{22} & \alpha_{23} & \alpha_{24} \\ \alpha_{31} & \alpha_{32} & \alpha_{33} & \alpha_{34} \end{pmatrix}$$

is obtained by the first matrix $\mathbf{A}$ in the following way. There exist a finite number of matrices $\mathbf{A} = \mathbf{A}_1, \mathbf{A}_2, \ldots, \mathbf{A}_s = \mathbf{A}'$, all of the same size as A, such that for each i, $2 \leq i \leq s$, A_i is obtained from A_{i-1} by an elementary row operation.† Since these elementary row operations leave the subspace corresponding to the rows unchanged, we conclude that if (6.3) holds, then the subspace $S(a', b', c')$ is equal to $S(a, b, c)$, where

$$\begin{aligned} a' &= \alpha_{11}e_1 + \alpha_{12}e_2 + \alpha_{13}e_3 + \alpha_{14}e_4 \\ b' &= \alpha_{21}e_1 + \alpha_{22}e_2 + \alpha_{23}e_3 + \alpha_{24}e_4 \\ c' &= \alpha_{31}e_1 + \alpha_{32}e_2 + \alpha_{33}e_3 + \alpha_{34}e_4, \end{aligned} \tag{6.4}$$

and $\{a, b, c\}$ are the vectors described by the rows of the original matrix

$$\begin{pmatrix} -3 & 2 & 1 & 4 \\ 4 & 1 & 0 & 2 \\ -10 & 3 & 2 & 6 \end{pmatrix}.$$

This means that all we have to do to find a basis for $S(a, b, c)$ is to apply elementary row operations to the matrix (6.2) until we find vectors $\{a', b', c'\}$ as in (6.4) where it is easy to test for linear independence of the nonzero vectors obtained.

One way to do this is called *gaussian elimination;* we apply elementary row operations to eliminate (or replace by zero) the coefficients of e_1 in all but one of the vectors, and then proceed to eliminate coefficients of the next basis vector involved, etc.

In our example we shall eliminate the coefficients of e_1 in the second and third rows. By replacing the second row by the second row plus $\frac{4}{3}$ of the first row, we have

$$\begin{pmatrix} -3 & 2 & 1 & 4 \\ 4 & 1 & 0 & 2 \\ -10 & 3 & 2 & 6 \end{pmatrix} \overset{\text{II}}{\sim} \begin{pmatrix} -3 & 2 & 1 & 4 \\ 0 & \frac{11}{3} & \frac{4}{3} & \frac{22}{3} \\ -10 & 3 & 2 & 6 \end{pmatrix}.$$

In the second matrix we replace the third row by the third row minus $\frac{10}{3}$ times the first row, giving

$$\begin{pmatrix} -3 & 2 & 1 & 4 \\ 0 & \frac{11}{3} & \frac{4}{3} & \frac{22}{3} \\ -10 & 3 & 2 & 6 \end{pmatrix} \overset{\text{II}}{\sim} \begin{pmatrix} -3 & 2 & 1 & 4 \\ 0 & \frac{11}{3} & \frac{4}{3} & \frac{22}{3} \\ 0 & \frac{-11}{3} & \frac{-4}{3} & \frac{-22}{3} \end{pmatrix}.$$

Now we can simply replace the third row by the third row plus the second row, obtaining

$$\begin{pmatrix} -3 & 2 & 1 & 4 \\ 0 & \frac{11}{3} & \frac{4}{3} & \frac{22}{3} \\ 0 & \frac{-11}{3} & \frac{-4}{3} & \frac{-22}{3} \end{pmatrix} \overset{\text{II}}{\sim} \begin{pmatrix} -3 & 2 & 1 & 4 \\ 0 & \frac{11}{3} & \frac{4}{3} & \frac{22}{3} \\ 0 & 0 & 0 & 0 \end{pmatrix}.$$

† Later we shall introduce a third type of elementary row operation, but it is not needed in this example.

Putting our work down in a more economical form (which the reader should use in doing problems of this sort), we have

$$\begin{pmatrix} -3 & 2 & 1 & 4 \\ 4 & 1 & 0 & 2 \\ -10 & 3 & 2 & 6 \end{pmatrix} \overset{\text{II}}{\sim} \begin{pmatrix} -3 & 2 & 1 & 4 \\ 0 & \frac{11}{3} & \frac{4}{3} & \frac{22}{3} \\ -10 & 3 & 2 & 6 \end{pmatrix}$$

$$\overset{\text{II}}{\sim} \begin{pmatrix} -3 & 2 & 1 & 4 \\ 0 & \frac{11}{3} & \frac{4}{3} & \frac{22}{3} \\ 0 & \frac{-11}{3} & \frac{-4}{3} & \frac{-22}{3} \end{pmatrix}$$

$$\overset{\text{II}}{\sim} \begin{pmatrix} -3 & 2 & 1 & 4 \\ 0 & \frac{11}{3} & \frac{4}{3} & \frac{22}{3} \\ 0 & 0 & 0 & 0 \end{pmatrix}.$$

From this we conclude that

$$S(a, b, c) = S(a', b')$$

where

$$\text{(6.5)} \qquad \begin{aligned} a' &= -3e_1 + 2e_2 + e_3 + 4e_4 \\ b' &= \tfrac{11}{3}e_2 + \tfrac{4}{3}e_3 + \tfrac{22}{3}e_4. \end{aligned}$$

It is now easy to check that a' and b' are linearly independent. In fact, suppose

$$\text{(6.6)} \qquad \lambda a' + \mu b' = 0.$$

Then, upon collecting the coefficients of e_1, e_2, e_3, e_4, we have

$$-3\lambda e_1 + (\cdots)e_2 + (\cdots)e_3 + (\cdots)e_4 = 0.$$

It follows that $\lambda = 0$, since $\{e_1, e_2, e_3, e_4\}$ are linearly independent. The equation (6.6) now becomes

$$\mu b' = 0,$$

and we have $\mu = 0$ since $b' \neq 0$.

We conclude that the vectors a' and b' form a basis for the subspace $S(a, b, c)$, because they are linearly independent, and generate the subspace.

A second application of this technique is to test vectors for linear dependence. In this case we can conclude that the original vectors $\{a, b, c\}$ are linearly dependent. Why? Because if they were linearly independent, the subspace $S(a, b, c)$ would have a basis of three vectors and another basis of two vectors, which is impossible by Theorem (5.3).

The problem of finding a relation of linear dependence among $\{a, b, c\}$ is not too hard to do at this point, but it will be postponed until

Section 9, since it is equivalent to the general problem of finding solutions of systems of homogeneous equations.

From this example we have already gained a great deal of insight into how subspaces of vector spaces are generated. We shall conclude this section by summarizing our results in the general situation. In studying the material, the reader will probably find it helpful to look at Examples A or C (at the end of the section) to see illustrations of the various points in the discussion.

Suppose V is a vector space over a field F, with a finite basis $\{a_1, \ldots, a_n\}$. Suppose $\{b_1, \ldots, b_m\}$ are vectors belonging to V, such that

$$\begin{aligned} b_1 &= \lambda_{11}a_1 + \lambda_{12}a_2 + \cdots + \lambda_{1n}a_n \\ b_2 &= \lambda_{21}a_1 + \lambda_{22}a_2 + \cdots + \lambda_{2n}a_n \\ &\cdots\cdots\cdots\cdots\cdots\cdots\cdots\cdots \\ b_m &= \lambda_{m1}a_1 + \lambda_{m2}a_2 + \cdots + \lambda_{mn}a_n. \end{aligned} \tag{6.7}$$

The *coefficient matrix* of the system of equations (6.7) is the matrix (see the definition given earlier in this section)

$$\mathbf{A} = \begin{pmatrix} \lambda_{11} & \lambda_{12} & \cdots & \lambda_{1n} \\ \lambda_{21} & \lambda_{22} & \cdots & \lambda_{2n} \\ \cdots & \cdots & \cdots & \cdots \\ \lambda_{m1} & \lambda_{m2} & \cdots & \lambda_{mn} \end{pmatrix}, \tag{6.8}$$

whose rows are the vectors in F_n,

$$\begin{aligned} r_1 &= \langle \lambda_{11}, \lambda_{12}, \ldots, \lambda_{1n} \rangle \qquad \text{(first row)} \\ &\cdots\cdots\cdots\cdots\cdots\cdots \\ r_m &= \langle \lambda_{m1}, \lambda_{m2}, \ldots, \lambda_{mn} \rangle \qquad (m\text{th row}) \end{aligned}$$

The *columns* of **A** are defined to be the vectors in F_m,

$$c_1 = \begin{pmatrix} \lambda_{11} \\ \lambda_{21} \\ \vdots \\ \lambda_{m1} \end{pmatrix}, \quad \ldots, \quad c_n = \begin{pmatrix} \lambda_{1n} \\ \lambda_{2n} \\ \vdots \\ \lambda_{mn} \end{pmatrix}$$

or we could write the first column as

$$c_1 = \langle \lambda_{11}, \lambda_{21}, \ldots, \lambda_{m1} \rangle;$$

the definition of m-tuple leaves some flexibility as to how the coefficients are displayed. We recall that the matrix **A** is called an *m-by-n matrix*, and the scalars $\{\lambda_{ij}\}$ are called the *entries* of **A**. The entry λ_{ij} belongs to the ith row and the jth column of **A**, and represents the coefficient of a_j in the expression of b_i as a linear combination of the basis vectors, in (6.7). As in the case of R_n, we use the notation $\mathbf{A} = 0$ to mean that all the entries of **A** are zero.

Example B. Let

$$\mathbf{A} = \begin{pmatrix} 2 & 1 & 0 \\ -3 & 1 & 2 \\ 1 & 1 & -1 \\ 2 & 1 & 4 \end{pmatrix} = \begin{pmatrix} \lambda_{11} & \lambda_{12} & \lambda_{13} \\ \lambda_{21} & \lambda_{22} & \lambda_{23} \\ \lambda_{31} & \lambda_{32} & \lambda_{33} \\ \lambda_{41} & \lambda_{42} & \lambda_{43} \end{pmatrix}.$$

Then **A** is a 4-by-3 matrix with rows

$$\begin{aligned} r_1 &= \langle 2, 1, 0\rangle \\ r_2 &= \langle -3, 1, 2\rangle \\ r_3 &= \langle 1, 1, -1\rangle \\ r_4 &= \langle 2, 1, 4\rangle. \end{aligned}$$

The element in the third row, first column is 1, for example. What is λ_{22}? λ_{33}? λ_{13}?

(6.9) DEFINITION. Let **A** be an *m-by-n* matrix with coefficients $\{\lambda_{ij}\}$. An *elementary row operation* applied to **A** is a rule that produces another *m-by-n* matrix from **A** by one of the following types of operations:

I. Interchanging two rows of **A**.
II. Replacing the ith row r_i of **A** by $r_i + \lambda r_j$, for some other row $r_j, j \neq i$, and $\lambda \in F$.
III. Replacing the ith row r_i of **A** by μr_i for some $\mu \neq 0$.

(6.10) DEFINITION. Let **A** be an *m-by-n* matrix. An *m-by-n* matrix **A**$'$ is said to be *row equivalent* to **A** if there exist *m-by-n* matrices $\mathbf{A}_0$, $\mathbf{A}_1, \ldots, \mathbf{A}_s$, such that

$$\mathbf{A}_0 = \mathbf{A}, \qquad \mathbf{A}_s = \mathbf{A}',$$

and for each $i, 1 \leq i \leq s$, $\mathbf{A}_i$ is obtained from $\mathbf{A}_{i-1}$ by applying an elementary row operation of type I, II, or III to $\mathbf{A}_{i-1}$. If **A**$'$ is row equivalent to **A**, we shall write $\mathbf{A}' \sim \mathbf{A}$.

The relation of row equivalence, $\mathbf{A}' \sim \mathbf{A}$, has the following properties (see the exercises):

$$\begin{aligned} &\mathbf{A} \sim \mathbf{A}; \\ &\mathbf{A} \sim \mathbf{A}' \text{ implies } \mathbf{A}' \sim \mathbf{A}; \\ &\mathbf{A} \sim \mathbf{A}' \text{ and } \mathbf{A}' \sim \mathbf{A}'' \text{ imply } \mathbf{A} \sim \mathbf{A}''. \end{aligned}$$

(6.11) THEOREM. *Let V be a vector space with a finite basis $\{a_1, \ldots, a_n\}$. Let $\{b_1, \ldots, b_m\}$ be vectors in V, and let*

$$\begin{aligned} b_1 &= \lambda_{11}a_1 + \cdots + \lambda_{1n}a_n \\ &\cdots\cdots\cdots\cdots\cdots\cdots\cdots\cdots . \\ b_m &= \lambda_{m1}a_1 + \cdots + \lambda_{mn}a_n \end{aligned}$$

Let

$$\mathbf{A}' = \begin{pmatrix} \lambda'_{11} & \cdots & \lambda'_{1n} \\ \cdots & \cdots & \cdots \\ \lambda'_{m1} & \cdots & \lambda'_{mn} \end{pmatrix}$$

be an m-by-n matrix which is row equivalent to the coefficient matrix.

$$\mathbf{A} = \begin{pmatrix} \lambda_{11} & \cdots & \lambda_{1n} \\ \cdots & \cdots & \cdots \\ \lambda_{m1} & \cdots & \lambda_{mn} \end{pmatrix}$$

of the $\{b_i\}$ *in terms of the basis vectors* $\{a_j\}$. *Then*

$$S(b_1, \ldots, b_m) = S(b'_1, \ldots, b'_m),$$

where

$$\begin{aligned} b'_1 &= \lambda'_{11}a_1 + \cdots + \lambda'_{1n}a_n \\ &\cdots\cdots\cdots\cdots\cdots\cdots \\ b'_m &= \lambda'_{m1}a_1 + \cdots + \lambda'_{mn}a_n. \end{aligned}$$

Proof. It is sufficient to prove the result for the case where $\mathbf{A}'$ is obtained from $\mathbf{A}$ by one elementary row operation. If the row operation is of type I or II, the conclusion of the theorem was checked in Example A, and we shall not repeat the details. Suppose now that the elementary row operation is of type III, so that the ith row r'_i of $\mathbf{A}'$ is given by

$$r'_i = \mu r_i,$$

for some $\mu \neq 0$. We have to prove that the subspaces generated by the vectors $\{b_1, \ldots, b_m\}$ and $\{b_1, \ldots, b_{i-1}, \mu b_i, b_{i+1}, \ldots, b_m\}$ are the same. To show that

(6.12) $$S(b_1, \ldots, b_m) \subset S(b_1, \ldots, b_{i-1}, \mu b_i, b_{i+1}, \ldots, b_m),$$

let

$$u = \beta_1 b_1 + \cdots + \beta_m b_m \in S(b_1, \ldots, b_m);$$

since $\mu \neq 0$, we can write

$$u = \beta_1 b_1 + \cdots + \beta_{i-1}b_{i-1} + \beta_i\mu^{-1}(\mu b_i) + \beta_{i+1}b_{i+1} + \cdots.$$

This proves the inclusion. The reverse inclusion is left to the reader.

We now come to the idea used in Example A for eliminating coefficients of $e_1, e_2, \ldots$ successively.

(6.13) Definition. An ordered collection of vectors $\{b_1, \ldots, b_p\}$ in F_n is said to be in *echelon form* if each $b_i \neq 0$ and if the position of the first nonzero entry in b_i is to the left of the position of the first nonzero entry in b_{i+1}, for $i = 1, 2, \ldots, p-1$. For example,

$$\langle 1, 0, -2, 3\rangle, \quad \langle 0, 1, 2, 0\rangle, \quad \langle 0, 0, 0, 1\rangle$$

are in echelon form, while

$$\langle 0, 1, 0, 0\rangle, \quad \langle 1, 0, 0, 0\rangle$$

are not.

The point of having vectors in echelon form is the following general fact, which was anticipated in Example A.

(6.14) LEMMA. *Let* **A** *be the coefficient matrix*

$$\begin{pmatrix} \lambda_{11} & \lambda_{12} & \cdots & \lambda_{1n} \\ \cdots & \cdots & \cdots & \cdots \\ \lambda_{m1} & \lambda_{m2} & \cdots & \lambda_{mn} \end{pmatrix}$$

of a set of vectors $\{b_1, \ldots, b_m\}$ *in terms of a set of basis vectors* $\{a_1, \ldots, a_n\}$ *of a vector space* V [*as in* (6.7)]. *Suppose that the rows* $r_1, \ldots, r_m$ *of the matrix* **A** *are in echelon form. Then the vectors* $\{b_1, \ldots, b_m\}$ *are linearly independent.*

Proof. We shall use induction on m. If $m = 1$, then by Definition (6.13), $b_1 \neq 0$, and $\{b_1\}$ is a linearly independent set. As an induction hypothesis, we assume that $m > 1$, and that the result holds for a matrix with $m - 1$ rows. Now let **A** be as in the statement of the lemma, and suppose that

(6.15) $$\beta_1 b_1 + \cdots + \beta_m b_m = 0, \qquad \beta_i \in F.$$

We show that $\beta_1 = 0$. Let λ_{1i} be the first nonzero entry of r_1. Then by the definition of echelon form, when (6.15) is expressed as a linear combination of the basis vectors $\{a_1, \ldots, a_n\}$, the coefficient of a_i is $\lambda_{1i}\beta_1$, which must be equal to zero. Since $\lambda_{1i} \neq 0$, we have $\beta_1 = 0$. Now consider the vectors $\{b_2, \ldots, b_m\}$. The coefficient matrix of these vectors has $m - 1$ rows, which are still in echelon form (why?). The relation (6.15) now has the form

$$\beta_2 b_2 + \cdots + \beta_m b_m,$$

and applying our induction hypothesis we have

$$\beta_2 = \cdots = \beta_m = 0.$$

This completes the proof of the lemma.

We come now to the main result of the section.

(6.16) THEOREM. *Let* **A** *be the coefficient matrix of a system of equations (6.7), expressing a set of vectors* $\{b_1, \ldots, b_m\}$ *as linear combinations of a given set of basis vectors of V. Then the following statements hold.*

(*i*) *There exists a matrix* $\mathbf{A}'$ *row equivalent to* **A**, *such that either* $\mathbf{A}' = 0$, *or there is a uniquely determined positive integer k, with* $1 \leq k \leq m$, *such that the first k rows of* $\mathbf{A}'$ *are in echelon form, and the remaining rows are all zero.*

(*ii*) *The vectors* $\{b_1', \ldots, b_k'\}$ *corresponding to the first k rows of* A' *form a basis for* $S(b_1, \ldots, b_k)$

(*iii*) *The original vectors* $\{b_1, \ldots, b_m\}$ *are linearly independent if and only if* $m = k$.

Proof. We first prove that if $\mathbf{A} \neq 0$, there exists a matrix $\mathbf{A}' \sim \mathbf{A}$ such that the first k rows of $\mathbf{A}'$ are in echelon form and the remaining rows are zero. The proof is by induction on the number of rows of **A**. If this number is one, then the result is clearly true, since a single nonzero vector is already in echelon form. Suppose now that $\mathbf{A} \neq 0$, and has $m > 1$ rows. By interchanging two rows, we may assume that the first row r_1 of **A** is different from zero, and has a nonzero entry as far to the left as possible. By adding multiples of the first row to the remaining rows, we obtain a matrix row equivalent to **A** of the following form:

$$\mathbf{A}_1 = \begin{pmatrix} 0 & \cdots & 0 & \lambda_{1i} & \cdots \\ 0 & \cdots & 0 & 0 & \cdots \\ \cdots & \cdots & \cdots & \cdots & \cdots \\ 0 & \cdots & 0 & 0 & \cdots \end{pmatrix}, \tag{6.17}$$

with $\lambda_{1i} \neq 0$. By the induction hypothesis, we may apply elementary row operations to the matrix consisting of the last $m - 1$ rows of $\mathbf{A}_1$, to obtain a matrix

$$\mathbf{A}' = \begin{pmatrix} 0 & \cdots & 0 & \lambda_{1i} & \cdots \\ 0 & \cdots & 0 & 0 & \cdots \\ 0 & \cdots & 0 & 0 & \cdots \end{pmatrix},$$

such that either all but the first row of $\mathbf{A}'$ consists of zeros, or rows $\{r_2', \ldots, r_k'\}$ of $\mathbf{A}'$ are in echelon form and the remaining rows are zero. It is then clear from the definition of echelon form that rows $\{r_1, \ldots, r_k\}$ of $\mathbf{A}'$ are in echelon form, and the first statement is proved.

Let $\{b_1', \ldots, b_k'\}$ be the vectors corresponding to the first k rows of $\mathbf{A}'$. By Theorem (6.11),

$$S(b_1, \ldots, b_m) = S(b_1', \ldots, b_k'),$$

and by Lemma (6.14), the vectors $\{b_1', \ldots, b_k'\}$ are linearly independent.

Therefore $\{b_1', \ldots, b_k'\}$ is a basis for $S(b_1, \ldots, b_m)$. Moreover, any other matrix satisfying the conditions satisfied by $\mathbf{A}'$ will have the property that its nonzero rows give a basis for $S(b_1, \ldots, b_m)$. It follows from Theorem (5.3) that the number of nonzero rows of $\mathbf{A}'$ is uniquely determined. At this point we have proved statements (*i*) and (*ii*) of the Theorem. Statement (*iii*) follows, as in Example A, by another application of Theorem (5.3). In fact, the vectors $\{b_1, \ldots, b_m\}$ are linearly independent if and only if m is the number of basis vectors of $S(b_1, \ldots, b_m)$. But the number of basis vectors in a basis of $S(b_1, \ldots, b_m)$ is equal to the number of nonzero rows in $\mathbf{A}'$, which is k. This completes the proof of the theorem.

Example C. We give one more example of how the techniques in this section are used.

Find a basis for the subspace of R_3 *generated by the vectors* $\langle 1, 3, 4\rangle$, $\langle 4, 0, 1\rangle$, $\langle 3, 1, 2\rangle$. *Test the vectors for linear dependence.*

Following our procedure, we express the vectors in terms of a basis of R_3, consisting of the unit vectors $\{e_1, e_2, e_3\}$:

$$\begin{aligned} b_1 &= \langle 1, 3, 4\rangle = e_1 + 3e_2 + 4e_3 \\ b_2 &= \langle 4, 0, 1\rangle = 4e_1 \qquad + e_3 \\ b_3 &= \langle 3, 1, 2\rangle = 3e_1 + e_2 + 2e_3. \end{aligned}$$

The matrix of coefficients is

$$\mathbf{A} = \begin{pmatrix} 1 & 3 & 4 \\ 4 & 0 & 1 \\ 3 & 1 & 2 \end{pmatrix}.$$

We proceed to find a matrix $\mathbf{A}' \sim \mathbf{A}$ whose first k-rows are in echelon form.

$$\mathbf{A} \overset{\text{II}}{\sim} \begin{pmatrix} 1 & 3 & 4 \\ 0 & -12 & -15 \\ 3 & 1 & 2 \end{pmatrix} \overset{\text{II}}{\sim} \begin{pmatrix} 1 & 3 & 4 \\ 0 & -12 & -15 \\ 0 & -8 & -10 \end{pmatrix}$$

$$\overset{\text{III}}{\sim} \begin{pmatrix} 1 & 3 & 4 \\ 0 & 1 & \frac{5}{4} \\ 0 & -8 & -10 \end{pmatrix} \overset{\text{II}}{\sim} \begin{pmatrix} 1 & 3 & 4 \\ 0 & 1 & \frac{5}{4} \\ 0 & 0 & 0 \end{pmatrix}.$$

Notice how the elementary row operation of type III was used to make the first nonzero entry in the second row equal to one. This step, while not essential, will often simplify the calculations.

Applying Theorem (6.16), we conclude that the vectors

$$\begin{aligned} b_1' &= e_1 + 3e_2 + 4e_3 \\ b_2' &= \qquad e_2 + \tfrac{5}{4}e_3 \end{aligned}$$

form a basis for $S(b_1, b_2, b_3)$, and that the original vectors $\{b_1, b_2, b_3\}$ are linearly dependent.

EXERCISES

1. Find matrices row equivalent to the following matrices, whose rows are in echelon form.

a. $\begin{pmatrix} 1 & 2 & 1 & 0 \\ 2 & 3 & 0 & -1 \\ 1 & 2 & 1 & -1 \end{pmatrix}$ **b.** $\begin{pmatrix} 1 & 1 & 2 \\ 0 & 2 & -1 \\ 1 & 2 & 4 \end{pmatrix}$

c. $\begin{pmatrix} 2 & -1 & 1 & 2 \\ 3 & 0 & 1 & 1 \\ 4 & 3 & -1 & 0 \end{pmatrix}$ **d.** $\begin{pmatrix} 1 & 1 & 1 \\ 0 & 1 & 1 \\ 1 & 1 & 0 \end{pmatrix}$.

2. Test the following sets of vectors for linear dependence. Assume the vectors all belong to R_n for the appropriate n.
   a. $\langle -1, 1\rangle, \langle 1, 2\rangle, \langle 1, 3\rangle$.
   b. $\langle 2, 1\rangle, \langle 1, 0\rangle, \langle -2, 1\rangle$.
   c. $\langle 1, 4, 3\rangle\ \langle 3, 0, 1\rangle, \langle 4, 1, 2\rangle$.
   d. $\langle 1, 1, 2\rangle, \langle 2, 1, 3\rangle, \langle 4, 0, -1\rangle, \langle -1, 0, 1\rangle$.
   e. $\langle 0, 1, 1, 2\rangle, \langle 3, 1, 5, 2\rangle, \langle -2, 1, 0, 1\rangle, \langle 1, 0, 3, -1\rangle$.
   f. $\langle 1, 1, 0, 0, 1\rangle, \langle -1, 1, 1, 0, 0\rangle, \langle 2, 1, 0, 1, 1\rangle, \langle 0, -1, -1, -1, 0\rangle$.
3. Find bases in echelon form for the vector spaces with the sets of vectors as generators given in parts (a) to (f) of Exercise 2.
4. Show that the relation of row equivalence of matrices has the properties
   a. $\mathbf{A} \sim \mathbf{A}$
   b. $\mathbf{A} \sim \mathbf{B}$ implies $\mathbf{B} \sim \mathbf{A}$
   c. $\mathbf{A} \sim \mathbf{B}$ and $\mathbf{B} \sim \mathbf{C}$ imply $\mathbf{A} \sim \mathbf{C}$.
5. Determine whether or not the following pairs of matrices are row equivalent.

a. $\begin{pmatrix} 1 & 1 & 0 & 1 \\ -1 & 2 & 1 & 0 \\ 1 & 4 & 1 & 2 \end{pmatrix}, \quad \begin{pmatrix} 0 & 6 & 2 & 2 \\ 0 & 3 & 1 & 1 \\ 2 & 2 & 0 & 2 \end{pmatrix}$.

b. $\begin{pmatrix} 1 & 1 & 0 & 1 \\ -1 & 2 & 1 & 0 \\ 1 & 4 & 1 & 2 \end{pmatrix}, \quad \begin{pmatrix} -2 & 1 & 1 & -1 \\ 0 & 3 & 1 & 1 \\ 1 & 0 & 1 & 1 \end{pmatrix}$.

6. Test some sets of polynomials in $P(R)$ for linear dependence, etc.

7. SOME GENERAL THEOREMS ABOUT FINITELY GENERATED VECTOR SPACES

In this section we prove some additional theorems about vector spaces that will be essential for the theory of systems of linear equations, and for other applications later in the book.

(7.1) LEMMA. *If* $\{a_1, \ldots, a_m\}$ *is linearly dependent and if* $\{a_1, \ldots, a_{m-1}\}$ *is linearly independent, then* a_m *is a linear combination of* $a_1, \ldots, a_{m-1}$.

Proof. By the hypothesis, we have

$$\lambda_1 a_1 + \cdots + \lambda_m a_m = 0,$$

where some $\lambda_i \neq 0$. If $\lambda_m = 0$, then some $\lambda_i \neq 0$ for $1 \leq i \leq m - 1$, and the equation of linear dependence becomes

$$\lambda_1 a_1 + \cdots + \lambda_{m-1} a_{m-1} = 0,$$

contrary to the assumption that $\{a_1, \ldots, a_{m-1}\}$ is linearly independent. Therefore $\lambda_m \neq 0$, and we have

$$a_m = \lambda_m^{-1}[(-\lambda_1)a_1 + \cdots + (-\lambda_{m-1})a_{m-1}]$$
$$\sum_{i=1}^{m-1} (-\lambda_m^{-1}\lambda_i)a_i$$

as we wished to prove.

To illustrate (7.1), let $a = \langle 0, 1\rangle$, $b = \langle 1, -1\rangle$, $c = \langle -2, 2\rangle$ in R_2. Then a, b, c are linearly dependent (why?). Moreover, a and b are linearly independent, and so by (7.1), c must be a linear combination of a and b. (Show that this is correct.) But b and c are not linearly independent, and in this case it is easily shown that a is *not* a linear combination of b and c. Therefore the hypothesis in (7.1) is essential

(7.2) THEOREM. *Every finitely generated vector space V has a basis.*

Proof. First we consider the case in which V consists of the zero vector alone. Then the zero vector spans V, but cannot be a basis (why?). In this case we agree that the empty set is a basis for V, so that the dimension of V is zero. Now let $V \neq \{0\}$ be a vector space with n generators. By Theorem (5.1) any set of $n + 1$ vectors in V is linearly dependent, and since a set consisting of a single nonzero vector is linearly independent, it follows that, for some integer $m \geq 1$, V contains linearly independent vectors $b_1, \ldots, b_m$ such that any set of $m + 1$ vectors in V is linearly dependent. We prove that $\{b_1, \ldots, b_m\}$ is a basis for V, and for this it is sufficient to show, for any vector $b \in V$, that $b \in S(b_1, \ldots, b_m)$. Because of the properties of the set $\{b_1, \ldots, b_m\}$, $\{b_1, \ldots, b_m, b\}$ is a linearly dependent set. Since $\{b_1, \ldots, b_m\}$ is linearly independent, Lemma (7.1) implies that $b \in S(b_1, \ldots, b_m)$, and the theorem is proved.

We have now shown that a finitely generated vector space is determined as soon as we know a basis for it—the space then consists of the set of all linear combinations of the basis vectors. The next two theorems give some idea of where to look for a basis if we know a set of generators for the vector space, or at least, some linearly independent vectors in it.

(7.3) THEOREM. *Let $V = S(a_1, \ldots, a_m)$ be a finitely generated vector space with generators $\{a_1, \ldots, a_m\}$. Then a basis for V can be selected from among the set of generators $\{a_1, \ldots, a_m\}$. In other words, a set of generators for a finitely generated vector space always contains a basis.*

Proof. The result is clear if $V = \{0\}$. Now let $V \neq \{0\}$. For some index r, where $1 \leq r \leq m$, we may assume, for a suitable ordering of $a_1, \ldots, a_m$, that $\{a_1, \ldots, a_r\}$ is linearly independent and that any larger set of the a_i's is linearly dependent. Then by Lemma (7.1), it follows that $a_{r+1}, \ldots, a_m$ all belong to $S(a_1, \ldots, a_r)$. Therefore $S(a_1, \ldots, a_m) = S(a_1, \ldots, a_r)$, and since $\{a_1, \ldots, a_r\}$ is linearly independent, we conclude that it is a basis of V.

(7.4) THEOREM. *Let $\{b_1, \ldots, b_q\}$ be a linearly independent set of vectors in a finitely generated vector space V. If $b_1, \ldots, b_q$ is not a basis of V, then there exist other vectors $b_{q+1}, \ldots, b_m$ in V such that $\{b_1, \ldots, b_m\}$ is a basis of V.*

Proof. By Theorem (7.2), V has a basis $\{a_1, \ldots, a_n\}$. By Theorem (5.1), we have $q \leq n$. If $q = n$, then Theorem (5.1) implies that for every i, the set

$$\{b_1, \ldots, b_n, a_i\}$$

is a linearly dependent set. By Lemma (7.1), $a_i \in S(b_1, \ldots, b_n)$, and it follows that

$$V = S(b_1, \ldots, b_n).$$

Therefore $\{b_1, \ldots, b_n\}$ is a basis of V, and there is nothing to prove in this case.

Next suppose $q = n - 1$. Then some $a_i \notin S(b_1, \ldots, b_q)$, and by Lemma (7.1) again, $\{b_1, \ldots, b_q, a_i\}$ is a linearly independent set of n elements. By the argument in the first paragraph, we conclude that $\{b_1, \ldots, b_q, a_i\}$ is a basis of V, and the theorem is proved in this case.

Now suppose $n - q > 1$, and by induction we can assume the theorem holds whenever the difference between dim V and the number of vectors $\{b_1, \ldots, b_q\}$ is less than $n - q - 1$. Since $n - q > 1$, some $a_i \notin S(b_1, \ldots, b_q)$. As before, $\{b_1, \ldots, b_q, a_i\}$ is a linearly independent set of $q + 1$ elements. Since the difference $n - (q + 1) = n - q - 1$, we can apply our induction hypothesis to supplement $\{b_1, \ldots, b_q, a_i\}$ by additional vectors to produce a basis. This completes the proof of the theorem.

We next consider certain operations on subspaces of vector space V that lead to new subspaces. If S, T are subspaces of vector space V, it follows at once, from the definition, that $S \cup T$ is not always a subspace, as the following example in R_2 shows (see Exercise 8 of Section 4):

$$S = \{\langle 0, \beta\rangle, \beta \in R\}, \qquad T = \{\langle \alpha, 0\rangle, \alpha \in R\}.$$

Because of this example, we define, for subspaces S and T of V,

$$S + T = \{s + t; s \in S, t \in T\}.$$

It is easy to show that $S + T$ is a subspace and is, in fact, the smallest subspace containing $S \cup T$. On the other hand, if S and T are subspaces, $S \cap T$ is always a subspace. It is natural to ask for the dimensions of $S \cap T$ and $S + T$, given the dimensions of S and T. The answer is provided by the following result, suggested by a counting procedure for finite sets: If A and B are finite sets, then the number of objects in $A \cup B$ is the sum of the numbers of objects in A and B, less the number of objects in the overlap $A \cap B$.

(7.5) THEOREM. *Let S and T be finitely generated subspaces of a vector space V. Then $S \cap T$ and $S + T$ are finitely generated subspaces, and we have*

$$\dim (S + T) + \dim (S \cap T) = \dim S + \dim T.$$

Proof. We shall give a sketch of the proof, leaving some of the details to the reader. Although it is not actually necessary, we consider first the special case $S \cap T = \{0\}$. Let $\{s_1, \ldots, s_d\}$ be a basis of S and $\{t_1, \ldots, t_e\}$ be a basis of T. Then it is easily checked that the vectors $s_1, \ldots, s_d$, $t_1, \ldots, t_e$ generate $S + T$, and (7.5) will follow in this case if we can show that these vectors are linearly independent. Suppose we have

$$\alpha_1 s_1 + \cdots + \alpha_d s_d + \beta_1 t_1 + \cdots + \beta_e t_e = 0.$$

Then

$$\alpha_1 s_1 + \cdots + \alpha_d s_d = -\beta_1 t - \cdots - \beta_e t_e \in S \cap T = \{0\}.$$

Since the s's and t's are linearly independent, we have $\alpha_1 = \cdots = \alpha_d = 0$ and $\beta_1 = \cdots = \beta_e = 0$.

Now suppose that $S \cap T \neq 0$. Then $S \cap T$ is finitely generated (why?). By Theorem (7.2), $S \cap T$ has a basis $\{u_1, \ldots, u_c\}$ where $c = \dim S \cap T$. By Theorem (7.4) we can find sets of vectors $\{v_1, \ldots, v_d\}$ and $\{w_1, \ldots, w_e\}$ such that

$$\{u_1, \ldots, u_c, v_1, \ldots, v_d\} \text{ is a basis for } S,$$
$$\{u_1, \ldots, u_c, w_1, \ldots, w_e\} \text{ is a basis for T}.$$

Then $S + T = S(u_1, \ldots, u_c, v_1, \ldots, v_d, w_1, \ldots, w_e)$, and Theorem (7.5) will be proved if we can show that these vectors are linearly independent. Suppose we have

(7.6)
$$\xi_1 u_1 + \cdots + \xi_c u_c + \lambda_1 v_1 + \cdots + \lambda_d v_d + \mu_1 w_1 + \cdots + \mu_e w_e = 0.$$

Then

$$\lambda_1 v_1 + \cdots + \lambda_d v_d = -(\sum \xi_i u_i) - (\sum \mu_i w_i) \in S \cap T$$

and hence there exist elements of F, $\zeta_1, \ldots, \zeta_c$, such that

$$\lambda_1 v_1 + \cdots + \lambda_d v_d = \zeta_1 u_1 + \cdots + \zeta_c u_c.$$

Since $\{v_1, \ldots, v_d, u_1, \ldots, u_c\}$ are linearly independent, we have $\lambda_1 = \cdots = \lambda_d = 0$. Then from (7.6) we have $\xi_1 = \cdots = \xi_c = \mu_1 = \cdots = \mu_e = 0$, and the proof is completed.

EXERCISES

1. Determine whether $\langle 1, 1, 1\rangle$ belongs to the subspace of R_3 generated by $\langle 1, 3, 4\rangle$, $\langle 4, 0, 1\rangle$, $\langle 3, 1, 2\rangle$. Explain your reasoning.
2. Determine whether $\langle 2, 0, -4, -2\rangle$ belongs to the subspace of R_4 generated by $\langle 0, 2, 1, -1\rangle$, $\langle 1, -1, 1, 0\rangle$, $\langle 2, 1, 0, -2\rangle$.
3. Prove that every subspace S of a finitely generated subspace T of a vector space V is finitely generated, and that $\dim S \leq \dim T$, with equality if and only if $S = T$.
4. Let S and T be two-dimensional subspaces of R_3. Prove that $\dim (S \cap T) \geq 1$.
5. Let

$$\begin{array}{lll} a_1 = \langle 2, 1, 0, -1\rangle & a_3 = \langle 1, -3, 2, 0\rangle & a_5 = \langle -2, 0, 6, 1\rangle \\ a_2 = \langle 4, 8, -4, -3\rangle & a_4 = \langle 1, 10, -6, -2\rangle & a_6 = \langle 3, -1, 2, 4\rangle \end{array}$$

and let

$$S = S(a_1, a_2, a_3, a_4), \qquad T = S(a_4, a_5, a_6).$$

Find $\dim S$, $\dim T$, $\dim (S + T)$, and, using Theorem (7.5), find $\dim (S \cap T)$.
6. Let F be the field of 2 elements,† and let V be a two-dimensional vector space over F. How many vectors are there in V? How many one-dimensional subspaces? How many different bases are there?

---

† See Exercise 4 of Section 2.

8. SYSTEMS OF LINEAR EQUATIONS

The rest of this chapter is devoted to the problem of solving systems of linear equations with real coefficients. This problem, as we have mentioned before, is one of the sources of linear algebra.

We shall describe a *system of m linear equations in n unknowns* by the notation

$$
\begin{aligned}
\alpha_{11}x_1 + \alpha_{12}x_2 + \cdots + \alpha_{1n}x_n &= \beta_1 \\
\alpha_{21}x_1 + \alpha_{22}x_2 + \cdots + \alpha_{2n}x_n &= \beta_2 \\
\cdots\cdots\cdots\cdots\cdots\cdots\cdots\cdots\cdots\cdots \\
\alpha_{m1}x_1 + \alpha_{m2}x_2 + \cdots + \alpha_{mn}x_n &= \beta_m
\end{aligned} \tag{8.1}
$$

where the α_{ij} and β_i are fixed real numbers and the $x_1, \ldots, x_n$ are the *unknowns*. The indexing is chosen such that for $1 \leq i \leq m$, the ith equation is

$$\alpha_{i1}x_1 + \cdots + \alpha_{in}x_n = \beta_i$$

where the first index appearing with α_{ij} stands for the equation in which α_{ij} appears and the second index j denotes the unknown x_j of which α_{ij} is the coefficient. Thus α_{21} is the coefficient of x_1 in the second equation, and so forth.

The matrix whose rows are the vectors

$$r_1 = \langle \alpha_{11}, \ldots, \alpha_{1n} \rangle, \quad \ldots, \quad r_m = \langle \alpha_{m1}, \ldots, \alpha_{mn} \rangle$$

is called the *coefficient matrix* of the system (8.1). (We recall that in Section 6 matrices were introduced in a slightly different situation.) As in Section 6, we use the notation

$$\begin{pmatrix} \alpha_{11} & \cdots & \alpha_{1n} \\ \alpha_{21} & \cdots & \alpha_{2n} \\ \cdots & \cdots & \cdots \\ \alpha_{m1} & \cdots & \alpha_{mn} \end{pmatrix}$$

for the matrix with rows $r_1, \ldots, r_m$ as above. The *entries* $\{\alpha_{ij}\}$ of the matrix are arranged in such a way that α_{ij} is the jth entry of the ith row. For example, α_{13} is the third entry in the first row, α_{21} the first entry in the second row, and so forth. We shall often use the more compact notation A or (α_{ij}) to denote matrices.

Although matrices were defined in terms of their rows, the notation suggests that with each m-by-n matrix we can associate two sets of vectors in the vector spaces R_m and R_n, respectively, namely, the *row vectors*

$$\{r_1, \ldots, r_m\} \subset R_n$$

where the ith row vector is

$$r_i = \langle \alpha_{i1}, \ldots, \alpha_{in} \rangle, \qquad 1 \le i \le m,$$

and the *column vectors*

$$\{c_1, \ldots, c_n\} \subset R_m$$

where the jth column vector is

$$c_j = \langle \alpha_{1j}, \alpha_{2j}, \ldots, \alpha_{mj} \rangle.$$

We should also notice that row and column vectors are special kinds of matrices, and we shall often write

$$\mathbf{r}_i = (\alpha_{i1}, \ldots, \alpha_{in}), \qquad \mathbf{c}_j = \begin{pmatrix} \alpha_{1j} \\ \alpha_{2j} \\ \vdots \\ \alpha_{mj} \end{pmatrix}.$$

The *row subspace* of the m-by-n matrix (α_{ij}) is the subspace $S(r_1, \ldots, r_m)$ of R_n, and the *column subspace* is the subspace $S(c_1, \ldots, c_n)$ of R_m.

A *solution* of the system (8.1) is an n-tuple of real numbers $\{\lambda_1, \ldots, \lambda_n\}$ such that

$$\begin{gathered} \alpha_{11}\lambda_1 + \cdots + \alpha_{1n}\lambda_n = \beta_1 \\ \cdots\cdots\cdots\cdots\cdots\cdots\cdots \\ \alpha_{m1}\lambda_1 + \cdots + \alpha_{mn}\lambda_n = \beta_m. \end{gathered}$$

In other words, the numbers $\{\lambda_i\}$ in the solution satisfy the equations (8.1) upon being substituted for the unknowns. We may identify a solution with a vector in R_n and may therefore speak of a *solution vector* of the system (8.1). Recalling the definition of the column vectors, we see that $x = \langle \lambda_1, \ldots, \lambda_n \rangle$ is a solution of the system (8.1) if and only if

(8.2) $$\lambda_1 c_1 + \lambda_2 c_2 + \cdots + \lambda_n c_n = b$$

where $b = \langle \beta_1, \ldots, \beta_n \rangle$, and we may describe the original system of equations by the more economical notation:

(8.3) $$x_1 c_1 + \cdots + x_n c_n = b.$$

A system of *homogeneous equations*, or a homogeneous system, is a system (8.3) in which the vector $b = 0$; if we allow the possibility $b \neq 0$, we speak of a *nonhomogeneous system*. If we have a homogeneous system,

(8.4) $$x_1 c_1 + \cdots + x_n c_n = 0.$$

then the zero vector $\langle 0, \ldots, 0 \rangle$ is always a solution vector, called the *trivial solution*. A solution different from $\langle 0, \ldots, 0 \rangle$ is called a *nontrivial solution*.

To clarify these concepts, consider the system

$$\begin{aligned} 3x_1 - x_2 + x_3 &= 1 \\ x_1 + x_2 - x_3 &= 2. \end{aligned}$$

The matrix of the system is

$$\begin{pmatrix} 3 & -1 & 1 \\ 1 & 1 & -1 \end{pmatrix};$$

the row vectors are

$$r_1 = \langle 3, -1, 1\rangle, \qquad r_2 = \langle 1, 1, -1\rangle;$$

and the column vectors are

$$c_1 = \langle 3, 1\rangle, \qquad c_2 = \langle -1, 1\rangle, \qquad c_3 = \langle 1, -1\rangle.$$

A solution vector is $\langle \frac{3}{4}, 0, -\frac{5}{4}\rangle$, as we check by substitution:

$$\begin{aligned} 3(\tfrac{3}{4}) - 0 - \tfrac{5}{4} &= 1 \\ \tfrac{3}{4} + 0 + \tfrac{5}{4} &= 2. \end{aligned}$$

In terms of the column vectors, the system is written

$$x_1c_1 + x_2c_2 + x_3c_3 = b,$$

where $b = \langle 1, 2\rangle$, and the solution vector $\langle \frac{3}{4}, 0, -\frac{5}{4}\rangle$ is expressed in the form

$$b = \tfrac{3}{4}c_1 - \tfrac{5}{4}c_3.$$

At this point the reader may wish to study the examples at the end of the section, before reading the proofs of the following theorems. The theorems explain in a general way what to expect from the solutions of a particular system of equations; the examples show how the solutions are actually found.

(8.5) THEOREM. *A nonhomogeneous system $x_1c_1 + \cdots + x_nc_n = b$ has a solution if and only if either of the following conditions is satisfied:*

1. *b belongs to the column space $S(c_1, \ldots, c_n)$.*
2. $\dim S(c_1, \ldots, c_n) = \dim S(c_1, \ldots, c_n, b)$.

Proof. The fact that the first statement is equivalent to the existence of a solution is immediate from (8.2). To prove that condition (2) is equivalent to condition (1), we proceed as follows. If $b \in S(c_1, \ldots, c_n)$, then clearly $S(c_1, \ldots, c_n, b) = S(c_1, \ldots, c_n)$ and condition (2) holds. Conversely, if condition (2) holds, then $b \in S(c_1, \ldots, c_n)$; otherwise b taken together with basis of $S(c_1, \ldots, c_n)$ would be a linearly independent set of $\dim S(c_1, \ldots, c_n) + 1$ elements, and we would contradict Theorem (5.1). This completes the proof of the theorem.

The result of Theorem (8.5) can be restated in a convenient way by using the following concepts.

(8.6) DEFINITION. Let (α_{ij}) be an m-by-n matrix with column vectors $\{c_1, \ldots, c_n\}$. The *rank* of the matrix is defined as the dimension of the column space $S(c_1, \ldots, c_n)$.

(8.7) DEFINITION. If $x_1c_1 + \cdots + x_nc_n = b$ is a nonhomogeneous system whose coefficient matrix has columns $c_1, \ldots, c_n$, then the m-by-$(n + 1)$ matrix with columns $\{c_1, \ldots, c_n, b\}$ is called the *augmented matrix* of the system.

The next result is immediate from our definitions and Theorem (8.5).

(8.8) THEOREM. *A nonhomogeneous system has a solution if and only if the rank of its coefficient matrix is equal to the rank of the augmented matrix.*

We have now settled in a theoretical way the question of whether a nonhomogeneous system has a solution or not. A method for computing the solutions in particular cases is given at the end of the section. If it does possess a solution, then we should ask to determine *all* solutions of the system. The key to this problem is furnished by the next theorem.

(8.9) THEOREM. *Suppose that a nonhomogeneous system*

(8.10)
$$x_1c_1 + \cdots + x_nc_n = b$$

has a solution x_0; then for all solutions x of the homogeneous system

(8.11)
$$x_1c_1 + \cdots + x_nc_n = 0,$$

$x_0 + x$ is a solution of (8.10) and all solutions of (8.10) can be expressed in this form.

Proof. Suppose first that $x = \langle\alpha_1, \ldots, \alpha_n\rangle$ is a solution of the homogeneous system and that $x_0 = \langle\alpha_1^{(0)}, \ldots, \alpha_n^{(0)}\rangle$ is a solution of (8.10). Then we have:

(8.12)
$$\alpha_1^{(0)}c_1 + \cdots + \alpha_n^{(0)}c_n = b$$

and

$$\alpha_1c_1 + \cdots + a_nc_n = 0.$$

Adding these equations, we obtain

$$(\alpha_1^{(0)} + \alpha_1)c_1 + \cdots + (\alpha_n^{(0)} + \alpha_n)c_n = b,$$

which asserts that $x_0 + x$ is a solution of (8.10).

Now let $y = \langle \beta_1, \ldots, \beta_n \rangle$ be an arbitrary solution of (8.10), so that we have

$$\beta_1 c_1 + \cdots + \beta_n c_n = b. \tag{8.13}$$

Subtracting (8.12) from (8.13), we obtain

$$(\beta_1 - \alpha_1^{(0)})c_1 + \cdots + (\beta_n - \alpha_n^{(0)})c_n = 0.$$

This asserts that $u = y - x_0$ is a solution of the homogeneous system, and we have

$$y = x_0 + u$$

as required. This completes the proof.

Thus the problem of finding all solutions of a nonhomogeneous system comes down to finding one solution of the nonhomogeneous system, and solving a homogeneous system. We shall show how to solve homogeneous systems in the next section.

We conclude this section with some examples of how to find solutions of nonhomogeneous systems, when they exist. We begin with the example discussed in Section 1.

Example A. Test the following system of equations for solvability, and find a solution if there is one:

$$\begin{aligned} x_1 + 2x_2 - 3x_3 + x_4 &= 1 \\ x_1 + x_2 + x_3 + x_4 &= 0. \end{aligned}$$

In this case, the matrix of the system is

$$\begin{pmatrix} 1 & 2 & -3 & 1 \\ 1 & 1 & 1 & 1 \end{pmatrix},$$

and the augmented matrix is

$$\begin{pmatrix} 1 & 2 & -3 & 1 & 1 \\ 1 & 1 & 1 & 1 & 0 \end{pmatrix}. \tag{8.14}$$

Rewriting the system in the form (8.3), we have

$$x_1c_1 + x_2c_2 + x_3c_3 + x_4c_4 = b,$$

where

$$c_1 = \langle 1, 1 \rangle, \quad c_2 = \langle 2, 1 \rangle, \quad c_3 = \langle -3, 1 \rangle, \quad c_4 = \langle 1, 1 \rangle, \quad b = \langle 1, 0 \rangle.$$

The rank of the matrix is

$$\dim S(c_1, c_2, c_3, c_4)$$

which is easily seen to be 2, since the vectors c_i are all contained in R_2; and putting the matrix

$$\begin{pmatrix} 1 & 1 \\ 2 & 1 \\ -3 & 1 \\ 1 & 1 \end{pmatrix}$$

in echelon form gives

$$\begin{pmatrix} 1 & 1 \\ 0 & -1 \\ 0 & 0 \\ 0 & 0 \end{pmatrix}.$$

It follows that $S(c_1, c_2, c_3, c_4) = R_2$, and hence $b \in S(c_1, c_2, c_3, c_4)$. By Theorem (8.5), we conclude that a solution exists.

To find a solution, it is more convenient to work with the rows of the augmented matrix (8.14). We shall prove that if a 2-by-5 matrix

$$\mathbf{A}' = \begin{pmatrix} \alpha_{11} & \alpha_{12} & \cdots & \alpha_{15} \\ \alpha_{21} & \alpha_{22} & \cdots & \alpha_{25} \end{pmatrix}$$

is row equivalent to **A**, then the system of equations

$$\begin{aligned} \alpha_{11}x_1 + \alpha_{12}x_2 + \alpha_{13}x_3 + \alpha_{14}x_4 &= \alpha_{15} \\ \alpha_{21}x_1 + \alpha_{22}x_2 + \alpha_{23}x_3 + \alpha_{24}x_4 &= \alpha_{25} \end{aligned}$$

associated with the new matrix **A′** has exactly the same solutions as the original system. When this occurs, we shall call the new system *equivalent* to the original system of equations. It is sufficient to check this statement in case **A′** is obtained from **A** by a single elementary row operation. For an elementary row operation of type I (interchanging two rows) or III (multiplying a row by a nonzero constant) the result is clear. In order to discuss an elementary row operation of type II, let us write the original equations in the form

$$\begin{aligned} L_1 &= 0 \\ L_2 &= 0, \end{aligned}$$

where $L_1 = x_1 + 2x_2 - 3x_3 + x_4 - 1$, $L_2 = x_1 + x_2 + x_3 + x_4$. We have to prove that if $L_1' = L_1 + \lambda L_2, L_2' = L_2$, then the solutions of $L_1 = 0$ and $L_2 = 0$ are the same as the solutions of

$$L_1' = L_1 + \lambda L_2 = 0, \qquad L_2' = L_2 = 0.$$

Certainly, $L_1 = 0$ and $L_2 = 0$ imply $L_1' = 0$ and $L_2' = 0$. On the other hand, if $L_1' = 0$ and $L_2' = 0$, then $L_2 = 0$, and $L_1 + \lambda L_2 = 0$, so that $L_1 = 0$. This completes the proof.

Finally we are ready to solve the original system. The idea is to apply elementary row operations to put the augmented matrix in echelon form (as in Section 6). We then solve, if possible, the last equation, corresponding to the last nonzero row of the matrix, for the variables which have not been eliminated by putting the augmented matrix in echelon form. The remaining values can then easily be determined because of the echelon form of the matrix. In our problem, we replace the equations

$$L_1 = 0, \qquad L_2 = 0$$

by $L_1 = 0$, $L_2 - L_1 = 0$, and the latter equations are

$$\begin{aligned} x_1 + 2x_2 - 3x_3 + x_4 &= 1 \\ - x_2 + 4x_3 \qquad &= -1.\dagger \end{aligned}$$

Solving the second equation for the variables x_2, x_3, x_4 which were not eliminated by putting the matrix in echelon form, we have, for example,

$$x_2 = 1, \qquad x_3 = x_4 = 0.$$

Substituting in the first equation we have $x_1 + 2 = 1$, $x_1 = -1$, and it follows from our discussion that

$$\langle -1, 1, 0, 0 \rangle$$

is a solution of our original system of equations.

Several points should be made about this computational method.

1. By working on the augmented matrix with elementary row operations, we have a procedure that is easy to check at each step, and also one that is suitable for large-scale problems where machine computation is used.
2. We can check that the solution found really works by substituting in the original equations.
3. The method can be applied to an arbitrary system, and will tell whether or not a solution exists, as well as producing one.

We conclude the section with two more examples.

Example B. Test for solvability and, if solvable, find a solution:

$$\begin{aligned} -3x_1 + x_2 + 4x_3 &= -5 \\ x_1 + x_2 + x_3 &= 2 \\ -2x_1 \qquad + x_3 &= -3 \\ x_1 + x_2 - 2x_3 &= 5. \end{aligned}$$

Putting the augmented matrix in row echelon form, we have

$$\begin{pmatrix} -3 & 1 & 4 & -5 \\ 1 & 1 & 1 & 2 \\ -2 & 0 & 1 & -3 \\ 1 & 1 & -2 & 5 \end{pmatrix} \underset{\sim}{\text{I}} \begin{pmatrix} 1 & 1 & 1 & 2 \\ -3 & 1 & 4 & -5 \\ -2 & 0 & 1 & -3 \\ 1 & 1 & -2 & 5 \end{pmatrix}$$

† In writing equations, we shall usually omit terms involving unknowns whose coefficients are zero.

$$
\overset{\text{II}}{\sim}
\begin{pmatrix}
1 & 1 & 1 & 2\\
0 & 4 & 7 & 1\\
0 & 2 & 3 & 1\\
0 & 0 & -3 & 3
\end{pmatrix}
\sim
\begin{pmatrix}
1 & 1 & 1 & 2\\
0 & 2 & 3 & 1\\
0 & 0 & 1 & -1\\
0 & 0 & -3 & 3
\end{pmatrix}
\sim
\begin{pmatrix}
1 & 1 & 1 & 2\\
0 & 2 & 3 & 1\\
0 & 0 & 1 & -1\\
0 & 0 & 0 & 0
\end{pmatrix}.
$$

The system of equations corresponding to the last matrix is

$$
\begin{aligned}
x_1 + x_2 + x_3 &= 2\\
2x_2 + 3x_3 &= 1\\
x_3 &= -1.
\end{aligned}
$$

Solving, we obtain the solution

$$\langle 1, 2, -1\rangle,$$

which by our discussion of Example A is also a solution of the original system.

Example C. Test for solvability and, if solvable, find a solution:

$$
\begin{aligned}
-3x_1 + x_2 + 4x_3 &= 1\\
x_1 + x_2 + x_3 &= 0\\
-2x_1 \qquad + x_3 &= -1\\
x_1 + x_2 - 2x_3 &= 0.
\end{aligned}
$$

Proceeding as in Example B, we have

$$
\begin{pmatrix}
-3 & 1 & 4 & 1\\
1 & 1 & 1 & 0\\
-2 & 0 & 1 & -1\\
1 & 1 & -2 & 0
\end{pmatrix}
\sim
\begin{pmatrix}
1 & 1 & 1 & 0\\
-3 & 1 & 4 & 1\\
-2 & 0 & 1 & -1\\
1 & 1 & -2 & 0
\end{pmatrix}
\sim
\begin{pmatrix}
1 & 1 & 1 & 0\\
0 & 4 & 7 & 1\\
0 & 2 & 3 & -1\\
0 & 0 & -3 & 0
\end{pmatrix}
\sim
\begin{pmatrix}
1 & 1 & 1 & 0\\
0 & 2 & 3 & -1\\
0 & 0 & 1 & 3\\
0 & 0 & -3 & 0
\end{pmatrix}
\sim
\begin{pmatrix}
1 & 1 & 1 & 0\\
0 & 2 & 3 & -1\\
0 & 0 & 1 & 3\\
0 & 0 & 0 & 9
\end{pmatrix}
$$

This time the equivalent system of equations is

$$\begin{aligned} x_1 + x_2 + x_3 &= 0 \\ 2x_2 + 3x_3 &= -1 \\ x_3 &= 3 \\ 0 \cdot x_3 &= 9. \end{aligned}$$

Clearly, the system has no solutions.

EXERCISES

1. Test for solvability of the following systems of equations, and if solvable, find a solution:

   a. $$\begin{aligned} x_1 + x_2 + x_3 &= 8. \\ x_1 + x_2 + x_4 &= 1. \\ x_1 + x_3 + x_4 &= 14. \\ x_2 + x_3 + x_4 &= 14. \end{aligned}$$

   b. $$\begin{aligned} x_1 + x_2 - x_3 &= 3. \\ x_1 - 3x_2 + 2x_3 &= 1. \\ 2x_1 - 2x_2 + x_3 &= 4. \end{aligned}$$

   c. $x_1 + x_2 - 5x_3 = -1.$

   d. $$\begin{aligned} 2x_1 + x_2 + 3x_3 - x_4 &= 1. \\ 3x_1 + x_2 - 2x_3 + x_4 &= 0. \\ 2x_1 + x_2 - x_3 + 2x_4 &= -1. \end{aligned}$$

   e. $$\begin{aligned} -x_1 + x_2 + x_4 &= 0. \\ x_2 + x_3 &= 1. \end{aligned}$$

   f. $$\begin{aligned} x_1 + 2x_2 + 4x_3 &= 1. \\ 2x_1 + x_2 + 5x_3 &= 0. \\ 3x_1 - x_2 + 5x_3 &= 0. \end{aligned}$$

   g. $$\begin{aligned} 3x_1 + 4x_2 &= -1. \\ -x_1 - x_2 &= 1. \\ x_1 - 2x_2 &= 0. \\ 2x_1 + 3x_2 &= 0. \end{aligned}$$

   h. $$\begin{aligned} 2x_1 + x_2 - x_3 &= 0. \\ x_1 - x_3 &= 0. \\ x_1 + x_2 + x_3 &= 1. \end{aligned}$$

   i. $$\begin{aligned} x_1 + 4x_2 + 3x_3 &= 1. \\ 3x_1 + x_3 &= 1. \\ 4x_1 + x_2 + 2x_3 &= 1. \end{aligned}$$

2. For what values of α does the following system of equations have a solution?

$$\begin{aligned} 3x_1 - x_2 + \alpha x_3 &= 1 \\ 3x_1 - x_2 + x_3 &= 5 \end{aligned}$$

3. Prove that a system of m homogeneous equations in $n > m$ unknowns always has a nontrivial solution.

4. Prove that a system of homogeneous equations $x_1c_1 + \cdots + x_nc_n = 0$ in n unknowns has a nontrivial solution if and only if the rank of the coefficient matrix is less than n.

9. SYSTEMS OF HOMOGENEOUS EQUATIONS

As in the earlier sections in this chapter, this section begins with some general facts about solutions of systems of homogeneous equations, whose purpose is partly to provide a language for talking about the problems in a precise way, and partly to obtain some theorems which enable us to predict what will happen in a particular case before becoming bogged down in numerical calculations. Later in the section an efficient computational method will be given for solving particular systems of equations.

(9.1) THEOREM. *The set S of all solution vectors of a homogeneous system $x_1c_1 + \cdots + x_nc_n = 0$ forms a subspace of R_n.*

Proof. Let $a = \langle\lambda_1, \ldots, \lambda_n\rangle$ and $b = \langle\mu_1, \ldots, \mu_n\rangle$ belong to S. Then we have

$$\lambda_1c_1 + \cdots + \lambda_nc_n = 0, \qquad \mu_1c_1 + \cdots + \mu_nc_n = 0.$$

Adding these equations, we obtain

$$(\lambda_1 + \mu_1)c_1 + \cdots + (\lambda_n + \mu_n)c_n = 0,$$

which shows that $a + b \in S$. If $\lambda \in R$, then we have also

$$\lambda(\lambda_1c_1 + \cdots + \lambda_nc_n) = (\lambda\lambda_1)c_1 + \cdots + (\lambda\lambda_n)c_n = 0$$

and $\lambda a \in S$. This completes the proof.

From this theorem and the results of the preceding sections, we will know all solutions of a homogeneous system as soon as we find a basis for the *solution space*, that is, the set of solutions of the system.

(9.2) THEOREM. *Let*

$$\begin{array}{c} \alpha_{11}x_1 + \cdots + \alpha_{1n}x_n = 0 \\ \cdots\cdots\cdots\cdots\cdots\cdots\cdots \\ \alpha_{m1}x_1 + \cdots + \alpha_{mn}x_n = 0 \end{array}$$

be a homogeneous system of m equations in n unknowns, with column vectors $c_1, \ldots, c_n$ arranged such that, for some r, $\{c_1, \ldots, c_r\}$ is a basis for the subspace $S(c_1, \ldots, c_n)$. Then for each i, if $r + 1 \le i \le n$, there exists a relation of linear dependence

$$\lambda_1^{(i)}c_1 + \cdots + \lambda_r^{(i)}c_r - c_i = 0, \qquad \lambda_j^{(i)} \in R.$$

Then for $r + 1 \leq i \leq n$,

$$u_i = \langle \underbrace{\lambda_1^{(i)}, \ldots, \lambda_r^{(i)}, 0, \ldots, 0, -1}_{i}, 0, \ldots, 0 \rangle$$

(where it is to be understood that the -1 appears in the ith position in u_i) is a solution of the system, and $\{u_{r+1}, \ldots, u_n\}$ is a basis for the solution space of the system.

Proof. By Theorem (7.3) it is indeed possible to select a basis for $S(c_1, \ldots, c_n)$ from among the vectors $c_1, \ldots, c_n$ themselves. Rearranging the indices so that these vectors occupy the $1, \ldots, r$th positions changes the solution space only in that a corresponding change of position has been made in the components of the solution vectors.

Since $\{c_1, \ldots, c_r\}$ forms a basis for $S(c_1, \ldots, c_n)$, each vector c_i, where $r + 1 \leq i \leq n$, is a linear combination of the basis vectors, and we have a relation of linear dependence

$$\lambda_1^{(i)} c_1 + \cdots + \lambda_r^{(i)} c_r - c_i = 0, \qquad \lambda_j^{(i)} \in R, \quad r + 1 \leq i \leq n,$$

as in the statement of the theorem.

Therefore, the vectors

$$u_i = \langle \lambda_1^{(i)}, \ldots, \lambda_r^{(i)}, \ldots, -1, \ldots, 0 \rangle, \qquad r + 1 \leq i \leq n,$$

with the -1 in the ith position of u_i, are solutions of the system. It remains to show that the $\{u_i\}$ are linearly independent and that they generate the solution space.

To show that they are linearly independent, suppose we have a possible relation of linear dependence:

(9.3) $$\mu_{r+1} u_{r+1} + \cdots + \mu_n u_n = 0, \qquad \text{for } \mu_i \in R.$$

The left side is a vector in R_n all of whose components are zero. For $r + 1 \leq i \leq n$, the ith component of (9.3) is $-\mu_i$ (why?), and it follows that $\mu_{r+1} = \cdots = \mu_n = 0$. Now let $x = \langle \alpha_1, \ldots, \alpha_n \rangle$ be an arbitrary solution of the original system. From the definition of $u_{r+1}, \ldots, u_n$, we have

$$x + \sum_{k=r+1}^{m} \alpha_k u_k = \langle \xi_1, \ldots, \xi_r, 0, \ldots, 0 \rangle$$

where $\xi_1, \ldots, \xi_r$ are some elements of R. Since the set of solutions is a subspace of R_n, the vector $y = \langle \xi_1, \ldots, \xi_r, 0, \ldots, 0 \rangle$ is a solution of the original system and we have, by (8.4),

$$\xi_1 c_1 + \cdots + \xi_r c_r + 0c_{r+1} + \cdots + 0c_n = 0.$$

Since $c_1, \ldots, c_r$ are linearly independent, we have $\xi_1 = \cdots = \xi_r = 0$ and hence

$$x = \sum_{k=r+1}^{m} (-\alpha_k)u_k \in S(u_{r+1}, \ldots, u_n).$$

This completes the proof of the theorem.

(9.4) COROLLARY. *The dimension of the solution space of a homogeneous system in n unknowns is $n - r$, where r is the rank of the coefficient matrix.*

This result is immediate from our definition of the rank as the dimension of the column space of the coefficient matrix.

We shall now apply our result to derive a useful and unexpected result about the rank of a matrix

$$\mathbf{A} = \begin{pmatrix} \alpha_{11} & \cdots & \alpha_{1n} \\ \cdots & \cdots & \cdots \\ \alpha_{m1} & \cdots & \alpha_{mn} \end{pmatrix}$$

with columns $c_1, \ldots, c_n$ and rows $r_1, \ldots, r_m$. Let us define the *row rank* of $\mathbf{A}$ as the dimension of the row space $S(r_1, \ldots, r_m)$. Some writers call the rank as we have defined it the "column rank" but, as the next theorem shows, the row rank and the column rank are always equal.

With the matrix **A**, let us consider the homogeneous system

(9.5)
$$\begin{aligned} \alpha_{11}x_1 + \cdots + \alpha_{1n}x_n &= 0 \\ \cdots\cdots&\cdots\cdots. \\ \alpha_{m1}x_1 + \cdots + \alpha_{mn}x_n &= 0 \end{aligned}$$

It is convenient for this proof and for some arguments in the next section to use the notation

$$r_i \cdot x = \alpha_{i1}x_1 + \cdots + \alpha_{in}x_n$$

for the two vectors r_i and x, so that the system (9.4) can be described also by the system of equations

(9.6)
$$\begin{aligned} r_1 \cdot x &= 0 \\ \cdots&\cdots. \\ r_m \cdot x &= 0 \end{aligned}$$

The "inner product" $r \cdot x$ has the property that

$$(\lambda r + \mu s) \cdot x = \lambda(r \cdot x) + \mu(s \cdot x), \qquad \text{for } \lambda \text{ and } \mu \in R, \text{ and for } r \text{ and } s \in R_n.$$

(9.7) THEOREM. *The row rank of an m-by-n matrix* **A** *is equal to the rank of* **A**.

Proof. We shall use the notations we have just introduced. Now, without changing the column rank we may assume that $\{r_1, \ldots, r_t\}$ forms a basis for the row space of **A**, where t is the row rank of **A**. Then it follows easily that the system of equations

$$\text{(9.8)} \qquad \begin{array}{c} r_1 \cdot x = 0 \\ \cdots\cdots\cdots \\ r_t \cdot x = 0 \end{array}$$

has the same solution space as (9.5). To see this, it is sufficient to prove that any solution x of (9.8) is a solution of (9.5). For $t + 1 \leq i \leq m$, we have

$$r_i = \xi_1 r_1 + \cdots + \xi_t r_t, \qquad \xi_j \in R.$$

Then, by the properties of the inner product $r \cdot x$,

$$r_i \cdot x = \sum_{k=1}^{t} \xi_k (r_k \cdot x) = 0$$

since x is a solution of (9.8), and our assertion is proved.

The columns of the matrix $\mathbf{A}'$ whose rows are $r_1, \ldots, r_t$ are in R_t; hence

$$\text{rank } \mathbf{A}' \leq t$$

and

$$n - \text{rank } \mathbf{A}' \geq n - t.$$

Since (9.6) and (9.8) have the same solution space, we have by Corollary (9.4)

$$n - \text{rank } \mathbf{A} = n - \text{rank } \mathbf{A}' \geq n - t,$$

and hence rank $\mathbf{A} \leq t$. Interchanging the rows and columns of **A**, we obtain an n-by-m matrix $\mathbf{A}^*$; repeating the argument with $\mathbf{A}^*$, we have rank $\mathbf{A}^* \leq$ row rank of $\mathbf{A}^*$. But rank $\mathbf{A}^* = t$ and row rank $\mathbf{A}^* =$ rank **A**. Combining our results, we have

$$\text{rank } \mathbf{A} = \text{row rank } \mathbf{A}$$

and Theorem (9.7) is proved.

For our first example, we complete the discussion of the system of equations discussed in Section 1.

Example A. Find a basis for the solution space of the system of homogeneous equations

$$\begin{aligned} x_1 + 2x_2 - 3x_3 + x_4 &= 0 \\ x_1 + x_2 + x_3 + x_4 &= 0. \end{aligned}$$

As in the previous section, we find an equivalent system of equations (i.e., having the same solution space) whose coefficient matrix is in echelon form. This time it is unnecessary to work with the augmented matrix since the additional column consists entirely of zeros. We have

$$\begin{pmatrix} 1 & 2 & -3 & 1 \\ 1 & 1 & 1 & 1 \end{pmatrix} \sim \begin{pmatrix} 1 & 2 & -3 & 1 \\ 0 & -1 & 4 & 0 \end{pmatrix}.$$

An equivalent system is therefore

$$\begin{aligned} x_1 + 2x_2 - 3x_3 + x_4 &= 0 \\ 0x_1 - x_2 + 4x_3 + 0x_4 &= 0. \end{aligned} \tag{9.9}$$

Since the rank of the coefficient matrix is 2 [by computing the row rank, for example, taking account of Theorem (9.7)], we know that the dimension of the solution space is $4 - 2 = 2$. The second equation of (9.9), viewed as an equation in $\{x_2, x_3, x_4\}$, has two linearly independent solutions given by

$$\langle 1, \tfrac{1}{4}, 0\rangle \quad \text{and} \quad \langle 0, 0, 1\rangle.$$

Substituting these in the first equation gives us two linearly independent solutions for the original system:

$$u_1 = \langle -\tfrac{5}{4}, 1, \tfrac{1}{4}, 0\rangle, \qquad u_2 = \langle -1, 0, 0, 1\rangle.$$

Example B. Find the general solution (i.e., all solutions) of the system of nonhomogeneous equations

$$\begin{aligned} x_1 + 2x_2 - 3x_3 + x_4 &= 1 \\ x_1 + x_2 + x_3 + x_4 &= 0. \end{aligned}$$

By Theorem (8.9), the general solution is given by

$$u = x_0 + x$$

where x_0 is a solution of the nonhomogeneous system and x ranges over the solutions of the homogeneous system. Applying the results of Example A of Section 8 and Example A of this section, we have for the general solution,

$$\begin{aligned} u &= \langle -1, 1, 0, 0\rangle + \lambda\langle -\tfrac{5}{4}, 1, \tfrac{1}{4}, 0\rangle + \mu\langle -1, 0, 0, 1\rangle \\ &= \langle -1 - \tfrac{5}{4}\lambda - \mu, 1 + \lambda, \tfrac{1}{4}\lambda, \mu\rangle. \end{aligned}$$

where λ and μ are arbitrary real numbers.

Example C. Find all solutions of the homogeneous system

$$\begin{aligned} x_1 + x_2 - 2x_3 + x_4 &= 0 \\ -x_1 - x_2 + x_3 + 3x_4 &= 0. \\ 2x_1 + 2x_2 + 5x_3 \qquad &= 0. \end{aligned}$$

Proceeding as usual, we have

$$\begin{pmatrix} 1 & 1 & -2 & 1 \\ -1 & -1 & 1 & 3 \\ 2 & 2 & 5 & 0 \end{pmatrix} \sim \begin{pmatrix} 1 & 1 & -2 & 1 \\ 0 & 0 & -1 & 4 \\ 0 & 0 & 9 & -2 \end{pmatrix}$$
$$\sim \begin{pmatrix} 1 & 1 & -2 & 1 \\ 0 & 0 & -1 & 4 \\ 0 & 0 & 0 & 1 \end{pmatrix}.$$

This time an equivalent system of equations is

$$\begin{aligned} x_1 + x_2 - 2x_3 + x_4 &= 0 \\ - x_3 + 4x_4 &= 0 \\ x_4 &= 0. \end{aligned}$$

The rank of the coefficient matrix (which can always be found by the methods of Section 6) is 3; so the dimension of the solution space is $4 - 3 = 1$. This time the last equation gives only the information $x_4 = 0$. Substituting this information in the second equation yields $x_3 = 0$. We get a basis for our solution space from the first equation, which yields

$$u = \langle 1, -1, 0, 0 \rangle$$

as the desired basis vector. (Remember always to substitute back in the original equations as a check.)

It should be fairly clear at this point that, in general, the computational procedure reduces the problem of finding all solutions of a homogeneous system to finding all solutions of one equation in several unknowns. This can be done always by a direct application of the method of the proof of Theorem (9.2).

Example D. Find all solutions of

$$x_1 + 2x_2 - x_3 + x_4 + x_5 = 0.$$

By Theorem (9.2) a basis for the solution space is given by

$$\begin{array}{ll} \langle 2, -1, 0, 0, 0 \rangle, & \langle -1, 0, -1, 0, 0 \rangle, \\ \langle 1, 0, 0, -1, 0 \rangle, & \langle 1, 0, 0, 0, -1 \rangle. \end{array}$$

Example E. The purpose of this last example is to show that the theory of vector spaces is really deeper and more far-reaching than the study of R_n and systems of linear equations. Let us return to the second problem discussed in Section 1. We want to find all solutions of the differential equation

$$y'' + m^2 y = 0.$$

As we observed, the set of solutions is a subspace S of $\mathscr{F}(R)$. The functions

$$y_1 = \sin mx, \qquad y_2 = \cos mx$$

are both in S. Moreover they are linearly independent, as we see *not* by the computational methods of row equivalence of matrices, etc., but by returning to the original definition and its consequences. The only way $\{\sin mx, \cos mx\}$ could be linearly dependent in $\mathscr{F}(R)$ is for $\sin mx = \lambda \cos mx$ for some real number λ. But this cannot occur, for example, because the functions $\sin mx$ and $\cos mx$ are never both zero at the same time. To prove that the functions $\sin mx$ and $\cos mx$ form a basis of S, we have to use the important result that the differential equation $y'' + m^2y = 0$ has a unique solution y satisfying a given set of initial conditions

$$y(0) = \alpha, \qquad y'(0) = \beta$$

where α, β are given real numbers, and that it is always possible to find a function of the form $\lambda \sin mx + \mu \cos mx$ satisfying such a set of initial conditions. Of course, it would take more time to prove these results.† The point of the discussion is to show that although the theory of vector spaces gives us a language to describe the solutions of a differential equation, the details of finding the solution are very different from the analogous problem (from the point of view of vector spaces) of solving a system of linear equations.

EXERCISES

1. Find a basis for the solution space of the system

$$\begin{aligned} 3x_1 - x_2 \qquad\quad + x_4 &= 0 \\ x_1 + x_2 + x_3 + x_4 &= 0. \end{aligned}$$

2. Find bases for the solution spaces of the homogeneous systems associated with the systems given in Exercise 1 of Section 8.
3. Describe all solutions of the system

$$\begin{aligned} -x_1 + 2x_2 + x_3 + 4x_4 &= 0 \\ 2x_1 + x_2 - x_3 + x_4 &= 1. \end{aligned}$$

4. In each of the problems in Exercise 2 of Section 6, find a relation of linear dependence (with nonzero coefficients) if one exists.
5. In plane analytic geometry, given two points, such as (3, 1) and (−1, 0), a method is given for finding an equation

$$Ax + By + C = 0$$

† See Section 34 for proofs of these results.

such that both points are solutions of the equation. Show that this problem is equivalent to solving the homogeneous system

$$A\begin{pmatrix}3\\-1\end{pmatrix} + B\begin{pmatrix}1\\0\end{pmatrix} + C\begin{pmatrix}1\\1\end{pmatrix} = 0,$$

to find a nontrivial solution. Prove that if $\langle A', B', C'\rangle$ is another nontrivial solution, then $\langle A', B', C'\rangle$ is a scalar multiple of the first solution $\langle A, B, C\rangle$.

6. With reference to Exercise 5 let (α, β) and (γ, δ) be distinct points in the plane. Prove that there exist real numbers A, B, C, not all zero, such that both points satisfy the equation

$$Ax + By + C = 0,$$

and that if both points also satisfy

$$A'x + B'y + C' = 0,$$

then $\langle A', B', C'\rangle = \lambda\langle A, B, C\rangle$ for some $\lambda \in R$. A *line* in R_2 is defined as the set of solutions of an equation $Ax + By + C = 0$. Show that the above result proves that two distinct points in R_2 lie on a unique line.

10. LINEAR MANIFOLDS

In this section we shall apply the results of Sections 5 to 9 to a discussion of the generalizations in R_n of lines and planes in two and three dimensions.

We define a *line* in R_n as a one-dimensional subspace $S(a)$ or more generally, a translate $b + S(a)$ of a one-dimensional subspace by some fixed vector b. Thus a vector p belongs to the line $b + S(a)$ if, for some $\lambda \in R$ (see Figure 2.7),

$$p = b + \lambda a.$$

By analogy we might then define a *plane* in R_n as a two-dimensional

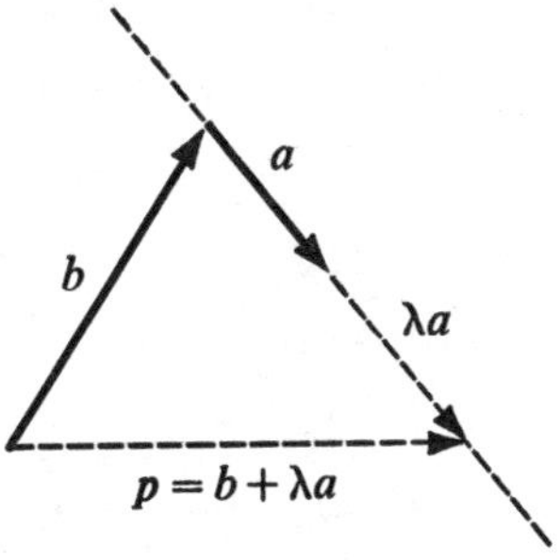

FIGURE 2.7

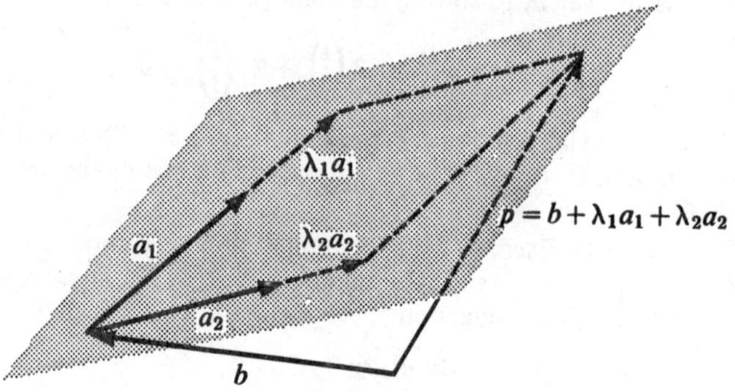

FIGURE 2.8

subspace $S \subset R_n$ or, more generally, a translate $b + S$ of a two-dimensional space S by a fixed vector b. If $\{a_1, a_2\}$ is a basis of S, then $p \in b + S$ if and only if (see Figure 2.8),

$$p = b + \lambda_1 a_1 + \lambda_2 a_2, \qquad \lambda_i \in R.$$

The general concept we are looking for is the following generalization of the concept of subspace.

(10.1) DEFINITION. A *linear manifold* V in R_n is the set of all vectors in $b + S$, where b is a fixed vector and S is a fixed subspace of R_n. Thus $p \in V$ if and only if $p = b + a$ for some $a \in S$. The subspace S is called the *directing space* of V. The *dimension* of V is the dimension of the subspace S.

(10.2) THEOREM. *Let $V = b + S$ be a linear manifold in R_n; then the directing space of V is the set of all vectors $p - q$ where $p, q \in V$.*

Proof. Let $p, q \in V$; then there exist vectors $a, a' \in S$ such that

$$p = b + a, \qquad q = b + a'.$$

Then

$$p - q = (b + a) - (b + a') = a - a' \in S.$$

On the other hand, if $a \in S$, then $a = b - (b - a)$, so that $a = p - q$ where $p = b$ and $q = b - a$ both belong to V. This completes the proof.

This theorem shows that the directing space of $V = b + S$ is determined independently of the vector b.

We know that lines and planes in R_2 and R_3 are also described as the sets of solutions of certain linear equations. The definition of a linear manifold $V = b + S$ suggests that a linear manifold can be described by equations in the general case. From Section 8 we know that vectors in $b + S$ have the form we obtained for the solutions of a nonhomogeneous system of linear equations, where b is the particular solution and S is the solution space of the associated homogeneous system.

In working out this idea we begin by indicating how to describe a subspace S by a system of linear equations.

(10.3) LEMMA. *Let S be an r-dimensional subspace of R_n; then there exists a set of $n - r$ homogeneous linear equations in n unknowns whose solution space is exactly S.*

Remark. No set of less than $n - r$ equations can have S as solution space. Why?

Proof of Lemma (10.3). Let $\{b_1, \ldots, b_r\}$ be a basis for S. The only system of equations which are possibly relevant to the problem is [in the notation of (9.6)]:

(10.4)
$$\begin{array}{c} b_1 \cdot x = 0 \\ \cdots\cdots\cdots. \\ b_r \cdot x = 0 \end{array}$$

But these obviously do not solve the problem, since it may happen (and usually does) that, say, $b_1 \cdot b_1 \neq 0$, so that b_1 is not, in general, a solution vector of (10.4). Let S^* be the solution space of (10.4). By (9.7) the rank of the matrix whose rows are $b_1, \ldots, b_r$ is r; hence S^* has dimension $n - r$, by (9.4). Let $\{c_1, \ldots, c_{n-r}\}$ be a basis for S^* and consider the system of equations

$$\begin{array}{c} c_1 \cdot x = 0 \\ \cdots\cdots\cdots\cdots. \\ c_{n-r} \cdot x = 0 \end{array}$$

By the same reasoning, the solution space S^{**} of this system has dimension r and, clearly, $S \subset S^{**}$ (why?). Since $\dim S = r$, we have by Exercise 3 of Section 7 the result that $S = S^{**}$, and the lemma is proved.

Remark. Note that our computational procedures in Sections 6 to 9 can be applied to carrying out each of the steps in the proof of Lemma (10.3). For example, let us find a system of equations whose solution space in R_4 has a basis consisting of the vectors $b_1 = \langle 1, 1, 0, 1\rangle$ and $b_2 = \langle 1, 0, 1, 1,\rangle$. Letting x denote $\langle x_1, x_2, x_3, x_4\rangle$, we form the system

(10.5)
$$\begin{array}{c} b_1 \cdot x = 0 \\ b_2 \cdot x = 0. \end{array}$$

The column vectors of the system are

$$c_1 = \langle 1, 1 \rangle, \qquad c_2 = \langle 1, 0 \rangle, \qquad c_3 = \langle 0, 1 \rangle, \qquad c_4 = \langle 1, 1 \rangle.$$

By inspection a basis for the solution space of the system (10.5) is

$$\begin{gathered} \langle 1, 0, 0, -1 \rangle \\ \langle -1, 1, 1, 0 \rangle, \end{gathered}$$

and according to Lemma (10.3), the desired system of equations is

$$\begin{aligned} x_1 \qquad\qquad - x_4 &= 0 \\ -x_1 + x_2 + x_3 \qquad &= 0. \end{aligned}$$

The next result completes our geometrical interpretation of systems of linear equations.

(10.6) THEOREM. *A necessary and sufficient condition for a set of vectors $V \subset R_n$ to form a linear manifold of dimension r is that V be the set of all solutions of a system of $n - r$ nonhomogeneous equations in n unknowns whose coefficient matrix has rank $n - r$.*

Proof. First let

$$x_1c_1 + \cdots + x_nc_n = b, \qquad c_i \in R_{n-r}$$

be a system of $n - r$ equations in n unknowns whose coefficient matrix has rank $n - r$. By Theorem (8.9) the set V of all solutions is a linear manifold $x_0 + S$ where S is the solution space of the homogeneous system $x_1c_1 + \cdots + x_nc_n = 0$. By Corollary (9.4), $\dim S = n - (n - r)$ and hence $\dim V = r$.

To prove the converse, let $V = y + S$ be a linear manifold of dimension r. By Lemma (10.3) there exists a system of $n - r$ equations whose solution space is S:

$$x_1c_1 + \cdots + x_nc_n = 0.$$

If $y = \langle \eta_1, \ldots, \eta_n \rangle$, let

$$b = \eta_1c_1 + \cdots + \eta_nc_n.$$

Then the nonhomogeneous system $x_1c_1 + \cdots + x_nc_n = b$ has for its solutions exactly the set $y + S = V$. This completes the proof of the theorem.

For example, let us find a system of equations for the linear manifold V whose directing space S has a basis

$$b_1 = \langle 1, 1, 0, 1 \rangle \qquad \text{and} \qquad b_2 = \langle 1, 0, 1, 1 \rangle,$$

and contains the vector $\langle 1, 2, 3, 4\rangle$. As we showed in the first example in this section, S is a solution space of the system

$$\begin{aligned} x_1 \qquad\qquad - x_4 &= 0 \\ -x_1 + x_2 + x_3 \qquad &= 0. \end{aligned}$$

As in the proof of Theorem (10.6), we substitute the vector $\langle 1, 2, 3, 4\rangle$ in the system to obtain

$$\begin{aligned} 1 \qquad\qquad - 4 &= -3 \\ -1 + 2 + 3 \qquad &= 4. \end{aligned}$$

Then the equations whose solutions form the linear manifold V are

$$\begin{aligned} x_1 \qquad\qquad - x_4 &= -3 \\ -x_1 + x_2 + x_3 \qquad &= 4, \end{aligned}$$

by Theorem (10.6).

EXERCISES

1. Find a set of homogeneous linear equations whose solution space is generated by the vectors:
   a. $\langle 2, 1, -3\rangle, \langle 1, -1, 0\rangle, \langle 1, 3, -4\rangle$.
   b. $\langle 2, 1, 1, -1\rangle, \langle -1, 1, 0, 4\rangle, \langle 1, 1, 2, 1\rangle$.
2. Find a set of homogeneous equations whose solution space S is generated by the vectors $\langle 3, -1, 1, 2\rangle$, $\langle 4, -1, -2, 3\rangle$, $\langle 10, -3, 0, 7\rangle$, $\langle -1, 1, -7, 0\rangle$. Find a system of nonhomogeneous equations whose set of solutions is the linear manifold with directing space S and which passes through $\langle 1, 1, 1, 1\rangle$.
3. A *hyperplane* is a linear manifold of dimension $n - 1$ in R_n. Prove that a linear manifold is a hyperplane if and only if it is the set of solutions of a single linear equation $\alpha_1 x_1 + \cdots + \alpha_n x_n = \beta$, with some $\alpha_i \neq 0$. Prove that a linear manifold of dimension r is the intersection of exactly $n - r$ hyperplanes.
4. A *line* is defined as a one-dimensional linear manifold in R_n. Prove that if p and q are distinct vectors belonging to a line L, then L consists of all vectors of the form
$$p + \lambda(q - p), \qquad \lambda \in R.$$
5. Prove that a line in R_n is the intersection of $n - 1$ hyperplanes. Find a system of linear equations whose set of solutions is the line passing through the points:
   a. $p = \langle 1, 1\rangle, q = \langle 2, -1\rangle$ in R_2.
   b. $p = \langle 1, 1, 2\rangle, q = \langle -1, 2, -1\rangle$ in R_3.
   c. $p = \langle 1, -1, 0, 2\rangle, q = \langle 2, 0, 1, 1\rangle$ in R_4.

6. Find two distinct vectors on the line belonging to the intersection of the hyperplanes:

$$x_1 + 2x_2 - x_3 = -1, \qquad 2x_1 + x_2 + 4x_3 = 2 \text{ in } R_3,$$
$$x_1 + x_2 = 0, \qquad x_2 - x_3 = 0, \qquad x_2 - 2x_4 = 0 \text{ in } R_4.$$

7. Prove that, if p and q are vectors belonging to a linear manifold V, then the line through p and q is contained in V.

8. Let S and T be subspaces of R_n, which are represented as the solution spaces of homogeneous systems

$$a_1 \cdot x = 0, \ldots, a_r \cdot x = 0$$

and

$$b_1 \cdot x = 0, \ldots, b_s \cdot x = 0,$$

respectively. Prove that $S \cap T$ is the solution space of the system

$$a_1 \cdot x = 0, \ldots, a_r \cdot x = 0, \qquad b_1 \cdot x = 0, \ldots, b_s \cdot x = 0.$$

Use this remark to find a basis for $S \cap T$, where S and T are as in Exercise 5 of Section 7.

9. Let $S_1 = S(e_1, e_2, e_3)$ in R_4, where $e_1 = \langle 1, 0, 0, 0 \rangle$, $e_2 = \langle 0, 1, 0, 0 \rangle$, and $e_3 = \langle 0, 0, 1, 0 \rangle$. Let $S_2 = S(a_1, a_2, a_3)$, where $a_1 = \langle 1, 1, 0, 1 \rangle$, $a_2 = \langle 2, -1, 3, -1 \rangle$, $a_3 = \langle -1, 0, 0, 2 \rangle$. Find $\dim (S_1 + S_2)$ and $\dim (S_1 \cap S_2)$. Find a basis for $S_1 \cap S_2$.

10. Find the point in R_3 where the line joining the points $\langle 1, -1, 0 \rangle$ and $\langle -2, 1, 1 \rangle$ pierces the plane

$$3x_1 - x_2 + x_3 - 1 = 0.$$

Chapter 3

Linear Transformations and Matrices

In order to compare different mathematical systems of the same type, it is essential to study the functions from one system to another which preserve the operations of the system. Thus in calculus we study functions f which preserve the limit operation: if $\lim_{i\to\infty} x_i = x$, then $\lim_{i\to\infty} f(x_i) = f(x)$. These are the continuous functions. For vector spaces, we investigate functions that preserve the vector space operations of addition and scalar multiplication. These are the linear transformations. In this chapter we develop the language of linear transformations and the connection between linear transformations and matrices. Deeper results about linear transformations appear later.

11. LINEAR TRANSFORMATIONS

Linear transformations are certain kinds of functions, and it is a good idea to begin this section by reviewing some of the definitions and terminology of functions on sets.

(11.1) Definition. Let X and Y be sets. A *function* $f: X \to Y$ (sometimes called a *mapping*) is a rule which assigns to each element of a set X a unique element of a set Y. We write $f(x)$ for the element of Y which the function f assigns to x. The set X is called the *domain* or *domain of definition* of f. Two functions f and f' are said to be *equal*, and we write

$f = f'$, if they have the same domain X and if, for all $x \in X, f(x) = f'(x)$. The function f is said to be *one-to-one* if $x_1 \neq x_2$ in X implies that $f(x_1) \neq f(x_2)$. [Note that this is equivalent to the statement that $f(x_1) = f(x_2)$ implies $x_1 = x_2$.] The function f is said to be *onto* Y if every $y \in Y$ can be expressed in the form $y = f(x)$ for some $x \in X$; we shall say that f is a function of X *into* Y when we want to allow the possibility that f is not onto Y. A one-to-one function f of a set X onto a set Y is called a *one-to-one correspondence* of X onto Y.

Examples of functions are familiar from calculus. We give some examples to show that the conditions of being onto or one-to-one are independent of each other. For example, the function $f: R \to R$ defined by

$$f(x) = x^2, \qquad x \in R,$$

is a function that is neither one-to-one nor onto (why?). The function $g: R \to R$ defined by

$$g(x) = e^x$$

is one-to-one [because $g'(x) > 0$ for all x, so that the function is strictly increasing in the sense that $x_1 < x_2$ implies $g(x_1) < g(x_2)$]. But g is not onto, since $g(x) > 0$ for all real numbers x. The function $h: R^* \to R$ defined by

$$h(x) = \log_e |x|, \qquad x \in R^* = \{x \in R, x \neq 0\}$$

is onto, but not one-to-one, since $h(x) = h(-x)$ for all x.

We are now ready to define linear transformations from one vector space to another. Throughout this section, F denotes an arbitrary field.

(11.2) DEFINITION. Let V and W be vector spaces over F. A *linear transformation of* V *into* W is a function $T: V \to W$ which assigns to each vector $v \in V$ a unique vector $w = T(v) \in W$ such that

$$\begin{aligned} T(v_1 + v_2) &= T(v_1) + T(v_2), && v_i \in V, \\ T(\alpha v) &= \alpha T(v), && \alpha \in F, \quad v \in V, \end{aligned}$$

We shall often use the notation Tv instead of $T(v)$, in order to avoid a forest of parentheses. We shall also use the arrow notation rather freely, in statements such as "let f be a function of $V \to W$" or "the function $f: V \to W$ carries $v_1 \to w_1$."

We first consider some examples of linear transformations.

Example A. The function $T: R_1 \to R_1$ defined by

$$T(x) = \alpha x$$

for some fixed $\alpha \in R$ is a linear transformation, as we can easily check from the definition. What is not obvious is that if $f: R_1 \to R_1$ is an arbitrary linear transformation of R_1 into R_1, then f is of the form given above. To see this, let $f: R_1 \to R_1$ be linear, and let $\alpha = f(1)$. Then for all x,

$$f(x) = f(x \cdot 1) = xf(1) = \alpha x,$$

using only the second part of the definition of a linear transformation.

The next example can be thought of as the obvious extension of Example A to the vector space R_n.

Example B. A system of linear equations

$$\begin{aligned} y_1 &= \alpha_{11}x_1 + \cdots + \alpha_{1n}x_n \\ &\cdots\cdots\cdots\cdots\cdots\cdots \\ y_m &= \alpha_{m1}x_1 + \cdots + \alpha_{mn}x_n \end{aligned} \tag{11.3}$$

with $\alpha_{ij} \in F$, can be regarded in two ways. In Chapter 2 we assumed $\{y_1, \ldots, y_m\}$ to be given, and asked how to solve for $\{x_1, \ldots, x_n\}$. This operation, however, is not a function, because the solution $\{x_1, \ldots, x_n\}$ is not uniquely determined by $\{y_1, \ldots, y_m\}$. For example, both $\langle 1, -1\rangle$ and $\langle 2, -2\rangle$ are solutions of the equation $x_1 + x_2 = y$, when $y = 0$. On the other hand, we may think of $\langle y_1, \ldots, y_m\rangle$ as a function of $\langle x_1, \ldots, x_n\rangle$, for a given coefficient matrix $\mathbf{A} = (\alpha_{ij})$. Thus the equation

$$x_1 + x_2 = y$$

defines a function of $R_2 \to R_1$, namely, the function $T: \langle \alpha_1, \alpha_2\rangle \to \langle \alpha_1 + \alpha_2\rangle$. For example $\langle 2, -3\rangle \to \langle -1\rangle$, $\langle 1, 1\rangle \to \langle 2\rangle$, $\langle 1, -1\rangle \to \langle 0\rangle$.

In general, the system (11.3) defines a function T from F_n into F_m, which assigns to each n-tuple $\langle x_1, \ldots, x_n\rangle$ in F_n the m-tuple $\langle y_1, \ldots, y_m\rangle$ in F_m. We leave it to the reader to verify that the function defined by a system (11.3), with a fixed coefficient matrix, is always a linear transformation of $F_n \to F_m$.

Example C. Let f be a continuous real valued function on the real numbers. Then the rule which assigns to f the integral $I(f) = \int_0^1 f(t)\,dt$ is a linear transformation from the vector space $C(R)$ into the vector space R_1. This fact is expressed by the rules

$$\begin{aligned} I(f_1 + f_2) &= I(f_1) = I(f_2), \\ I(\alpha f) &= \alpha I(f), \end{aligned}$$

for $f_1, f_2, f \in C(R)$ and $\alpha \in R$.

Example D. Let $P(R)$ be the vector space of polynomial functions defined on R. Then the derivative function D, which assigns to each polynomial function its derivative, is a linear transformation of $P(R) \to P(R)$. We have to check that the derivative of a polynomial function is a polynomial function, and that the conditions

$$D(f_1 + f_2) = D(f_1) + D(f_2)$$
$$D(\alpha f) = \alpha(Df)$$

hold for all polynomial functions f_1, f_2, f and scalars α.

These illustrations show that some of the operations studied in the theory of linear equations, and in calculus, are all examples of linear transformations.

The importance of linear transformations comes partly from the fact that they can be combined by certain algebraic operations, and the resulting algebraic system has many properties not to be found in the other algebraic systems we have studied so far.

(11.4) DEFINITION. Let F be an arbitrary field and let V and W be vector spaces over F. Let S and T be linear transformations of $V \to W$. Define a mapping $S + T$ of $V \to W$ by the rule

$$(S + T)v = S(v) + T(v), \qquad v \in V.$$

Then $S + T$ is a mapping of $V \to W$ called the *sum* of the linear transformations S and T.

(11.5) DEFINITION. Let M, N, P be vector spaces over F and let S be a linear transformation of $M \to N$ and T a linear transformation of $N \to P$; then the mapping $TS: M \to P$ defined by

$$(TS)(m) = T[S(m)]$$

is a mapping of $M \to P$ called the *product* of the linear transformations T and S. If $\alpha \in F$, $S \in L(M, N)$, then αS is defined to be the mapping of $M \to N$ such that $(\alpha S)(m) = \alpha(S(m))$, for all $m \in M$.

(11.6) THEOREM. *Let V, W be vector spaces over F and let $L(V, W)$ denote the set of all linear transformations of V into W; then $L(V, W)$ is a vector space over F, with respect to the operations*

$$S + T, \qquad S, T \in L(V, W),$$

and

$$\alpha S, \qquad \alpha \in F, \quad S \in L(V, W).$$

Proof. We check first that $S + T$ and αS actually belong to $L(V, W)$; the argument is simply a thorough workout with the axioms for a vector space. Let $v_1, v_2 \in V$; then

$$\begin{aligned}(S + T)(v_1 + v_2) &= S(v_1 + v_2) + T(v_1 + v_2)\\ &= [S(v_1) + S(v_2)] + [T(v_1) + T(v_2)]\\ &= [S(v_1) + T(v_1)] + [S(v_2) + T(v_2)]\\ &= (S + T)(v_1) + (S + T)(v_2).\end{aligned}$$

For $\xi \in F$ we have

$$(S + T)(\xi v) = S(\xi v) + T(\xi v) = \xi[S(v)] + \xi[T(v)]$$
$$= \xi[S(v) + T(v)] = \xi[(S + T)(v)].$$

Turning to the mapping αS, we have

$$(\alpha S)(v_1 + v_2) = \alpha[S(v_1) + S(v_2)] = \alpha S(v_1) + \alpha S(v_2)$$

and, for $\xi \in F$,

$$\begin{aligned}(\alpha S)(\xi v) &= \alpha[S(\xi v)] = \alpha[\xi S(v)] \\ &= (\alpha\xi)S(v) \quad = (\xi\alpha)S(v) \\ &= \xi[(\alpha S)(v)],\end{aligned}$$

since F satisfies the commutative law for multiplication.

The linear transformation 0, which sends each $v \in V$ into the zero vector, satisfies the condition that†

$$T + 0 = T, \qquad T \in L(V, W)$$

and the transformation $-T$, defined by

$$(-T)v = -T(v)$$

satisfies the condition that

$$T + (-T) = 0.$$

The verification of the other axioms is left to the reader.

The set of linear transformations of a vector space V into itself admits three operations: addition, scalar multiplication, and multiplication of linear transformations. The next result shows what algebraic properties these operations have.

(11.7) THEOREM. *Let V be a vector space over F, and let $S, T \in L(V, V)$ and $\alpha \in F$;* then

$$S + T, \qquad ST, \qquad \alpha S$$

are all elements of $L(V, V)$. With respect to the operations $S + T$ and αS, $L(V, V)$ is a vector space. Moreover, $L(V, V)$ has the further properties

$$S(TU) = (ST)U \qquad \textit{(associative law);}$$

there is a linear transformation 1 *called the* identity transformation *on V such that*

$$1v = v, \qquad v \in V$$

and

$$1T = T1 = T, \qquad T \in L(V, V).$$

† The symbol 0 is given still another meaning, but the context will always indicate which meaning is intended.

Finally, we have the distributive laws

$$S(T + U) = ST + SU, \qquad (S + T)U = SU + TU.$$

Proof. Because the properties of $L(V, V)$ relative to the operations $S + T$ and αS were already described in the preceding theorem, it is sufficient to consider the properties of ST. First we check that ST is a linear transformation. For $v_1, v_2 \in V$ and $\alpha \in F$ we have

$$\begin{aligned}(ST)(v_1 + v_2) &= S[T(v_1 + v_2)] = S[T(v_1) + T(v_2)] \\ &= S[T(v_1)] + S[T(v_2)] = (ST)v_1 + (ST)v_2\end{aligned}$$

and

$$\begin{aligned}(ST)(\alpha v_1) &= S[T(\alpha v_1)] = S[\alpha T(v_1)] \\ &= \alpha\{S[T(v_1)]\} = \alpha[ST(v_1)].\end{aligned}$$

This completes the proof that $ST \in L(V, V)$. The same argument shows that TS in Definition (11.5) is a linear transformation.

Now let S, T, U be elements of $L(V, V)$. The associative and distributive laws are verified by checking that the transformations $S(TU)$ and $(ST)U$, for example, have the same effect on an arbitrary vector in V. Let $v \in V$; then

$$[S(TU)](v) = S[(TU)(v)] = S\{T[U(v)]\}$$

while

$$[(ST)U](v) = ST[U(v)] = S\{T[U(v)]\}.$$

Similarly,

$$\begin{aligned}[S(T + U)](v) &= S[(T + U)(v)] = S[T(v) + U(v)] \\ &= S[T(v)] + S[U(v)] = (ST + SU)(v)\end{aligned}$$

and

$$\begin{aligned}[(S + T)U](v) &= (S + T)U(v) = S[U(v)] + T[U(v)] \\ &= (SU)(v) + (TU)(v) = (SU + TU)(v).\end{aligned}$$

The properties of the identity transformation 1 are immediate. This completes the proof of the theorem.

The previous result leads to the following general concept.

(11.8) Definition. A *ring* $\mathscr{R}$ is a mathematical system consisting of a nonempty set $\mathscr{R} = \{a, b, \ldots\}$ together with two operations, addition and multiplication, each of which assigns to a pair of elements a and b in $\mathscr{R}$ other elements of $\mathscr{R}$, denoted by $a + b$ in the case of addition and ab in the case of multiplication, such that the following conditions hold for all a, b, c in $\mathscr{R}$.

1. $a + b = b + a$.
2. $(a + b) + c = a + (b + c)$.
3. There is an element 0 such that $a + 0 = a$ for all $a \in \mathscr{R}$ and an element 1 such that $a1 = 1a = a$ for all a.
4. For each $a \in \mathscr{R}$ there is an element $-a$ such that $a + (-a) = 0$.
5. $(ab)c = a(bc)$.
6. $a(b + c) = ab + ac$, $(a + b)c = ac + bc$.
7. If the commutative law for multiplication ($ab = ba$, for $a, b \in \mathscr{R}$) holds, then $\mathscr{R}$ is called a *commutative ring*.

Any field is a commutative ring; the integers Z form a commutative ring which is not a field. In the exercises at the end of this section, the reader is asked to show that the commutative law for multiplication does not hold in general for the linear transformations in $L(V, V)$. Theorem (11.7) now can be stated more concisely:

(11.7′) THEOREM. *$L(V, V)$ is a ring.*

It is worth checking to what extent the proofs of the ring axioms for $L(V, V)$ depend on the assumption that the elements are *linear* transformations. To make this question more precise, let $M(V, V)$ be the set of *all* functions $T: V \to V$ and define $S + T$ and ST as for linear transformations. It can easily be verified that all the axioms for a ring hold for $M(V, V)$, with the exception of the one distributive law

$$S(T + U) = ST + SU,$$

which actually fails for suitably chosen S, T, and U in $M(V, V)$.

The mapping $T: \langle \alpha, \beta \rangle \to \langle \beta, 0 \rangle$, for $\alpha, \beta \in R$, is a linear transformation of R_2 such that $T \neq 0$ and $T^2 = 0$. It is impossible for T to have a reciprocal $\hat{T}$ such that $T\hat{T} = 1$, since $T\hat{T} = 1$ implies

$$T(T\hat{T}) = T \cdot 1 = T,$$

while, because of the associative law,

$$T(T\hat{T}) = T^2\hat{T} = 0 \cdot \hat{T} = 0,$$

which produces the contradiction $T = 0$. Because of this phenomenon, it is necessary to make the following definition.

(11.9) DEFINITION. A linear transformation $T \in L(V, V)$ is said to be *invertible* (or *nonsingular*) if there exists a linear transformation $T^{-1} \in L(V, V)$ (called the *inverse* of T) such that

$$TT^{-1} = T^{-1}T = 1.$$

An exercise at the end of this section shows that $TU = 1$ does not imply always that $UT = 1$; so if T' is to be shown the inverse of T, both the equations $TT' = 1$ and $T'T = 1$ must be checked.

Some other properties of invertible transformations can be best understood from the viewpoint of the following definition.

(11.10) DEFINITION. A *group* G is a mathematical system consisting of a nonempty set G, together with one operation which assigns to each pair of elements S, T in G a third element ST in G, such that the following conditions are satisfied.

1. $(ST)U = S(TU)$, for $S, T, U \in G$.
2. There is an element $1 \in G$ such that $S1 = 1S = S$ for all $S \in G$.
3. For each $S \in G$ there is an element $S^{-1} \in G$ such that $SS^{-1} = S^{-1}S = 1$.

(11.11) THEOREM. *Let G be the set of invertible linear transformations in $L(V, V)$; then G is a group, with respect to the product operation defined in* (11.5).

Proof. In view of the definition of invertible linear transformation and what has already been proved about $L(V, V)$, it is only necessary to check that if $S, T \in G$ then $ST \in G$. We have

$$(ST)T^{-1}S^{-1} = S \cdot 1S^{-1} = 1,$$
$$T^{-1}S^{-1}(ST) = T^{-1} \cdot 1T = 1.$$

Hence $ST \in G$ and $T^{-1}S^{-1}$ is an inverse of ST.

The next theorem, on the uniqueness of T^{-1}, etc., holds for groups in general.

(11.12) THEOREM. *Let G be an arbitrary group; then the equations*

$$AX = B, \qquad XA = B$$

have unique solutions $A^{-1}B$ and BA^{-1}, respectively. In particular, $AX = 1$ implies that $X = A^{-1}$. Similarly, $XA = 1$ implies $X = A^{-1}$.

Proof. We have

$$A(A^{-1}B) = (AA^{-1})B = 1B = B$$

and

$$(BA^{-1})A = B(A^{-1}A) = B \cdot 1 = B,$$

proving that solutions of the equations do exist. For the uniqueness, suppose that

$$AX' = B.$$

Then $A^{-1}(AX') = A^{-1}B$ and, since

$$A^{-1}(AX') = (A^{-1}A)X' = 1 \cdot X' = X',$$

we have $X' = A^{-1}B$. Similarly, $X'A = B$ implies $X' = BA^{-1}$. This completes the proof of the theorem.

Other important examples of groups will be considered in the next chapter.

We come next to a method for testing whether a linear transformation is invertible or not.

(11.13) THEOREM. *A linear transformation $T \in L(V, V)$ is invertible if and only if T is one-to-one and onto.*

Proof. Suppose first that T is invertible. Then there exists a linear transformation T^{-1} such that

$$TT^{-1} = T^{-1}T = 1.$$

Now suppose $T(v_1) = T(v_2)$ for vectors $v_1, v_2 \in V$. This relation implies that

$$T^{-1}(T(v_1)) = T^{-1}(T(v_2)),$$

and it follows that $v_1 = v_2$, since $T^{-1}T = 1$. At this point we have proved that T is one-to-one. To show that T is onto, let $v \in V$. Then

$$v = 1(v) = (TT^{-1})v = T(T^{-1}(v))$$

and we have shown that $v = T(u)$, where $u = T^{-1}(v)$. Therefore T is onto.

Conversely, suppose T is one-to-one and onto. We have to show that T is invertible, and this involves defining the linear transformation T^{-1}. Let $v \in V$. Since T is onto, there exists a vector $w \in V$ such that $T(w) = v$, and since T is one-to-one, there is only one such vector w. Therefore the rule

$$T^{-1}(v) = w, \qquad \text{where } T(w) = v,$$

defines a function $T^{-1}: V \to V$. We first have to show that T^{-1} is a linear transformation. Let $v_1, v_2 \in V$, and find w_1, w_2 such that

$$T(w_1) = v_1, \qquad T(w_2) = v_2.$$

Then since T is linear, we have

$$T(w_1 + w_2) = v_1 + v_2.$$

Therefore, by the definition of T^{-1},

$$T^{-1}(v_1 + v_2) = w_1 + w_2 = T^{-1}(v_1) + T^{-1}(v_2).$$

Similarly, let $T(w) = v$, and let $\alpha \in F$. Then $T(\alpha w) = \alpha T(w) = \alpha v$, and we have

$$T^{-1}(\alpha v) = \alpha w = \alpha T^{-1}(v),$$

completing the proof that $T^{-1} \in L(V, V)$. Finally, to check that $TT^{-1} = T^{-1}T = 1$, we have, for $v \in V$, and w such that $T(w) = v$,

$$TT^{-1}(v) = T(w) = v = 1v,$$

and

$$T^{-1}T(w) = T^{-1}(v) = w = 1w.$$

Since these equations hold for all vectors v and w, we conclude that T is invertible, and the theorem is proved.

(11.14) DEFINITION. A linear transformation $T: V \rightarrow W$ is called an *isomorphism* if T is a one-to-one mapping of V onto W. If there exists an isomorphism $T: V \rightarrow W$, we say that the vector spaces V and W are *isomorphic.*

If two vectors spaces are isomorphic, then they have the same structure as vector spaces, and every fact that holds for one vector space can be translated into a fact about the other. In fact, this is the reason for using the word "isomorphic," derived from Greek words meaning "of the same form." For example, if one space has dimension 5, so does the other, since the given isomorphism will carry a basis of one space onto a basis of the other. In case there is one isomorphism $T: V \rightarrow W$, there are many other isomorphisms different from T, and in translating properties of V to properties of W, we have to be careful to indicate which isomorphism we are using.

Example E. Suppose V is a vector space over R of dimension 3. We shall prove that there exists an isomorphism $T: V \rightarrow R_3$. Since $\dim V = 3$, V has a basis $\{v_1, v_2, v_3\}$. Then every vector $v \in V$ can be expressed in the form

$$v = \alpha_1 v_1 + \alpha_2 v_2 + \alpha_3 v_3$$

with coefficients $\alpha_i \in R$. Since $\{v_1, v_2, v_3\}$ are linearly independent, the coefficients $\{\alpha_1, \alpha_2, \alpha_3\}$ are uniquely determined by the vector v (why?). Therefore, the rule

$$T(v) = \langle \alpha_1, \alpha_2, \alpha_3 \rangle$$

defines a function $T: V \rightarrow R_3$. We have to check that T is linear, and one-to-one, and onto. Suppose v is given as above, and let $\alpha \in R$. Then

$$\alpha v = \alpha(\alpha_1 v_1 + \alpha_2 v_2 + \alpha_3 v_3) = \alpha\alpha_1 v_1 + \alpha\alpha_2 v_2 + \alpha\alpha_3 v_3.$$

Therefore,

$$T(\alpha v) = \langle \alpha\alpha_1, \alpha\alpha_2, \alpha\alpha_3 \rangle = \alpha\langle \alpha_1, \alpha_2, \alpha_3 \rangle = \alpha T(v).$$

Similarly, $T(v + v') = T(v) + T(v')$, since the coefficients of $v + v'$ with respect to the given basis are the sums of the coefficients of v and v', respectively. The fact that every vector in V is a linear combination of $\{v_1, v_2, v_3\}$ shows that T is onto. Finally, suppose $T(v) = T(v') = \langle \alpha_1, \alpha_2, \alpha_3 \rangle$. Then $v = \alpha_1 v_1 + \alpha_2 v_2 + \alpha_3 v_3 = v'$, and T is one-to-one.

Of course, there is nothing special about dimension 3; the same argument will show that a vector space of dimension n over R is isomorphic to R_n.

Example F. Let T and U be linear transformations from $R_2 \to R_2$ given by systems of equations

$$T: \begin{array}{l} y_1 = -x_1 + 2x_2 \\ y_2 = 0 \end{array} \qquad U: \begin{array}{l} y_1 = x_1 \\ y_2 = x_1 \end{array}.$$

Find systems of equations defining $T + U$, TU.

We have $T(\langle \alpha_1, \alpha_2 \rangle) = \langle -\alpha_1 + 2\alpha_2, 0 \rangle$, while $U(\langle \beta_1, \beta_2 \rangle) = \langle \beta_1, \beta_1 \rangle$. Therefore,

$$\begin{aligned} (T + U)(\langle \alpha_1, \alpha_2 \rangle) &= T\langle \alpha_1, \alpha_2 \rangle + U\langle \alpha_1, \alpha_2 \rangle \\ &= \langle -\alpha_1 + 2\alpha_2, 0 \rangle + \langle \alpha_1, \alpha_1 \rangle \\ &= \langle 2\alpha_2, \alpha_1 \rangle. \end{aligned}$$

Therefore a system of equations defining $T + U$ is given by

$$\begin{aligned} y_1 &= 2x_2 \\ y_2 &= x_1. \end{aligned}$$

To find TU, we have

$$\begin{aligned} TU\langle \alpha_1, \alpha_2 \rangle &= T(U\langle \alpha_1, \alpha_2 \rangle) = T\langle \alpha_1, \alpha_1 \rangle \\ &= \langle -\alpha_1 + 2\alpha_1, 0 \rangle = \langle \alpha_1, 0 \rangle, \end{aligned}$$

and a system of equations defining TU is given by

$$\begin{aligned} y_1 &= x_1 \\ y_2 &= 0. \end{aligned}$$

Example G. Prove that in order for a linear transformation T defined by a system of equations

$$y_i = \sum_{j=1}^{n} \alpha_{ij} x_j, \qquad i = 1, \ldots, m,$$

to be one-to-one, it is necessary and sufficient that the homogeneous system

$$\sum_{j=1}^{n} \alpha_{ij} x_j = 0, \qquad i = 1, \ldots, m$$

have only the trivial solution.

First suppose the linear transformation T is one-to-one. Then $T(0) = 0$, and if $v = \langle \alpha_1, \ldots, \alpha_n \rangle$ is a solution of the homogeneous system, then $T(v) = 0$. Since T is one-to-one, $v = 0$ is the only solution of the homogeneous system. Conversely, suppose the homogeneous system has only the trivial solution, and let $T(v) = T(v')$, where $v = \langle \alpha_1, \ldots, \alpha_n \rangle$, $v' = \langle \alpha_1', \ldots, \alpha_n' \rangle$. Then $v - v'$ is a solution of the homogeneous system, and $v - v' = 0$.

Example H. Test the linear transformation $T: R_3 \to R_2$ given by the system of equations

$$\begin{aligned} y_1 &= x_1 - 2x_2 + x_3 \\ y_2 &= x_1 \qquad\quad + x^3 \end{aligned}$$

to determine whether or not T is one-to-one.

The dimension of the solution space of the homogeneous system

$$\begin{aligned} x_1 - 2x_2 + x_3 &= 0 \\ x_1 \qquad\quad + x_3 &= 0 \end{aligned}$$

is $3 - \text{rank}\,(\mathbf{A})$, where $\mathbf{A}$ is the coefficient matrix. Since $\text{rank}\,(\mathbf{A}) = 2$, the dimension is one, and there do exist nontrivial solutions to the homogeneous system. Therefore T is not one-to-one.

Example I. Consider the linear transformation $T: R_n \to R_m$ defined by the system of equations

$$y_i = \sum_{j=1}^{n} \alpha_{ij} x_j, \qquad i = 1, \ldots, m.$$

Show that a vector $w \in R_m$ is the image $T(v)$ of some vector $v \in R_n$ if and only if w is a linear combination of the column vectors of the matrix of the system.

In Chapter 2, we showed that if $w = \langle y_1, \ldots, y_n \rangle$, then the system of equations can be written in the form

$$w = x_1 c_1 + \cdots + x_n c_n,$$

where $c_1, \ldots, c_n$ are the column vectors of the system. This statement solves the problem.

Example J. Test the linear transformation $T: R_3 \to R_2$ given in Example H to decide whether it is onto.

By the discussion in Example I, T is onto if and only if every vector in R_2 is a linear combination of the column vectors of the matrix

$$\mathbf{A} = \begin{pmatrix} 1 & -2 & 1 \\ 1 & 0 & 1 \end{pmatrix}.$$

This is certainly the case, since $\mathbf{A}$ has rank 2, and we conclude that the transformation is onto.

EXERCISES

1. Which of the following mappings of $R_2 \to R_2$ are linear transformations?
   a. $\langle x_1, x_2\rangle \to \langle y_1, y_2\rangle$, where $y_1 = 3x_1 - x_2 + 1$ and $y_2 = -x_1 + 2x_2$.
   b. $\langle x_1, x_2\rangle \to \langle y_1, y_2\rangle$, where $y_1 = 3x_1 + x_2^2$ and $y_2 = -x_1$.
   c. $\langle x_1, x_2\rangle \to \langle x_2, x_1\rangle$.
   d. $\langle x_1, x_2\rangle \to \langle x_1 + x_2, x_2\rangle$.
   e. $\langle x_1, x_2\rangle \to \langle 2x_1, x_2\rangle$.
2. Verify that the function $T: F_n \to F_m$ described in Example A is a linear transformation.
3. Let T be the linear transformation of $F_2 \to F_2$ defined by the system

$$\begin{aligned} y_1 &= -3x_1 + x_2 \\ y_2 &= \quad\; x_1 - x_2 \end{aligned}$$

   and let U be the linear transformation defined by the system

$$\begin{aligned} y_1 &= x_1 + x_2 \\ y_2 &= x_1 \end{aligned}$$

   Find a system of linear equations defining the linear transformations

$$2T, \qquad T - U, \qquad T^2, TU, UT, T^2 + 2U.$$

   Is $TU = UT$?
4. Let D be the linear transformation of the polynomial functions $P[R]$ defined by $Df =$ derivative of f. Let M be the mapping of $P[R] \to P[R]$ defined by $(Mf)(x) = xf(x)$. Is M a linear transformation of $P[R]$? Find the transformations $M + D$, DM, MD. Is $MD = DM$?
5. Let T be a linear transformation of V into W. Prove that $T(0) = 0$ and that $T(-v) = -T(v)$ for all $v \in V$. If V_1 is a subspace of V, prove that $T(V_1)$, which consists of all vectors $\{Ts, s \in V_1\}$, is a subspace of W.
6. Using the methods of Examples G and H, test the linear transformations defined by the following systems of equations to determine whether they are one-to-one.
   a. $$\begin{aligned} y_1 &= 3x_1 - x_2 \\ y_2 &= \; x_1 + x_2 \end{aligned}$$
   b. $$\begin{aligned} y_1 &= 3x_1 - x_2 + x_3 \\ y_2 &= -x_1 \qquad\;\; + 2x_3 \end{aligned}$$
   c. $$\begin{aligned} y_1 &= x_1 + 2x_2 \\ y_2 &= x_1 - x_2 \\ y_3 &= -2x_1 \end{aligned}$$
   d. $$\begin{aligned} y_1 &= x_1 \qquad\quad - x_3 \\ y_2 &= x_1 + x_2 \\ y_3 &= -x_1 - 3x_2 - 2x_3 \end{aligned}$$
   e. $$\begin{aligned} y_1 &= x_1 + 2x_2 + x_3 \\ y_2 &= x_1 + x_2 \\ y_3 &= \qquad x_2 + x_3. \end{aligned}$$

7. Using Example I, prove that a linear transformation defined by a system of equations carries R_n onto R_m if and only if the rank of the coefficient matrix of the system is m.
8. Using Example I, test the linear transformations given in Exercise 6 to decide which ones map R_n onto R_m.
9. Let T be a linear transformation of R_n into R_n defined by a system of linear equations

$$y_i = \sum_{j=1}^{n} \alpha_{ij}x_j, \qquad i = 1, 2, \ldots, n.$$

Prove that the following statements about T are equivalent.
   a. T is one-to-one.
   b. T is onto.
   c. T is an isomorphism of F_n onto F_n.
   d. T is an invertible linear transformation.
10. Let I be the linear transformation of $P[R] \to P[R]$ defined by

$$I(f) = \alpha_0 x + \frac{\alpha_1 x^2}{2} + \cdots + \frac{\alpha_k x^{k+1}}{k+1},$$

for a polynomial function $f(x) = \alpha_0 + \alpha_1 x + \cdots + \alpha_k x^k$. Let D be the derivative operation in $P[R]$. Show that $DI = 1$, but that neither D nor I are isomorphisms. Is D one-to-one? Onto? Answer the same questions for I.

12. ADDITION AND MULTIPLICATION OF MATRICES

In Section 11 we defined the operations of addition, multiplication by scalars, and multiplication of linear transformations. In this section the corresponding operations are defined for matrices. In Section 13 we shall relate linear transformations of arbitrary finite dimensional vector spaces with matrices; Section 12 can be regarded as an introduction to that discussion.

First of all there is no difficulty in defining the vector space operations on matrices; we simply treat m-by-n matrices as $m \cdot n$-tuples.

(12.1) DEFINITION. Let $\mathbf{A} = (\alpha_{ij})$ and $\mathbf{B} = (\beta_{ij})$ be two m-by-n matrices with coefficients in a field F. The sum $\mathbf{A} + \mathbf{B}$ is defined to be the m-by-n matrix whose (i, j) entry is $\alpha_{ij} + \beta_{ij}$. If $\alpha \in F$, then $\alpha\mathbf{A}$ is the m-by-n matrix whose (i, j) entry is $\alpha\alpha_{ij}$.

We can now state the following result.

(12.2) THEOREM. *The set of all m-by-n matrices with coefficients in F forms a vector space over F, with respect to the operations given in Definition* (12.1).

The proof is the same as the verification that F_n is a vector space and is omitted.

The definition of matrix multiplication is not as obvious. In order to start, let us consider a problem similar to one discussed in the problems in Section 11.

Let $T: R_2 \to R_3$ be the linear transformation

$$\begin{aligned} y_1 &= x_1 + x_2 \\ y_2 &= -x_1 + x_2 \\ y_3 &= x_1 \end{aligned} \tag{12.3}$$

and let $U: R_3 \to R_3$ be the linear transformation

$$\begin{aligned} y_1 &= -x_1 \phantom{{}+x_2} + x_3 \\ y_2 &= x_1 + x_2 \\ y_3 &= -x_1 \phantom{{}+x_2} + 2x_3 . \end{aligned}$$

The matrices associated with these systems of equations are

$$T \leftrightarrow \mathbf{A} = \begin{pmatrix} 1 & 1 \\ -1 & 1 \\ 1 & 0 \end{pmatrix}, \qquad U \leftrightarrow \mathbf{B} = \begin{pmatrix} -1 & 0 & 1 \\ 1 & 1 & 0 \\ -1 & 0 & 2 \end{pmatrix},$$

The linear transformation UT maps R_2 into R_3. Let us work out the system of equations defining it.

If

$$T(x) = y, \qquad U(y) = z$$

then

$$UT(x) = U(y) = z.$$

Rewriting the system for U so that it carries $\{y_1, y_2, y_3\} \to \{z_1, z_2, z_3\}$, we have the system

$$U: \quad \begin{aligned} z_1 &= -y_1 \phantom{{}+y_2} + y_3 \\ z_2 &= y_1 + y_2 \\ z_3 &= -y_1 \phantom{{}+y_2} + 2y_3 \end{aligned} \tag{12.4}$$

Now we can substitute for $\{y_1, y_2, y_3\}$ using (12.3) to find the system associated with UT:

$$\begin{aligned} z_1 &= -(x_1 + x_2) + x_1 &&= -x_2 \\ z_2 &= (x_1 + x_2) + (-x_1 + x_2) &&= 2x_2 \\ z_3 &= -(x_1 + x_2) + 2x_1 &&= x_1 - x_2 \end{aligned} \tag{12.5}$$

which is associated with the matrix

$$\mathbf{C} = \begin{pmatrix} 0 & -1 \\ 0 & 2 \\ 1 & -1 \end{pmatrix}.$$

It is natural to define the matrix **C** to be the product of the matrices **B** and **A**. Then the product UT of linear transformations will correspond to the product **BA** of the corresponding matrices.

In order to see how the product $\mathbf{C} = \mathbf{BA}$ is formed, let us write the equations (12.3) and (12.4) in the form

$$\begin{aligned} y_1 &= \alpha_{11}x_1 + \alpha_{12}x_2 \\ y_2 &= \alpha_{21}x_1 + \alpha_{22}x_2 \\ y_3 &= \alpha_{31}x_1 + \alpha_{32}x_2 \end{aligned}$$

and

$$\begin{aligned} z_1 &= \beta_{11}y_1 + \beta_{12}y_2 + \beta_{13}y_3 \\ z_2 &= \beta_{21}y_1 + \beta_{22}y_2 + \beta_{23}y_3 \\ z_3 &= \beta_{31}y_1 + \beta_{32}y_2 + \beta_{33}y_3 \end{aligned}$$

respectively. Then (12.5) becomes

$$\begin{aligned} z_1 &= \beta_{11}(\alpha_{11}x_1 + \alpha_{12}x_2) + \beta_{12}(\alpha_{21}x_1 + \alpha_{22}x_2) + \beta_{13}(\alpha_{31}x_1 + \alpha_{32}x_2) \\ z_2 &= \beta_{21}(\alpha_{11}x_1 + \alpha_{12}x_2) + \beta_{22}(\alpha_{21}x_1 + \alpha_{22}x_2) + \beta_{23}(\alpha_{31}x_1 + \alpha_{32}x_2) \\ z_3 &= \beta_{31}(\alpha_{11}x_1 + \alpha_{12}x_2) + \beta_{32}(\alpha_{21}x_1 + \alpha_{22}x_2) + \beta_{33}(\alpha_{31}x_1 + \alpha_{32}x_2). \end{aligned}$$

Then the entry in the (1, 1) position of the product matrix **BA** may be obtained by taking the first row of **B** and the first column of **A**, multiplying corresponding elements together, and adding. The other entries are seen to be obtained by the same process.

We can now make a formal definition.

(12.6) DEFINITION. Let $\mathbf{B} = (\beta_{ij})$ be a q-by-n matrix (q rows and n columns) with coefficients in F, $\mathbf{A} = (\alpha_{kl})$ an n-by-m matrix. Then the product matrix $\mathbf{C} = \mathbf{BA}$ is defined by the q-by-m matrix whose (i, j) entry γ_{ij} is given by

$$\gamma_{ij} = \beta_{i1}\alpha_{1j} + \cdots + \beta_{in}\alpha_{nm} = \sum_{k=1}^{n} \beta_{ik}\alpha_{kj},$$

for $1 \leq i \leq q$, $1 \leq j \leq m$. In general, multiplication of a q-by-r matrix **B** and an s-by-t matrix **A** is defined if and only if $r = s$, and in that case results in a q-by-t matrix **BA** according to the rule we have given.

The process of multiplying two matrices together is much simpler than the formula given in the definition may lead one to think.

In order to multiply two matrices **BA**, proceed as follows:

1. Check that the number of columns in **B** is equal to the number of rows of **A**. If this is not the case, then the product **BA** is not defined.
2. Assuming (1) is satisfied, the product matrix **BA** will have as many rows as **B** and as many columns as **A**. To find the entry in the first row, first column of **BA**, take the first row of **B** and the first column

of **A**, multiply corresponding elements and add (this is just what the formula in the definition says!). To find the entry of **BA** in the first row, second column, take the first row of **B**, the second column of **A**, multiply corresponding elements, and add. In general, to find the entry in the ith row, jth column of **BA**, take the ith row of **B**, the jth column of **A**, multiply corresponding elements and add.

Example A. Multiply the following matrices with real coefficients.

(i)
$$\begin{pmatrix} -1 & 0 & 1 \\ 1 & 1 & 1 \\ -1 & 0 & 2 \end{pmatrix}\begin{pmatrix} 1 & 1 \\ -1 & 1 \\ 1 & 0 \end{pmatrix} = \begin{pmatrix} \gamma_{11} & \gamma_{12} \\ \gamma_{21} & \gamma_{22} \\ \gamma_{31} & \gamma_{32} \end{pmatrix}.$$

where

$$\gamma_{11} = (-1, 0, 1)\begin{pmatrix} 1 \\ -1 \\ 1 \end{pmatrix} = 0$$

$$\gamma_{12} = (-1, 0, 1)\begin{pmatrix} 1 \\ 1 \\ 0 \end{pmatrix} = -1$$

$$\gamma_{21} = (1, 1, 1)\begin{pmatrix} 1 \\ -1 \\ 1 \end{pmatrix} = 1$$

$$\gamma_{22} = (1, 1, 1)\begin{pmatrix} 1 \\ 1 \\ 0 \end{pmatrix} = 2$$

etc. We have written, for example,

$$(1, 1, 1)\begin{pmatrix} 1 \\ 1 \\ 0 \end{pmatrix}$$

to stand for multiplying corresponding elements and adding. Thus

$$(1, 1, 1)\begin{pmatrix} 1 \\ 1 \\ 0 \end{pmatrix} = 1 \cdot 1 + 1 \cdot 1 + 1 \cdot 0 = 2.$$

Here are some more examples.

(ii)
$$\begin{pmatrix} 1 & 1 \\ -1 & 1 \\ 1 & 0 \end{pmatrix}\begin{pmatrix} -1 & 0 & 1 \\ 1 & 1 & 0 \\ -1 & 0 & 2 \end{pmatrix} \quad \text{not defined}$$

(iii)
$$\begin{pmatrix} 2 & 1 \\ 0 & -1 \end{pmatrix}\begin{pmatrix} 1 \\ -1 \end{pmatrix} = \begin{pmatrix} 1 \\ 1 \end{pmatrix}$$

(iv)
$$(1 \quad -1)\begin{pmatrix} 2 & 1 \\ 0 & -1 \end{pmatrix} = (2 \quad 2).$$

The next result is perhaps the most useful fact about matrix multiplication.

(12.7) THEOREM. *The associative law for matrix multiplication holds. More precisely, let* **A**, **B**, **C** *be three matrices with coefficients in a field F, and suppose that the products* **AB** *and* **BC** *are defined. Then the products* (**AB**)**C** *and* **A**(**BC**) *are defined, and we have*

$$(\mathbf{AB})\mathbf{C} = \mathbf{A}(\mathbf{BC}).$$

Proof. Let $\mathbf{A} = (\alpha_{ij})$ be an m-by-n matrix, $\mathbf{B} = (\beta_{ij})$ an n-by-p matrix, and $\mathbf{C} = (\gamma_{ij})$ a p-by-q matrix. Then the (i, j) entry of $(\mathbf{AB})\mathbf{C}$ is $\sum_{s=1}^{p} (\sum_{r=1}^{n} \alpha_{ir}\beta_{rs})\gamma_{sj}$. The (i, j) entry of $\mathbf{A}(\mathbf{BC})$ is $\sum_{t=1}^{n} \alpha_{it} (\sum_{u=1}^{p} \beta_{tu}\gamma_{uj})$, and it is not difficult to check that these expressions are equal.

Matrix multiplication can be used to rewrite systems of equations. For example, the system

$$\begin{aligned} 2x_1 - x_2 &= 1 \\ x_1 + x_2 &= 0 \end{aligned}$$

can be written in matrix form

$$\begin{pmatrix} 2 & -1 \\ 1 & 1 \end{pmatrix} \begin{pmatrix} x_1 \\ x_2 \end{pmatrix} = \begin{pmatrix} 1 \\ 0 \end{pmatrix}.$$

More generally, a system

(12.8)
$$\begin{aligned} \alpha_{11}x_1 + \cdots + \alpha_{1n}x_n &= \beta_1 \\ \cdots\cdots&\cdots\cdots \\ \alpha_{m1}x_m + \cdots + \alpha_{mn}x_n &= \beta_m \end{aligned}$$

of m equations in n unknowns can be written in matrix form

$$\mathbf{A}x = b$$

where $\mathbf{A} = (\alpha_{ij})$, and

$$x = \begin{pmatrix} x_1 \\ \vdots \\ x_n \end{pmatrix}, \qquad b = \begin{pmatrix} \beta_1 \\ \vdots \\ \beta_m \end{pmatrix}.$$

The linear transformation defined by the system (12.8) can be described as follows:

(12.9)
$$T(\mathbf{x}) = \mathbf{A} \cdot \mathbf{x},$$

where $\mathbf{x}$ is the column vector with entries $x_1, \ldots, x_n$, regarded as an n-by-1 matrix, and $\mathbf{A} \cdot \mathbf{x}$ denotes matrix multiplication.

These observations take the pain out of working with linear transformations defined by systems of equations.

Example B. Let $T\colon R_n \to R_m$ be a linear transformation defined by a system of equations (12.8) with coefficient matrix **A**, and let $U\colon R_m \to R_p$ be defined by another system of equations with coefficient matrix **B**. Then the product linear transformation $UT\colon R_n \to R_p$, can be computed using the formula (12.9), as follows. Let **x** be a column vector, representing an element of R_n. Then

$$UT(\mathbf{x}) = U(T(\mathbf{x})) = U(\mathbf{Ax}) = \mathbf{B}(\mathbf{Ax}),$$

where the products on the right-hand side are all in terms of matrix multiplication. By the associative law for matrix multiplication,

$$UT(\mathbf{x}) = \mathbf{B}(\mathbf{Ax}) = (\mathbf{BA})\mathbf{x},$$

which expresses the fact that the product of two linear transformations is represented by a system of equations whose matrix is the product (in the same order) of the coefficient matrices corresponding to the linear transformations.

For a numerical example, let T and U be defined by systems of equations as follows:

$$T\colon\ \begin{aligned} y_1 &= 3x_1 - x_2 \\ y_2 &= -x_1 + x_2 \end{aligned} \qquad U\colon\ \begin{aligned} y_1 &= -x_1 - x_2 \\ y_2 &= x_1 \end{aligned}$$

Then, letting

$$\mathbf{x} = \begin{pmatrix} x_1 \\ x_2 \end{pmatrix},$$

we have

$$\begin{aligned} T(\mathbf{x}) &= \begin{pmatrix} 3 & -1 \\ -1 & 1 \end{pmatrix}\begin{pmatrix} x_1 \\ x_2 \end{pmatrix}, \\ UT(\mathbf{x}) &= \begin{pmatrix} -1 & -1 \\ 1 & 0 \end{pmatrix}\begin{pmatrix} 3 & -1 \\ -1 & 1 \end{pmatrix}\begin{pmatrix} x_1 \\ x_2 \end{pmatrix} \\ &= \begin{pmatrix} -2 & 0 \\ 3 & -1 \end{pmatrix}\begin{pmatrix} x_1 \\ x_2 \end{pmatrix}. \end{aligned}$$

Thus a system of equations defining UT is given by

$$\begin{aligned} y_1 &= -2x_1 \\ y_2 &= \ \ 3x_1 - x_2. \end{aligned}$$

In the remainder of this section we give some other applications of matrix multiplication.

(12.10) DEFINITION. The n-by-n identity matrix **I** is the matrix

$$\mathbf{I} = \begin{pmatrix} 1 & 0 & \cdots & 0 \\ 0 & 1 & & \vdots \\ \vdots & & \ddots & \vdots \\ 0 & \cdots & 0 & 1 \end{pmatrix}$$

with zeros except in the (1, 1), (2, 2), . . ., (n, n) positions where the entries are all equal to one.

Then one checks that

$$\mathbf{AI} = \mathbf{IA} = \mathbf{A}$$

for all n-by-n matrices **A**. An n-by-n matrix **A** is said to be *invertible*, if there exists a matrix **B** such that $\mathbf{AB} = \mathbf{BA} = \mathbf{I}$.

Using the associative law for matrix multiplication, the proof of Theorem (11.12) can be applied word for word to matrices; it shows that if **A** is invertible, then the matrix **B** such that $\mathbf{AB} = \mathbf{BA} = \mathbf{I}$ is uniquely determined, and will be called the *inverse* of **A** (notation $\mathbf{A}^{-1}$). The problem is, how do we compute $\mathbf{A}^{-1}$?

Example C. Finding the inverse of a matrix using elementary row operations. Some details in the discussion below are omitted and are left to the reader.

Let **I** be the n-by-n identity matrix. Define an *elementary matrix* as any one of the matrices $\mathbf{P}_{ij}$, $\mathbf{B}_{ij}(\lambda)$, $\mathbf{D}_i(\mu)$ for all $i, j = 1, \ldots, n$, and $\lambda, \mu \in F$. These matrices are defined as follows.

1. $\mathbf{P}_{ij}$ is obtained from **I** by interchanging the ith and jth rows.
2. $\mathbf{B}_{ij}(\lambda)$ is obtained from **I** by adding λ times the jth row of **I** to the ith row.
3. $\mathbf{D}_i(\mu)$ is obtained from **I** by multiplying the ith row of I by μ.

For example, in the set of 2-by-2 matrices we have

$$\mathbf{P}_{12} = \begin{pmatrix} 0 & 1 \\ 1 & 0 \end{pmatrix}, \quad \mathbf{B}_{12}(\lambda) = \begin{pmatrix} 1 & \lambda \\ 0 & 1 \end{pmatrix}, \quad \mathbf{D}_2(\mu) = \begin{pmatrix} 1 & 0 \\ 0 & \mu \end{pmatrix}.$$

Now let **A** be an arbitrary n-by-n matrix.

1. $\mathbf{P}_{ij}\mathbf{A}$ is obtained from **A** by interchanging the ith and jth rows of **A**.
2. $\mathbf{B}_{ij}(\lambda)\mathbf{A}$ is obtained from **A** by adding λ times the jth row of **A** to the ith row.
3. $D_i(\mu)\mathbf{A}$ is obtained from **A** by multiplying the ith row of **A** by μ.

As in Chapter 2, the operations on **A** described above are called *elementary operations* and may be referred to as types 1, 2, and 3, respectively.

Now let **A** be an arbitrary n-by-n matrix. By Theorem (6.16) there exists a sequence of elementary operations of types 1, 2, or 3 which reduces **A** to a matrix whose rows are in echelon form, and the number of nonzero rows is the rank of **A**.

Now suppose **A** is invertible. Then $\mathbf{A} \cdot \mathbf{x} = 0$ implies $\mathbf{x} = 0$, and so the dimension of the solution space of the homogeneous system of equations with coefficient matrix **A** is zero. By Corollary (9.4), **A** has rank n. Therefore the matrix whose rows are in echelon form, which we obtained in the first part of the argument, will have the form

$$\begin{pmatrix} \alpha_{11} & \alpha_{12} & \cdots & \alpha_{1n} \\ 0 & \alpha_{22} & \cdots & \alpha_{2n} \\ \cdots & \cdots & \cdots & \cdots \\ 0 & 0 & 0\ \ 0 & \alpha_{nn} \end{pmatrix},$$

with $\alpha_{11}, \ldots, \alpha_{nn}$ all different from zero. Applying elementary operations of type 3, we may assume that $\alpha_{11}, \ldots, \alpha_{nn}$ are all equal to one. Then by adding a suitable multiple of the second row to the first, we can make $\alpha_{12} = 0$, without changing α_{11}. Similarly, we can add multiples of the third row to the first and second to make $\alpha_{13} = \alpha_{23} = 0$, without affecting α_{11} or α_{22}. Continuing in this way, we reduce **A** to the identity matrix.

The preceding discussion shows that there exist elementary matrices $\mathbf{E}_1, \ldots, \mathbf{E}_s$ of types 1, 2, and 3 such that

$$\mathbf{E}_s\mathbf{E}_{s-1} \cdots \mathbf{E}_1\mathbf{A} = \mathbf{I}.$$

By the uniqueness of the inverse of a matrix,

$$\mathbf{A}^{-1} = \mathbf{E}_s\mathbf{E}_{s-1} \cdots \mathbf{E}_1\mathbf{I};$$

so that if the elementary operations given by $\mathbf{E}_1, \ldots, \mathbf{E}_s$ reduce **A** to **I**, the same sequence of elementary operations applied to **I** will yield $\mathbf{A}^{-1}$.

As a numerical example, let us test for invertibility, and if invertible, find $\mathbf{A}^{-1}$, for the matrix

$$\mathbf{A} = \begin{pmatrix} 2 & -1 \\ 3 & 0 \end{pmatrix}.$$

We do the work in two columns; in one column we apply elementary row operations to reduce **A** to the identity matrix, and in the other column we apply the same elementary row operations to *I*.

$$\mathbf{A} \sim \begin{pmatrix} 2 & -1 \\ 0 & \frac{3}{2} \end{pmatrix} \qquad \mathbf{I} \sim \begin{pmatrix} 1 & 0 \\ -\frac{3}{2} & 1 \end{pmatrix}$$

$$\sim \begin{pmatrix} 1 & -\frac{1}{2} \\ 0 & 1 \end{pmatrix} \qquad \sim \begin{pmatrix} \frac{1}{2} & 0 \\ -1 & \frac{2}{3} \end{pmatrix}$$

$$\sim \begin{pmatrix} 1 & 0 \\ 0 & 1 \end{pmatrix} \qquad \sim \begin{pmatrix} 0 & \frac{1}{3} \\ -1 & \frac{2}{3} \end{pmatrix}.$$

We conclude that

$$\mathbf{A}^{-1} = \begin{pmatrix} 0 & \frac{1}{3} \\ -1 & \frac{2}{3} \end{pmatrix},$$

and can easily check that this is the case.

The interpretation of matrix multiplication in terms of systems of equations leads to the following application of matrix multiplication to partial differentiation. We shall restrict ourselves to a particular case of the general situation.

Example D. *The chain rule for partial derivatives.* Let

$$T: R_2 \to R_3$$

be a function such that if $x = \langle x_1, x_2 \rangle$, then $T(x) = y = \langle y_1, y_2, y_3 \rangle$ where

$$\begin{aligned} y_1 &= F_1(x_1, x_2) \\ y_2 &= F_2(x_1, x_2) \\ y_3 &= F_3(x_1, x_2), \end{aligned}$$

and F_1, F_2, F_3 are functions such that the partial derivatives

$$\frac{\partial y_i}{\partial x_j} = \frac{\partial F_i}{\partial x_j}$$

exist, and are continuous for $1 \leq j \leq 2$, and $1 \leq i \leq 3$, at a point x.

We define the *jacobian matrix* of the function T at the point x to be the 3×2 matrix

$$\mathbf{J}(T)_x = \begin{pmatrix} \dfrac{\partial y_1}{\partial x_1} & \dfrac{\partial y_1}{\partial x_2} \\ \dfrac{\partial y_2}{\partial x_1} & \dfrac{\partial y_2}{\partial x_2} \\ \dfrac{\partial y_3}{\partial x_1} & \dfrac{\partial y_3}{\partial x_2} \end{pmatrix}_x,$$

where the notation $(.)_x$ means that the derivatives are evaluated at x. Now suppose $U: R_3 \to R_3$ is a function such that $U(y) = z$, where $z = \langle z_1, z_2, z_3 \rangle$, and

$$z_i = z_i(y_1, y_2, y_3),$$

for $1 \leq i \leq 3$. Suppose that the partial derivatives $\partial z_i/\partial y_j$ also exist and are continuous at the point y and form the jacobian matrix

$$\mathbf{J}(U)_y = \begin{pmatrix} \dfrac{\partial z_1}{\partial y_1} & \dfrac{\partial z_1}{\partial y_2} & \dfrac{\partial z_1}{\partial y_3} \\ \dfrac{\partial z_2}{\partial y_1} & \dfrac{\partial z_2}{\partial y_2} & \dfrac{\partial z_2}{\partial y_3} \\ \dfrac{\partial z_3}{\partial y_1} & \dfrac{\partial z_3}{\partial y_2} & \dfrac{\partial z_3}{\partial y_3} \end{pmatrix}_y.$$

The function $UT: R_2 \to R_3$ is given by $(UT)(x) = U(T(x)) = U(y) = z$ and has the jacobian matrix

$$\mathbf{J}(UT)_x = \begin{pmatrix} \dfrac{\partial z_1}{\partial x_1} & \dfrac{\partial z_1}{\partial x_2} \\ \dfrac{\partial z_2}{\partial x_1} & \dfrac{\partial z_2}{\partial x_2} \\ \dfrac{\partial z_3}{\partial x_1} & \dfrac{\partial z_3}{\partial x_2} \end{pmatrix}_x$$

if the partial derivatives exist. In any book on the calculus of functions of several variables it is proved that the partial derivatives $\partial z_i/\partial x_j$ do exist and are given by *chain rules* such as

$$\frac{\partial z_1}{\partial x_1} = \frac{\partial z_1}{\partial y_1}\frac{\partial y_1}{\partial x_1} + \frac{\partial z_1}{\partial y_2}\frac{\partial y_2}{\partial x_1} + \frac{\partial z_1}{\partial y_3}\frac{\partial y_3}{\partial x_1}.$$

The point is that these chain rules can all be expressed (and remembered!) in the simple form

$$\mathbf{J}(UT)_x = \mathbf{J}(U)_{y=T(x)} \cdot \mathbf{J}(T)_x, \tag{12.11}$$

where the multiplication is matrix multiplication of the jacobian matrices.

For example, let $T\colon R_1 \to R_3$ be given by

$$y = T(x) = \langle \sin x, \cos x, x \rangle,$$

and $U\colon R_3 \to R_2$, be defined by

$$U(y) = z = \langle z_1, z_2 \rangle$$

where $z_1 = y_1 y_2$, $z_2 = y_2 y_3$. We have

$$\mathbf{J}(T)_x = \begin{pmatrix} \dfrac{dy_1}{dx} \\[2mm] \dfrac{dy_2}{dx} \\[2mm] \dfrac{dy_3}{dx} \end{pmatrix} = \begin{pmatrix} \cos x \\ -\sin x \\ 1 \end{pmatrix}.$$

Moreover,

$$\mathbf{J}(U)_y = \begin{pmatrix} \dfrac{\partial z_1}{\partial y_1} & \dfrac{\partial z_1}{\partial y_2} & \dfrac{\partial z_1}{\partial y_3} \\[2mm] \dfrac{\partial z_2}{\partial y_1} & \dfrac{\partial z_2}{\partial y_2} & \dfrac{\partial z_2}{\partial y_3} \end{pmatrix} = \begin{pmatrix} y_2 & y_1 & 0 \\ 0 & y_3 & y_2 \end{pmatrix}.$$

Applying the formula (12.11), we obtain

$$\begin{aligned} \mathbf{J}(UT)_x &= \begin{pmatrix} y_2 & y_1 & 0 \\ 0 & y_3 & y_2 \end{pmatrix} \begin{pmatrix} \cos x \\ -\sin x \\ 1 \end{pmatrix} \\ &= (y_2 \cos x - y_1 \sin x, -y_3 \sin x + y_2) \\ &= \left(\frac{dz_1}{dx}, \frac{dz_2}{dx}\right). \end{aligned}$$

Thus, for example,

$$\frac{dz_1}{dx} = y_2 \cos x - y_1 \sin x = (\cos x)^2 - (\sin x)^2.$$

EXERCISES

The numerical problems refer to matrices with real entries.

1. Compute the following matrix products.

$$\begin{pmatrix} -1 & 2 \\ -1 & 0 \end{pmatrix}\begin{pmatrix} 1 & 1 \\ 0 & 1 \end{pmatrix}, \quad \begin{pmatrix} -1 & 2 & 3 \\ 1 & 1 & 1 \end{pmatrix}\begin{pmatrix} 1 & 1 \\ 1 & 0 \\ 2 & -1 \end{pmatrix},$$

$$\begin{pmatrix} 0 & 1 & 1 \\ -1 & 1 & 0 \end{pmatrix}\begin{pmatrix} 1 \\ 2 \\ 0 \end{pmatrix}, \quad (1 \;\; 1 \;\; 2)\begin{pmatrix} -1 \\ 0 \\ 1 \end{pmatrix},$$

$$\begin{pmatrix} 0 & 1 \\ 0 & 0 \end{pmatrix}\begin{pmatrix} 0 & 1 \\ 0 & 0 \end{pmatrix}, \quad \begin{pmatrix} 1 & 0 & 0 \\ 0 & -1 & 0 \\ 0 & 0 & 2 \end{pmatrix}\begin{pmatrix} 1 & 1 & 0 \\ -1 & 2 & 1 \\ 1 & 1 & 3 \end{pmatrix},$$

$$\begin{pmatrix} 1 & 1 & 0 \\ -1 & 2 & 1 \\ 1 & 1 & 3 \end{pmatrix}\begin{pmatrix} 1 & 0 & 0 \\ 0 & -1 & 0 \\ 0 & 0 & 2 \end{pmatrix},$$

2. Does matrix multiplication of n-by-n matrices satisfy the commutative law, $\mathbf{AB} = \mathbf{BA}$?

3. Letting $\mathbf{x}$ denote a column vector of appropriate size, solve the matrix equation $\mathbf{Ax} = \mathbf{b}$, in the following cases. Note that by (12.8)–(12.9), the matrix equation $\mathbf{Ax} = \mathbf{b}$ is equivalent to a system of linear equations.

a. $\mathbf{A} = \begin{pmatrix} 1 & 1 \\ -1 & 0 \end{pmatrix}, \quad \mathbf{b} = \begin{pmatrix} 2 \\ 1 \end{pmatrix}.$

b. $\mathbf{A} = \begin{pmatrix} 1 & 1 & 0 \\ -1 & 1 & 2 \end{pmatrix}, \quad \mathbf{b} = \begin{pmatrix} 0 \\ 0 \end{pmatrix}.$

c. $\mathbf{A} = \begin{pmatrix} 1 & 0 & 0 & 0 & 0 \\ 0 & 1 & 0 & 1 & 0 \\ -1 & 0 & 0 & 0 & 1 \end{pmatrix}, \quad \mathbf{b} = \begin{pmatrix} 1 \\ 0 \\ -1 \end{pmatrix}.$

d. $\mathbf{A} = \begin{pmatrix} 1 & 1 & 0 & 1 \\ 0 & 1 & -1 & 0 \end{pmatrix}, \quad \mathbf{b} = \begin{pmatrix} 1 \\ 2 \end{pmatrix}.$

4. Define a *diagonal matrix* $\mathbf{D}$ over F to be an n-by-n matrix whose (i, j) entry is zero unless $i = j$. We usually write

$$\mathbf{D} = \begin{pmatrix} \delta_1 & & 0 \\ & \ddots & \\ 0 & & \delta_n \end{pmatrix}$$

to denote a diagonal matrix. Figure out the effect of multiplying an arbitrary matrix on the left and on the right by a diagonal matrix $\mathbf{D}$.

5. Two matrices $\mathbf{A}$ and $\mathbf{B}$ are said to *commute* if $\mathbf{AB} = \mathbf{BA}$. Prove that the only n-by-n matrices which commute with all the n-by-n diagonal matrices over F are the diagonal matrices themselves.

6. Verify the statements made in Example C about the effect of multiplying an arbitrary n-by-n matrix $\mathbf{A}$ on the left by the elementary matrices $\mathbf{P}_{ij}$, $\mathbf{B}_{ij}(\lambda)$, and $\mathbf{D}_i(\mu)$.

7. Test the following matrices for invertibility, and if invertible, find the inverse by the method of Example C.

a. $\begin{pmatrix} 2 & -1 \\ -2 & 1 \end{pmatrix}$ b. $\begin{pmatrix} 2 & 1 \\ 1 & 1 \end{pmatrix}$

c. $\begin{pmatrix} 3 & 1 & 0 \\ 1 & 2 & 1 \\ 0 & -1 & 2 \end{pmatrix}$ d. $\begin{pmatrix} 1 & 1 & 0 \\ 0 & 1 & 1 \\ -1 & 1 & 0 \end{pmatrix}$

e. $\begin{pmatrix} 1 & 1 & 1 & 0 \\ 0 & 1 & 1 & 1 \\ 1 & 0 & 1 & 1 \\ 1 & 1 & 0 & 1 \end{pmatrix}$,

8. Show that the equation $\mathbf{Ax} = \mathbf{b}$, where $\mathbf{A}$ is an n-by-n matrix, and $\mathbf{x}$ and $\mathbf{b}$ column vectors, has a unique solution if and only if $\mathbf{A}$ is an invertible matrix. In case $\mathbf{A}$ is invertible, show that the solution of the equation is given by $\mathbf{x} = \mathbf{A}^{-1}\mathbf{b}$.

13. LINEAR TRANSFORMATIONS AND MATRICES

Throughout the rest of this chapter, V and W denote finite-dimensional vector spaces over F. The first result asserts that a linear transformation is completely determined if we know its effect on a set of basis elements and that, conversely, we may define a linear transformation by assigning arbitrary images for a set of basis elements.

(13.1) THEOREM. *Let $\{v_1, \ldots, v_n\}$ be a basis of V over F. If S and T are elements of $L(V, W)$ such that $S(v_i) = T(v_i)$, $1 \le i \le n$, then $S = T$. Moreover, let $w_1, \ldots, w_n$ be arbitrary vectors in W. Then there exists one and only one linear transformation $T \in L(V, W)$ such that $T(v_i) = w_i$.*

Proof. Let $v = \sum_1^n \xi_i v_i$. Then $S(v_i) = T(v_i)$, $1 \le i \le n$, implies that

$$S(v) = S\left(\sum \xi_i v_i\right) = \sum \xi_i S(v_i) = \sum \xi_i T(v_i) = T(v).$$

Since this holds for all $v \in V$, we have $S = T$, and the first part of the theorem is proved.

To prove the second part, let $w_1, \ldots, w_n$ be given and define a mapping $T: V \to W$ by setting

$$T\left(\sum \xi_i v_i\right) = \sum \xi_i w_i, \qquad \xi_i \in F.$$

Since $\{v_1, \ldots, v_n\}$ is a basis of V, $\sum \xi_i v_i = \sum \eta_i v_i$ implies $\xi_i = \eta_i$, $1 \le i \le n$, and hence $T(\sum \xi_i v_i) = T(\sum \eta_i v_i)$, and we have shown that T is a function. It is immediate from the definition that T is a linear transformation of $V \to W$ such that $T(v_i) = w_i$, $1 \le i \le n$, and the uniqueness of T is clear by the first part of the theorem. This completes the proof.

Now consider a fixed basis $\{v_1, \ldots, v_n\}$ of V over F and for simplicity let $T \in L(V, V)$. Then for each i, $T(v_i)$ is a linear combination of $v_1, \ldots, v_n$, and the coefficients can be used to define the rows or columns of an n-by-n matrix which together with the basis $\{v_1, \ldots, v_n\}$ determines completely the linear transformation T, because of the preceding theorem. The question whether we should let $T(v_i)$ give the rows or columns of the matrix corresponding to T is answered by requiring that the matrix of a product of two transformations be the product of their corresponding matrices.

For example, let V be a two-dimensional vector space over F with basis $\{v_1, v_2\}$. Let S and T in $L(V, V)$ be defined by

$$\begin{aligned} S(v_1) &= -v_1 + 2v_2, & T(v_1) &= 2v_1 + 3v_2, \\ S(v_2) &= v_1 + v_2, & T(v_2) &= -v_2. \end{aligned}$$

Then ST is the linear transformation given by

$$\begin{aligned} ST(v_1) &= S(2v_1 + 3v_2) = 2(-v_1 + 2v_2) + 3(v_1 + v_2) = v_1 + 7v_2, \\ ST(v_2) &= S(-v_2) = -(v_1 + v_2) = -v_1 - v_2. \end{aligned}$$

The matrices corresponding to S, T, ST, if we let $S(v_i)$ correspond to the *rows* of the matrix of S, etc., are respectively

$$\begin{pmatrix} -1 & 2 \\ 1 & 1 \end{pmatrix}, \quad \begin{pmatrix} 2 & 3 \\ 0 & -1 \end{pmatrix}, \quad \begin{pmatrix} 1 & 7 \\ -1 & -1 \end{pmatrix}$$

and we have

$$\begin{pmatrix} -1 & 2 \\ 1 & 1 \end{pmatrix} \cdot \begin{pmatrix} 2 & 3 \\ 0 & -1 \end{pmatrix} \neq \begin{pmatrix} 1 & 7 \\ -1 & -1 \end{pmatrix}.$$

Let us see if we have better luck by letting $S(v_i)$ correspond to the columns of the matrix of S, etc. In this case the matrices corresponding to S, T, ST are respectively

$$\begin{pmatrix} -1 & 1 \\ 2 & 1 \end{pmatrix}, \quad \begin{pmatrix} 2 & 0 \\ 3 & -1 \end{pmatrix}, \quad \begin{pmatrix} 1 & -1 \\ 7 & -1 \end{pmatrix}$$

and this time it is true that

$$\begin{pmatrix} -1 & 1 \\ 2 & 1 \end{pmatrix} \cdot \begin{pmatrix} 2 & 0 \\ 3 & -1 \end{pmatrix} = \begin{pmatrix} 1 & -1 \\ 7 & -1 \end{pmatrix}.$$

All this suggests the following definition.

(13.2) DEFINITION. Let $\{v_1, \ldots, v_n\}$ be a basis of V and let $T \in L(V, V)$. The *matrix of T with respect to the basis*† $\{v_1, \ldots, v_n\}$ of V is the n-by-n

† The matrix of T depends not only on the *set* of basis vectors $\{v_1, \ldots, v_n\}$, but on the order in which they are given. The order is usually clear from the context, but in complicated situations we shall sometimes say *ordered basis*, to emphasize a particular order we have in mind.

matrix whose ith column, for $1 \leq i \leq n$, is the set of coefficients obtained when $T(v_i)$ is expressed as a linear combination of $v_1, \ldots, v_n$. Thus the matrix (α_{rs}) of T is described by the equations

$$T(v_i) = \sum_{j=1}^{n} \alpha_{ji} v_j = \alpha_{1i} v_1 + \cdots + \alpha_{ni} v_n .$$

To give another example, let T be the linear transformation of a three-dimensional vector space with basis $\{v_1, v_2, v_3\}$ such that

$$\begin{aligned} T(v_1) &= 2v_1 - 3v_3 \\ T(v_2) &= v_2 + 5v_3 \\ T(v_3) &= v_1 - v_2 . \end{aligned}$$

Then the matrix of T with respect to the basis $\{v_1, v_2, v_3\}$ is

$$\begin{pmatrix} 2 & 0 & 1 \\ 0 & 1 & -1 \\ -3 & 5 & 0 \end{pmatrix}.$$

Let us check whether Definition (13.2) is consistent with the interpretation of the matrix of a linear transformation given by a system of equations. For example, let T be defined by the system

$$\begin{aligned} y_1 &= 3x_1 - x_2 \\ y_2 &= x_1 + 2x_2 . \end{aligned}$$

Let

$$e_1 = \begin{pmatrix} 1 \\ 0 \end{pmatrix}, \qquad e_2 = \begin{pmatrix} 0 \\ 1 \end{pmatrix};$$

then e_1 and e_2 form a basis for R_2, and we can compute the matrix of T with respect to this basis according to Definition (13.2). Using the results of the last example in the preceding section, we have

$$T(e_1) = \begin{pmatrix} 3 & -1 \\ 1 & 2 \end{pmatrix} \begin{pmatrix} 1 \\ 0 \end{pmatrix} = \begin{pmatrix} 3 \\ 1 \end{pmatrix} = 3e_1 + e_2$$

$$T(e_2) = \begin{pmatrix} 3 & -1 \\ 1 & 2 \end{pmatrix} \begin{pmatrix} 0 \\ 1 \end{pmatrix} = \begin{pmatrix} -1 \\ 2 \end{pmatrix} = -e_1 + 2e_2 .$$

Thus the matrix of T is

$$\begin{pmatrix} 3 & -1 \\ 1 & 2 \end{pmatrix}.$$

The reader can easily check that for a general system of n equations in n unknowns with matrix **A**, the matrix of the corresponding linear transformation with respect to the basis $\{e_1, \ldots, e_n\}$ defined as above, is **A**, according to Definition (13.2).

We now give some general results on the connection between linear transformations and matrices.

We recall that if **A**, **B** are n-by-n matrices with coefficients (α_{ij}) and (β_{ij}), respectively, their sum and product are defined by

$$(\alpha_{ij}) + (\beta_{ij}) = (\alpha_{ij} + \beta_{ij}),$$

$$(\alpha_{ij})(\beta_{ij}) = (\gamma_{ij}), \qquad \gamma_{ij} = \sum_{k=1}^{n} \alpha_{ik}\beta_{kj}.$$

In treating a matrix as an n^2-tuple, it is also natural to define

$$\alpha(\alpha_{ij}) = (\alpha\alpha_{ij}), \qquad \alpha \in F.$$

We come now to the result that links the algebraic structure of $L(V, V)$ introduced in the preceding section with the algebraic structure of the set $M_n(F)$ of all n-by-n matrices with coefficients in F.

(13.3) THEOREM. *Let $\{v_1, \ldots, v_n\}$ be a fixed basis of V over F. The mapping $T \to (\alpha_{ij})$ which assigns to each linear transformation T its matrix (α_{ij}) with respect to the basis $\{v_1, \ldots, v_n\}$ is a one-to-one mapping of $L(V, V)$ onto $M_n(F)$ such that, if $T \to (\alpha_{ij})$ and $S \to (\beta_{ij})$, then*

$$T + S \to (\alpha_{ij}) + (\beta_{ij}),$$
$$TS \to (\alpha_{ij})(\beta_{ij}),$$
$$\alpha T \to \alpha(\alpha_{ij}).$$

Proof. The fact that the mapping is one-to-one and onto is clear by Theorem (13.1). The fact that $T + S$ and αT map onto the desired matrices is clear from the definition and, of course, the result on TS should be true because this property motivated our definition of the matrix corresponding to T. However, we should check the details. We have

$$Tv_i = \sum_{j=1}^{n} \alpha_{ji}v_j, \qquad Sv_i = \sum_{j=1}^{n} \beta_{ji}v_j.$$

Then
$$(TS)v_i = T(Sv_i) = T\left(\sum_{j=1}^{n} \beta_{ji}v_j\right)$$
$$= \sum_{j=1}^{n} \beta_{ji}T(v_j) = \sum_{j=1}^{n} \beta_{ji} \sum_{k=1}^{n} \alpha_{kj}v_k$$
$$= \sum_{k=1}^{n} \left(\sum_{j=1}^{n} \alpha_{kj}\beta_{ji}\right)v_k.$$

Thus the (k, i) entry of the matrix of TS is $\sum_{j=1}^{n} \alpha_{kj}\beta_{ji}$, which is also the (k, i) entry of the product $(\alpha_{ij})(\beta_{ij})$. This completes the proof.

(13.4) COROLLARY. *Let* **A**, **B**, **C** *be n-by-n matrices with coefficients in F; then*

$$\mathbf{A}(\mathbf{BC}) = (\mathbf{AB})\mathbf{C}$$
$$\mathbf{A}(\mathbf{B} + \mathbf{C}) = \mathbf{AB} + \mathbf{AC}, \qquad (\mathbf{A} + \mathbf{B})\mathbf{C} = \mathbf{AC} + \mathbf{BC}.$$

The proof is immediate by Theorems (13.3) and (11.6) and does not require any computation at all.

Some remarks on (13.3) are appropriate at this point. First, Theorem (13.3) is simply the assertion that $L(V, V)$ is isomorphic with $M_n(F)$ as a ring and as a vector space over F. Theorem (13.3) asserts that all computations in $L(V, V)$ can equally well be carried out in $M_n(F)$. Experience shows that frequently the shortest and most elegant solution of a problem comes by working in $L(V, V)$, but the reader will find that there are times when calculations with matrices cannot be avoided. The preceding theorem also has the complication that the correspondence between $L(V, V)$ and $M_n(F)$ depends on the choice of a basis in V. Our next task is to work out this connection explicitly.

Let $\{v_1, \ldots, v_n\}$ be a basis of V over F and let $\{w_1, \ldots, w_n\}$ be a set of vectors in V. Then we have

(13.5) $$w_i = \sum_{j=1}^{n} \mu_{ji} v_j, \qquad 1 \le i \le n.$$

We assert that $\{w_1, \ldots, w_n\}$ is another basis of V if and only if the matrix (μ_{ij}) is invertible [see Definition (12.10)]. To see this, suppose first that $\{w_1, \ldots, w_n\}$ is a basis. Then we can express each

$$v_i = \sum_{j=1}^{n} \eta_{ji} w_j$$

where (η_{ij}) is an n-by-n matrix. Substituting in (13.5), we obtain

$$w_i = \sum_{j=1}^{n} \mu_{ji} v_j = \sum_{j=1}^{n} \mu_{ji} \left(\sum_{k=1}^{n} \eta_{kj} w_k \right) = \sum_{k=1}^{n} \left(\sum_{j=1}^{n} \eta_{kj} \mu_{ji} \right) w_k .$$

Since $\{w_1, \ldots, w_n\}$ are linearly independent, we obtain

$$\sum_{j=1}^{n} \eta_{kj} \mu_{ji} = \begin{cases} 1 & \text{if } i = k \\ 0 & \text{if } i \neq k \end{cases}$$

and we have proved that $(\eta_{ij})(\mu_{ij}) = \mathbf{I}$. Similarly, $(\mu_{ij})(\eta_{ij}) = \mathbf{I}$ and we have shown that (μ_{ij}) is an invertible matrix. We leave as an exercise the proof that if (μ_{ij}) is invertible then $\{w_1, \ldots, w_n\}$ is a basis.

(13.6) THEOREM. *Let $\{v_1, \ldots, v_n\}$ be a basis of V and let $\{w_1, \ldots, w_n\}$ be another basis such that*

$$w_i = \sum_{j=1}^{n} \mu_{ji} v_j$$

where (μ_{rs}) is an invertible matrix. Let $T \in L(V, V)$, and let (α_{ij}) and

(α'_{ij}) be the matrices of T with respect to the bases $\{v_1, \ldots, v_n\}$ and $\{w_1, \ldots, w_n\}$, respectively. Then we have

$$(\mu_{ij})(\alpha'_{ij}) = (\alpha_{ij})(\mu_{ij})$$

or

$$(\alpha'_{ij}) = (\mu_{ij})^{-1}(\alpha_{ij})(\mu_{ij}).$$

Proof. We have

$$\begin{aligned} T(w_i) &= \sum_{j=1}^{n} \alpha'_{ji} w_j = \sum_{j=1}^{n} \alpha'_{ji} \sum_{k=1}^{n} \mu_{kj} v_k \\ &= \sum_{k=1}^{n} \left(\sum_{j=1}^{n} \mu_{kj} \alpha'_{ji} \right) v_k, \end{aligned}$$

while on the other hand we have

$$\begin{aligned} T(w_i) &= T\left(\sum_{j=1}^{n} \mu_{ji} v_j \right) = \sum_{j=1}^{n} \mu_{ji} \left(\sum_{k=1}^{n} \alpha_{kj} v_k \right) \\ &= \sum_{k=1}^{n} \left(\sum_{j=1}^{n} \alpha_{kj} \mu_{ji} \right) v_k. \end{aligned}$$

Therefore

$$\sum_{j=1}^{n} \mu_{kj} \alpha'_{ji} = \sum_{j=1}^{n} \alpha_{kj} \mu_{ji}, \qquad 1 \le i,\ k \le n,$$

and the theorem is proved.

(13.7) DEFINITION. Two matrices **A** and **B** in $M_n(F)$ are *similar* if there exists an invertible matrix **X** in $M_n(F)$ such that

$$\mathbf{B} = \mathbf{X}^{-1}\mathbf{A}\mathbf{X}.$$

Theorem (13.6) asserts that, if **A** and **B** are matrices of $T \in L(V, V)$ with respect to different bases, then **A** and **B** are similar. The reader may verify that the converse of this statement is also true. Thus if **A** and **B** are similar matrices, then **A** and **B** can always be viewed as matrices of a single linear transformation with respect to different bases.

In order to remember how Theorem (13.6) works in particular cases, it may be helpful to restate it in another way.

(13.6)′ THEOREM. *Let V be a vector space over F, with*

Original basis $\{v_1, \ldots, v_n\}$.

Let $T \in L(V, V)$ have the matrix **A** *with respect to the original basis. Now suppose V has the*

New basis $\{w_1, \ldots, w_n\}$,

and that the matrix of T with respect to the new basis is **B**. *Then*

$$\mathbf{B} = \mathbf{S}^{-1}\mathbf{A}\mathbf{S},$$

where **S** *is the matrix expressing the new basis in terms of the original one:*

$$\mathbf{S} = (\mu_{ij}),\ w_i = \sum_{j=1}^{n} \mu_{ji} v_j .$$

Example A. As an illustration of the preceding theorem, let $T \in L(R_2, R_2)$ be defined by:

$$T(e_1) = e_1 - e_2$$
$$T(e_2) = e_1 + 2e_2,$$

where $e_1 = \langle 1, 0 \rangle$, $e_2 = \langle 0, 1 \rangle$. The matrix **A** of T with respect to the basis $\{e_1, e_2\}$ is

$$\mathbf{A} = \begin{pmatrix} 1 & 1 \\ -1 & 2 \end{pmatrix}.$$

Let us find the matrix **B** of T with respect to the basis $\{u_1 = e_1 + e_2, u_2 = e_1 - e_2\}$. We have

$$T(u_1) = T(e_1) + T(e_2) = 2e_1 + e_2,$$
$$T(u_2) = T(e_1) - T(e_2) = -3e_2.$$

In order to express $T(u_1)$ and $T(u_2)$ as a linear combination of u_1 and u_2, we have to solve for e_1 and e_2 in terms of u_1 and u_2. We obtain

$$e_1 = \tfrac{1}{2}(u_1 + u_2), \qquad e_2 = \tfrac{1}{2}(u_1 - u_2),$$

and

$$T(u_1) = \tfrac{3}{2}u_1 + \tfrac{1}{2}u_2,$$
$$T(u_2) = -\tfrac{3}{2}u_1 + \tfrac{3}{2}u_2.$$

The matrix **B** of T with respect to the basis $\{u_1, u_2\}$ is

$$\mathbf{B} = \begin{pmatrix} \frac{3}{2} & -\frac{3}{2} \\ \frac{1}{2} & \frac{3}{2} \end{pmatrix}.$$

To find an invertible matrix **S** such that $\mathbf{S}^{-1}\mathbf{A}\mathbf{S} = \mathbf{B}$, we use the equations expressing the new basis $\{u_1, u_2\}$ in terms of the original basis $\{e_1, e_2\}$. For **S** we obtain the matrix

$$\mathbf{S} = \begin{pmatrix} 1 & 1 \\ 1 & -1 \end{pmatrix}.$$

To conclude this section, we apply our basic results about basis and dimension from Section 7, together with the results of this section, to obtain some useful theorems about linear transformations. First of all, let V and W be finite dimensional vector spaces over F, and let $T \in L(V, W)$. Then the subsets

$$T(V) = \{w \in W \mid w = T(v) \text{ for some } v \in V\}$$

and

$$n(T) = \{v \in V \mid T(v) = 0\}$$

are easily shown to be subspaces of W and V, respectively. (The proof of this fact is left as an exercise for the reader.) We shall see that these subspaces have a great deal to do with the behavior of T. First let us make a formal definition.

(13.8) DEFINITION. Let $T \in L(V, W)$, and let $T(V)$ and $n(T)$ be the subspaces of W and V, respectively, defined above. Then the subspace $T(V)$ of W is called the *range* of T, and its dimension is called the *rank* of T. The subspace $n(T)$ of V is called the *null space* of T, and its dimension is called the *nullity* of T.

The next result is certainly the most useful and important theorem about linear transformations we have had so far.

(13.9) FUNDAMENTAL THEOREM. *Let* $T \in L(V, W)$. *Then* $\dim T(V) + \dim n(T) = \dim V$. *In other words, the sum of the rank of T and the nullity of T is equal to the dimension of V.*

Proof. Let $\{v_1, \ldots, v_k\}$ be a basis for the null space $n(T)$ [with our convention that this set is empty if $n(T) = 0$]. By Theorem (7.4), there exist vectors $\{v_{k+1}, \ldots, v_n\}$ in V such that $\{v_1, \ldots, v_k, \ldots, v_n\}$ is a basis for V. It is sufficient to prove that $\{T(v_{k+1}), \ldots, T(v_n)\}$ is a basis for $T(V)$. We have first the fact that

$$T\left(\sum_{i=1}^{n} \xi_i v_i\right) = \sum_{i=k+1}^{n} \xi_i T(v_i), \qquad \xi_i \in F,$$

because $T(v_i) = 0$ for $1 \le i \le k$. Finally, suppose we have

$$\sum_{i=k+1}^{n} \eta_i T(v_i) = 0$$

for some $\eta_i \in F$. Then

$$T\left(\sum_{i=k+1}^{n} \eta_i v_i\right) = 0,$$

and $\sum_{i=k+1}^{n} \eta_i v_i \in n(T)$. Because of the way the basis $\{v_i\}$ for V was chosen, it follows that all the $\{\eta_i\}$ are zero, and the theorem is proved.

Our final theorem of this section is the following important application of the fundamental theorem (13.9).

(13.10) THEOREM. *Let* $T \in L(V, V)$ *for some finite dimensional space* V *over* F. *Then the following statements are equivalent:*

1. T is invertible.
2. T is one-to-one.
3. T is onto.

Proof. We prove that (1) implies (2), (2) implies (3), and (3) implies (1). First we assume (1). Then there exists $T^{-1} \in L(V, V)$ such that $TT^{-1} = T^{-1}T = 1$. Suppose $T(v_1) = T(v_2)$. Applying T^{-1} we obtain $T^{-1}T(v_1) = T^{-1}T(v_2)$, and $v_1 = v_2$. Thus (1) implies (2).

Next assume (2). Then the null space $n(T) = 0$. By Theorem (13.9) we have $\dim T(V) = n$, and it follows that $T(V) = V$, and that T is onto.

Finally, assume T is onto. By Theorem (13.9) again, T is also one-to-one. It follows that if $\{v_1, \ldots, v_n\}$ is a basis for V, then $\{T(v_1), \ldots, T(v_n)\}$ is also a basis. By Theorem (13.1), there exists a linear transformation U such that $UT(v_i) = v_i$, $i = 1, \ldots, n$. By (13.1) again, we have $UT = 1$. On the other hand, $TU(Tv_i) = Tv_i$ for $i = 1, \ldots, n$, and since $\{Tv_i, \ldots, Tv_n\}$ is a basis, we have $TU = 1$. Therefore T is invertible, and the theorem is proved.

EXERCISES

In all the Exercises, all vector spaces involved are assumed to have finite bases, and the field F can be taken to be the field of real numbers in the numerical problems.

1. Let $S, T, U \in L(V, V)$ be given by

$$S(u_1) = u_1 - u_2, \quad T(u_1) = u_2, \quad U(u_1) = 2u_1$$
$$S(u_2) = u_1, \quad T(u_2) = u_1, \quad U(u_2) = -2u_2$$

where $\{u_1, u_2\}$ is a basis for V. Find the matrices of S, T, U with respect to the basis $\{u_1, u_2\}$ and with respect to the new basis $\{w_1, w_2\}$ where

$$w_1 = 3u_1 - u_2$$
$$w_2 = u_1 + u_2.$$

Find invertible matrices $\mathbf{X}$ in each case such that $\mathbf{X}^{-1}\mathbf{A}\mathbf{X} = \mathbf{A}'$ where $\mathbf{A}$ is the matrix of the transformation with respect to the old basis, and $\mathbf{A}'$ the matrix with respect to the new basis.

2. Let S, T, U be linear transformations such that (letting $\{u_1, u_2\}$ or $\{u_1, u_2, u_3\}$ be bases of the vector spaces)

$$S(u_1) = u_1 + u_2,$$
$$S(u_2) = -u_1 - u_2,$$
$$T(u_1 = u_1 - u_2,$$
$$T(u_2) = 2u_2,$$
$$U(u_1) = u_1 + u_2 - u_3$$
$$U(u_2) = u_2 - 3u_3$$
$$U(u_3) = -u_1 - 3u_2 - 2u_3.$$

   a. Find the rank and nullity of S, T, U.
   b. Which of these linear transformations are invertible?

c. Find bases for the null spaces of S, T, and U.

d. Find bases for the range spaces of S, T, and U.

3. Let V be the space of all polynomial functions $f(x) = \alpha_0 + \alpha_1 x + \cdots + \alpha_k x^k$, with real coefficients α_i and k fixed. Show that the derivative transformation D defined in Example D of Section 11, maps V into V. What is the matrix of D with respect to the basis $1, x, \ldots, x^k$ of V? What is the rank of $D: V \to V$? What is the nullity?

4. Prove that $\dim T(V) = \text{rank } \mathbf{A}$, where $\mathbf{A}$ is the matrix of $T \in L(V, V)$ with respect to an arbitrary basis of V.

5. Let V be an n-dimensional vector space over R, and let $f: V \to R_1$ be a nonzero linear transformation from V to the one-dimensional space R_1. Prove that $\dim n(f) = n - 1$.

6. Let V, f be as in Exercise 5, and let α be a fixed real number. Prove that the set

$$\{v \in V \mid f(v) = \alpha\}$$

is a hyperplane in V, i.e., a linear manifold of dimension $n - 1$ in V, as defined in Section 10.

7. Let $T \in L(V, V)$. Prove that $T^2 = 0$ if and only if $T(V) \subset n(T)$.

8. Give an example of a linear transformation $T: V \to V$ which shows that it can happen that $T(V) \cap n(T) \neq 0$.

9. Let $T \in L(V, V)$. Prove that there exists a nonzero linear transformation $S \in L(V, V)$ such that $TS = 0$ if and only if there exists a nonzero vector $v \in V$ such that $T(v) = 0$.

10. Let $S, T \in L(V, V)$ be such that $ST = 1$. Prove that $TS = 1$. (Exercise 10, Section 11, shows that this result is not true unless V is finite dimensional.)

11. Let $T \in L(V, W)$, where $\dim V = m$ and $\dim W = n$. Let $\{v_1, \ldots, v_m\}$ be a basis of V and $\{w_1, \ldots, w_n\}$ a basis for W. Define the matrix $\mathbf{A}$ of T with respect to the pair of bases $\{v_i\}$ and $\{w_j\}$ to be the n-by-m matrix $\mathbf{A} = (\alpha_{ij})$, where

$$Tv_i = \sum_{j=1}^{n} \alpha_{ji} w_j, \qquad 1 \leq i \leq m, \quad 1 \leq j \leq n.$$

The vector spaces V and W are isomorphic via the basis $\{v_i\}$ and $\{w_j\}$ to the spaces F_m and F_n, respectively (see Section 11, Example E). Show that if $\mathbf{x} \in F_m$ is the column vector corresponding to the vector $x \in V$ via the isomorphism, then $\mathbf{Ax}$ is the column vector in F_n corresponding to T_x. In other words, the correspondence between linear transformations and matrices is such that the action of T on a vector x is realized by matrix multiplication $\mathbf{Ax}$.

Chapter 4

Vector Spaces with an Inner Product

This chapter begins with an optional section on symmetry of plane figures, which shows how some natural geometrical questions lead to the problem of studying linear transformations that preserve length. The concept of length in a general vector space over the real numbers is introduced in the next section, where it is shown how length is related to an inner product. The language of orthonormal bases and orthogonal transformations is developed with some examples from geometry and analysis. Beside the fact that the real numbers form a field, we shall use heavily in this chapter the theory of inequalities and absolute value, and the fact that every real number $a \geq 0$ has a unique nonnegative square root $\sqrt{a}$.

14. THE CONCEPT OF SYMMETRY

The word *symmetry* has rich associations for most of us. A person familiar with sculpture and painting knows the importance of symmetry to the artist and how the distinctive features of certain types of architecture and ornaments results from the use of symmetry. A geologist knows that crystals are classified according to the symmetry properties they possess. A naturalist knows the many appearances of symmetry in the shapes of plants, shells, and fish. The chemist knows that the symmetry properties of molecules are related to their chemical properties. In this section we shall consider the mathematical concept of symmetry, which will provide some worthwhile insight into all of the preceding examples.

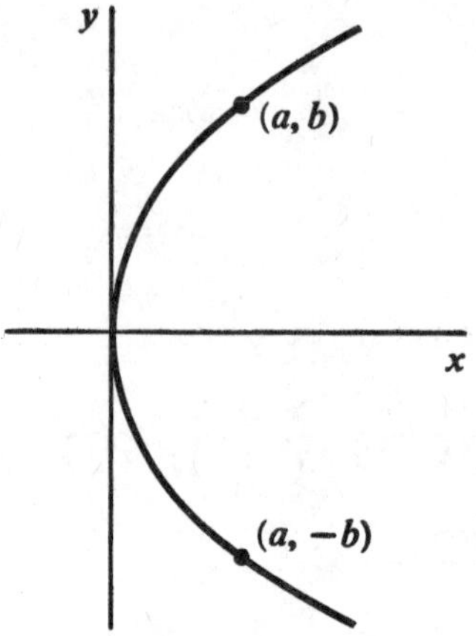

FIGURE 4.1

Let us begin with a simple example from analytic geometry, Figure 4.1. What does it mean to say the graph of the parabola $y^2 = x$ is symmetric about the x axis? One way of looking at it is to say that if we fold the plane along the x axis the two halves of the curve $y^2 = x$ fit together. But this is a little too vague. A more precise description comes from the observation that the operation of plane-folding defines a transformation T, of the points in the plane, which assigns to a point (a, b) the point $(a, -b)$ which meets it after the plane is folded. The symmetry of the parabola is now described by asserting that if (a, b) is a point on the parabola so is the transformed point $(a, -b)$. This sort of symmetry is called *bilateral symmetry.*

Now consider the example of a triod, Figure 4.2. What sort of symmetry does it possess? It clearly has bilateral symmetry about the three lines joining the center with the points a, b, and c. The triod has also a new sort of symmetry, *rotational symmetry.* If we rotate points in

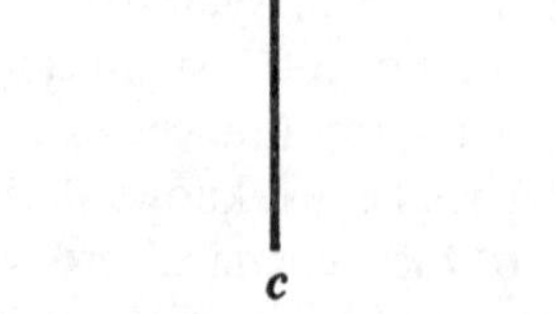

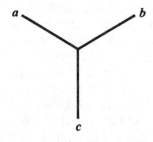

FIGURE 4.2

the plane through an angle of 120°, leaving the center fixed, then the triod is carried onto itself. Does the triod have the same symmetry properties as the winged triod of Figure 4.3? Clearly not; the winged triod has only rotational symmetry and does not possess the bilateral symmetry of the triod. Thus we see that the symmetry properties of figures may serve to distinguish them.

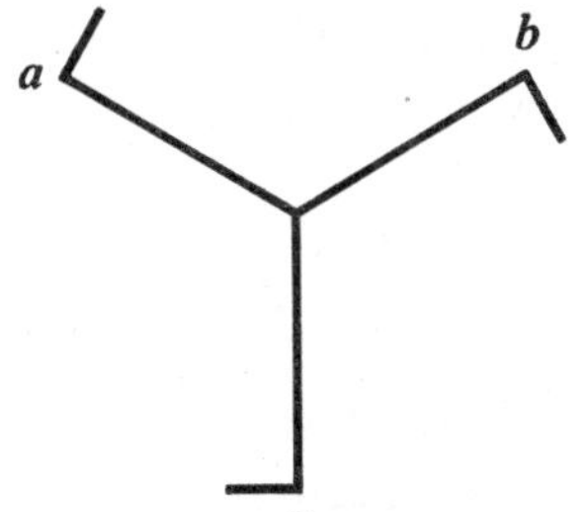

FIGURE 4.3

We have not yet arrived at a precise definition of symmetry. Another example suggests the underlying idea. The circle with center 0, Figure 4.4, possesses all the kinds of symmetry we have discussed so far. However, we can look at the symmetry of the circle from another viewpoint. The circle is carried onto itself by any transformation T which preserves distance and which leaves the center fixed, for the circle consists of precisely those points p whose distances from 0 are a fixed constant r and, if T is a distance preserving transformation such that $T(0) = 0$, then the distance of $T(p)$ from 0 will again be r and $T(p)$ is on the circle. This suggests the following definition.

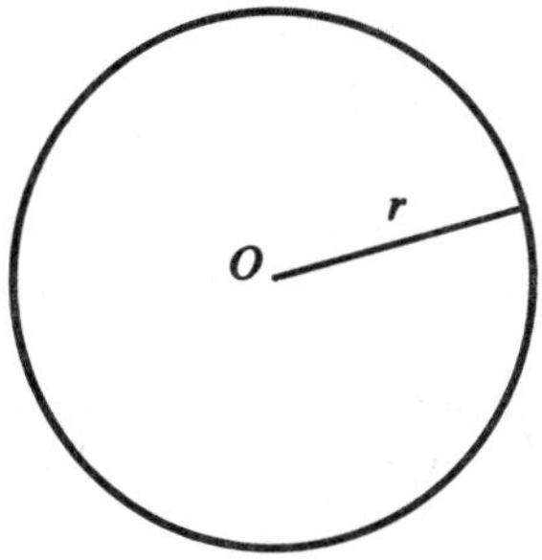

FIGURE 4.4

(14.1) DEFINITION. By a *figure* X we mean a set of points in the plane. A *symmetry* of a figure X is a transformation T of the plane into itself such that:

(*i*) $T(X) = X$; that is, T sends every point in X onto another point in X, and every point in X is the image of some point in X under T.
(*ii*) T preserves distance; that is, if $d(p, q)$ denotes the distance between the points p and q, then

$$d[T(p), T(q)] = d(p, q)$$

for all points p and q.

(14.2) THEOREM. *The set G of all symmetries of a figure X form a group G, called the* symmetry group *of the figure [see Definition (11.10)].*

Proof. We show first that if $S, T \in G$ then the product ST defined by

$$ST(p) = S[T(p)]$$

belongs to G. We have $ST(X) = S[T(X)] = S(X) = X$ and $d(ST(p), ST(q)) = d(S[T(p)], S[T(q)]) = d(T(p), T(q)) = d(p, q)$ and hence $ST \in G$. We have to check also that

$$S(TU) = (ST)U, \qquad \text{for } S, T, U \in G,$$

which is a consequence of the fact that both mappings send $p \to S\{T[U(p)]\}$. The transformation 1 such that $1(p) = p$ for all p belongs, clearly, to G and satisfies

$$S \cdot 1 = 1 \cdot S = S, \qquad S \in G.$$

Finally, if $S \in G$, there exists a symmetry S^{-1} of X which unwinds whatever S has done (the reader can give a more rigorous proof) such that

$$SS^{-1} = S^{-1}S = 1.$$

This completes the proof of the theorem.

The vague question, "What kinds of symmetry can a figure possess?" is now replaced by the precise mathematical question, "What are the possible symmetry groups?" We shall investigate a simple case of this problem in this section. First we look at some examples of symmetry groups.

The symmetry group of the triod, Figure 4.5, consists of the three bilateral symmetries about the arms of the triod and three rotations through angles of 120°, 240°, and 360°. By means of the group operation we show that the symmetries of the triod all can be built up from two of the symmetries. For example, let R be the counterclockwise rotation

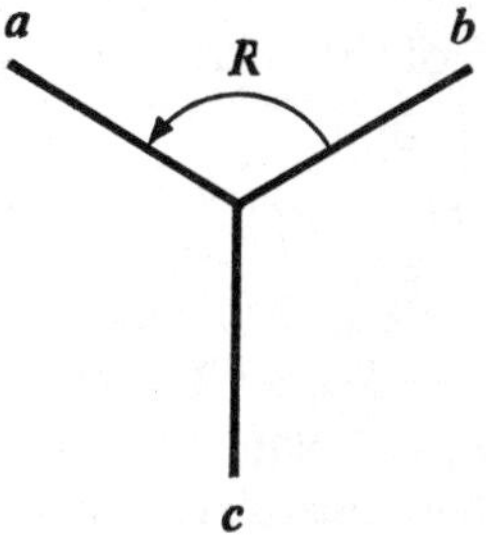

FIGURE 4.5

through an angle of 120° and let S be the bilateral symmetry about the arm a. The reader may verify that the symmetry group G of the triod consists of the symmetries

$$\{1, R, R^2, S, SR, SR^2\}.$$

We may also verify that $S^2 = 1$, $R^3 = 1$, and $SR = R^{-1}S$ and that these rules suffice to multiply arbitrary elements of G.

The symmetry group of the winged triod is easily seen to consist of exactly the symmetries

$$\{1, R, R^2\}.$$

More generally, let X be the n-armed figure (Figure 4.6), S the bilateral

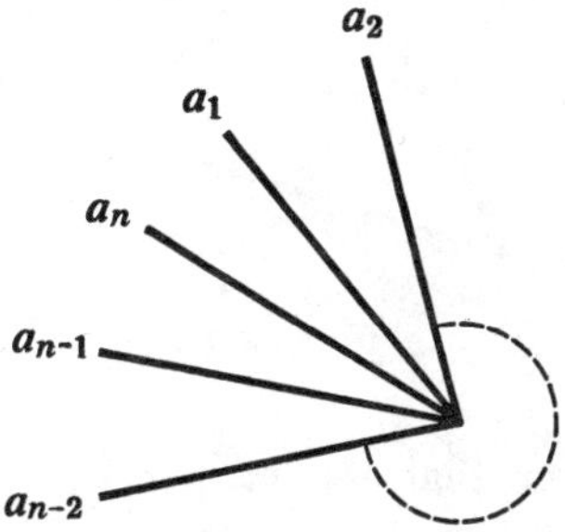

FIGURE 4.6

symmetry about the arm a_1, and R the rotation carrying $a_2 \to a_1$. Then the group of X consists exactly of the symmetries

$$\{1, R, R^2, \ldots, R^{n-1}, S, SR, \ldots, SR^{n-1}\} \tag{14.3}$$

and these symmetries are multiplied according to the rules

$$R^n = 1, \qquad S^2 = 1, \qquad SR = R^{-1}S.$$

The symmetry group of the corresponding winged figure consists of precisely the rotations

(14.4) $$\{1, R, R^2, \ldots, R^{n-1}\}.$$

The group (14.3) is called the *dihedral group* D_n; the group (14.4) is called the *cyclic group* C_n. We shall sketch a proof that these are the only *finite* symmetry groups of plane figures.

We require first some general remarks. It will be convenient to identify the plane with the vector space R_2 of all pairs of real numbers $\langle\alpha, \beta\rangle$. If p is a vector $\langle\alpha, \beta\rangle$, then the *length* $\|p\|$ is defined by

$$\|p\| = \sqrt{\alpha^2 + \beta^2}$$

and the distance from p to q (see Figure 4.7) is then given by

$$d(p, q) = \|p - q\|.$$

We require also the fact from plane analytic geometry that $p = \langle\alpha, \beta\rangle$ is perpendicular to $q = \langle\gamma, \delta\rangle$ (notation: $p \perp q$) if and only if

$$\alpha\gamma + \beta\delta = 0.$$

(14.5) *A distance-preserving transformation T of R_2 which leaves the zero element fixed is a linear transformation.*

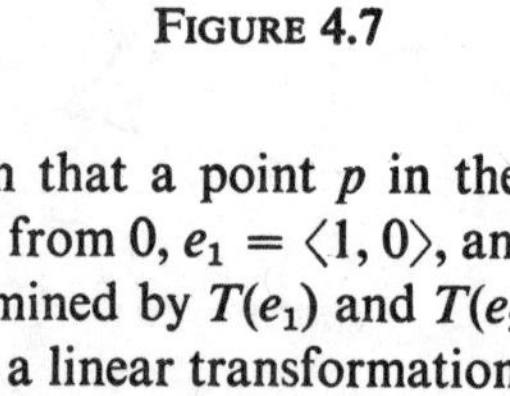

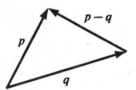

FIGURE 4.7

Proof. It can be shown that a point p in the plane is completely determined by its distances from 0, $e_1 = \langle 1, 0\rangle$, and $e_2 = \langle 0, 1\rangle$. Therefore $T(p)$ is completely determined by $T(e_1)$ and $T(e_2)$, since $T(0) = 0$. From Chapter 3 we can define a linear transformation $\tilde{T}$ such that

$$T(e_1) = \tilde{T}(e_1), \qquad T(e_2) = \tilde{T}(e_2).$$

Since both T and $\tilde{T}$ are determined by their action on e_1 and e_2, we have $T = \tilde{T}$.

We prove next the following characterization of a distance preserving transformation.

(14.6) *A linear transformation T preserves distances if and only if:*

(14.7) $$\|T(e_1)\| = \|T(e_2)\| = 1 \quad \textit{and} \quad T(e_1) \perp T(e_2).$$

Proof. If T preserves distances, then it is clear that (14.7) holds. Conversely, suppose T is a linear transformation such that (14.7) holds. To prove that T preserves distances it is sufficient to prove that, for all vectors p,

$$\|T(p)\| = \|p\|,$$

for then

$$d[T(p), T(q)] = \|T(p) - T(q)\| = \|T(p - q)\| = \|p - q\| = d(p, q).$$

Now let $p = \xi e_1 + \eta e_2$ and let $T(e_1) = \langle\alpha, \beta\rangle$, $T(e_2) = \langle\gamma, \delta\rangle$. Then (14.7) implies that

$$\alpha^2 + \beta^2 = \gamma^2 + \delta^2 = 1, \qquad \alpha\gamma + \beta\delta = 0.$$

Then

$$\begin{aligned}\|T(p)\|^2 &= \|\xi T(e_1) + \eta T(e_2)\|^2 = \|\langle\xi\alpha + \eta\gamma, \xi\beta + \eta\delta\rangle\|^2 \\ &= (\xi\alpha + \eta\gamma)^2 + (\xi\beta + \eta\delta)^2 \\ &= \xi^2\alpha^2 + 2\xi\alpha\eta\gamma + \eta^2\gamma^2 + \xi^2\beta^2 + 2\xi\beta\eta\delta + \eta^2\delta^2 \\ &= \xi^2(\alpha^2 + \beta^2) + \eta^2(\gamma^2 + \delta^2) + 2\xi\eta(\alpha\gamma + \beta\delta) \\ &= \xi^2 + \eta^2 = \|p\|^2.\end{aligned}$$

This completes the proof of (14.6).

Now let T be a distance-preserving transformation and leaving the origin fixed, and let $T(e_1) = \langle\alpha, \beta\rangle$. Then $T(e_2)$ is a point on the unit circle whose radius vector is perpendicular to $T(e_1)$, Figure 4.8. It follows that either

$$T(e_2) = \langle -\beta, \alpha\rangle$$

or

$$T(e_2) = \langle\beta, -\alpha\rangle.$$

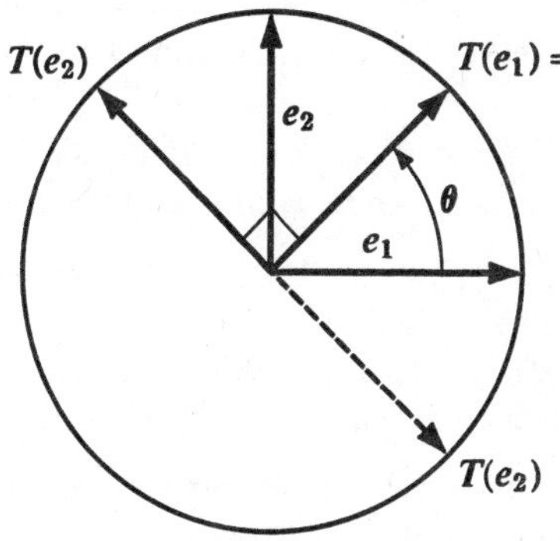

FIGURE 4.8

In the former case the matrix of T with respect to $\{e_1, e_2\}$ is

$$\mathbf{A} = \begin{pmatrix} \alpha & -\beta \\ \beta & \alpha \end{pmatrix},$$

while in the latter case the matrix is

$$\mathbf{A} = \begin{pmatrix} \alpha & \beta \\ \beta & -\alpha \end{pmatrix}.$$

In the former case T is a rotation through an angle θ such that $\cos \theta = \alpha$, while in the latter case T is a bilateral symmetry about the line making an angle $\frac{1}{2}\theta$ with e_1. Notice that in the second case the matrix of T^2 is

$$\begin{pmatrix} \alpha & \beta \\ \beta & -\alpha \end{pmatrix}\begin{pmatrix} \alpha & \beta \\ \beta & -\alpha \end{pmatrix} = \begin{pmatrix} \alpha^2 + \beta^2 & 0 \\ 0 & \alpha^2 + \beta^2 \end{pmatrix} = \mathbf{I}$$

so that $T^2 = 1$. The second kind of transformation will be called a *reflection*, the first a *rotation*.

The next result shows how reflections and rotations combine.

(14.8) *Let S and T be distance-preserving linear transformations; then:*

1. *If S, T are rotations, then ST is a rotation.*
2. *If S, T are reflections, ST is a rotation.*
3. *If one of S, T is a rotation and the other is a reflection, then ST is a reflection.*

The proof of (14.8) is left to the Exercises.

Now we are ready to determine the possible finite symmetry groups G of plane figures. We assume that all the elements of G are linear transformations (and so leave the origin fixed). To determine a group means to show that it is *isomorphic* with a group that has previously been constructed, in the sense that a one-to-one correspondence exists between the two groups which preserves the multiplication laws of the two groups.

(14.9) THEOREM. *Let G be a finite group of distance-preserving linear transformations. Then G is isomorphic with one of the following:*

1. *The cyclic group $C_n = \{1, R, R^2, \ldots, R^{n-1}\}$, $R^n = 1$, consisting of the powers of a single rotation.*
2. *The dihedral group $D_n = \{1, R, \ldots, R^{n-1}, S, SR, \ldots, SR^{n-1}\}$ where R is a rotation, S a reflection, and $R^n = 1$, $S^2 = 1$, $SR = R^{-1}S$.*

Proof. We may suppose $G \neq \{1\}$. First suppose all the elements of G are rotations and let R be a rotation in G through the least positive angle. Consider the powers of R, $\{1, R, R^2, \ldots\}$. These exhaust G, as will be

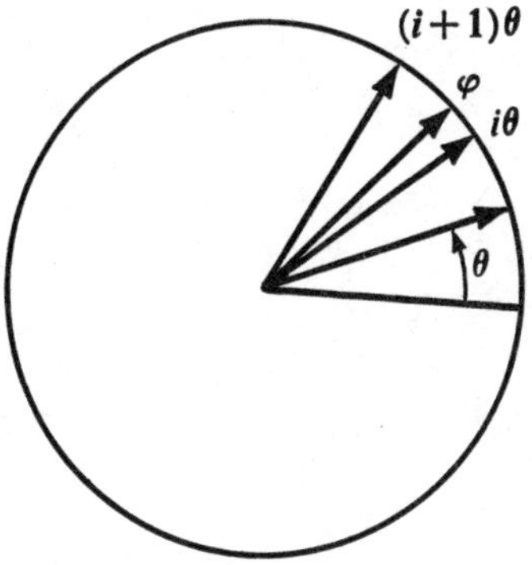

FIGURE 4.9

seen. Suppose $T \in G$ is a rotation different from all the powers of R. If φ is the angle of T and θ the angle of R, then, for some i, $i\theta < \varphi < (i+1)\theta$, Figure 4.9. Then G contains TR^{-i} which is a rotation in G through an angle $\varphi - i\theta$ smaller than θ, contrary to our assumption. Therefore G consists of the powers of R and is a cyclic group.

Now suppose G contains a reflection S. Let H be the set of all rotations contained in G. Then H itself is a group and, by the first part of the proof, there exists a rotation R in H such that

$$H = \{1, R, R^2, \ldots, R^{n-1}\}, \qquad R^n = 1.$$

Now let $X \in G$; either $X \in H$, or X is a reflection. In the latter case SX is a rotation. Hence $SX = R^i$ for some i, and since $S^2 = 1$ we have

$$X = S(SX) = SR^i;$$

and we have proved that

$$G = \{1, R, \ldots, R^{n-1}, S, SR, \ldots, SR^{n-1}\}.$$

Finally, SR is a reflection; hence $(SR)^2 = 1$ or $SRSR = 1$, and we have $SR = R^{-1}S$. It follows that G is isomorphic with a dihedral group, and the theorem is proved.

For further discussion of this topic, the reader is urged to consult the books of Weyl and of Coxeter listed in the Bibliography.

EXERCISES

All exercises in this set refer to vectors in R_2. Figures should be drawn in order to see the point of most of the exercises.

1. For vectors $a = \langle \alpha_1, \alpha_2 \rangle$, $b = \langle \beta_1, \beta_2 \rangle$, define their inner product $(a, b) = \alpha_1\beta_1 + \alpha_2\beta_2$. Show that the inner product satisfies the following rules:
   - **a.** $(a, b) = (b, a)$
   - **b.** $(a, b + c) = (a, b) + (a, c)$
   - **c.** $(\lambda a, b) = \lambda(a, b)$, for all $\lambda \in R$.
   - **d.** $\|a\| = \sqrt{(a, a)}$
   - **e.** $a \perp b$ if and only if $(a, b) = 0$.
   - **f.** $a \perp b$ if and only if $\|a + b\| = \|a - b\|$. (Draw a figure to illustrate this statement.)
2. A *line* L in R_2 is defined to be the set of all vectors for the form $p + x$, where p is a fixed vector, and x ranges over some one-dimensional subspace S of R_2. Thus if $S = S(a)$, the line L consists of all vectors of the form $p + \lambda a$, where $\lambda \in R$. We shall use the notation $p + S$ for the line L described above.†
   - **a.** Let $p + S$ and $q + S$ be two lines with the same one-dimensional subspace S. Show that $p + S$ and $q + S$ either coincide or have no vectors in common. In the latter case, we say that the lines are *parallel.*
   - **b.** Show that there is one and only one line L containing two distinct vectors p and q, and that L consists of all vectors of the form $p + \lambda(q - p)$, $\lambda \in R$.
   - **c.** Show that three distinct vectors p, q, r are collinear if and only if $S(q - p) = S(q - r)$.
   - **d.** Show that two distinct lines are either parallel or intersect in a unique vector.
3. Let L be the line containing two distinct vectors p and q, and let r be a vector not on L. Show that a vector u on L such that $(u - r) \perp (q - p)$ is a solution of the simultaneous equations

$$(u - r, q - p) = 0,$$
$$u = p + \lambda(q - p), \qquad \lambda \in R.$$

   Show that there is a unique value of λ for which these equations are satisfied. Derive a formula for the perpendicular distance from a point to a line. Test your formula on some numerical examples.
4. Let

$$\mathbf{A} = \begin{pmatrix} \alpha_{11} & \alpha_{12} \\ \alpha_{21} & \alpha_{22} \end{pmatrix}$$

   by a 2-by-2 matrix with entries from R. Define the *determinant* of $\mathbf{A}$, $D(\mathbf{A})$, by the formula

$$D(\mathbf{A}) = \alpha_{11}\alpha_{22} - \alpha_{12}\alpha_{21}.$$

---

† This definition of a line in R_2 is identical with the definition given in Section 10; no results from Section 10 are needed to do any of these problems, however.

a. Show that if **A** is the matrix of a distance preserving linear transformation T of R_2, then

$$D(\mathbf{A}) = \pm 1,$$

and that $D(\mathbf{A}) = +1$ if and only if **A** is a rotation, while $D(\mathbf{A}) = -1$ if and only if **A** is a reflection.

b. Prove that for any 2-by-2 matrices **A** and **B**, $D(\mathbf{AB}) = D(\mathbf{A})D(\mathbf{B})$.

c. Derive the statements (1), (2), (3) in (14.8).

15. INNER PRODUCTS

Let V be a vector space over the real numbers R.

(15.1) DEFINITION. An *inner product* on V is a function which assigns to each pair of vectors u, v in V a real number (u, v) such that the following conditions are satisfied.

(*i*) (u, v) is a *bilinear function*; that is,

$$\begin{aligned}(u + v, w) &= (u, w) + (v, w),\\ (u, v + w) &= (u, v) + (u, w),\\ (\alpha u, v) &= (u, \alpha v) = \alpha(u, v),\end{aligned}$$

for all $u, v, w \in V$ and $\alpha \in R$.

(*ii*) The function (u, v) is *symmetric*; that is,

$$(u, v) = (v, u), \qquad u, v \in V.$$

(*iii*) The function is *positive definite*; that is,

$$(u, u) \geq 0$$

and

$$(u, u) = 0 \quad \text{if and only if} \quad u = 0.$$

Examples. (i) Let $\{e_1, \ldots, e_n\}$ be a basis for V over R and let

$$(u, v) = \sum_{i=1}^{n} \xi_i \eta_i$$

where $u = \sum \xi_i e_i$, $v = \sum \eta_i e_i$.

(ii) Let V be a subspace of the vector space $C(R)$ of continuous functions on the real numbers, and define

$$(f, g) = \int_0^1 f(t)g(t)\,dt, \qquad f, g \in V.$$

In both cases the reader may verify that the functions defined are actually inner products.

In the case of R_2 the inner product defined in Example (*i*) is known to be connected with the angle between the vectors u and v; indeed, if u and v have length 1, then $(u, v) = \cos \theta$ where θ is the angle between u and v. Thus $|(u, v)| \leq 1$ if both u and v have length 1. Our next task is to verify that this same inequality holds in general.

(15.2) DEFINITION. Let (u, v) be a fixed inner product on V. Define the *length* $\|u\|$ of a vector $u \in V$ by

$$\|u\| = \sqrt{(u, u)}.$$

Note that by part (3) of Definition (15.1) we have

$$\|u\| \geq 0, \qquad \|u\| = 0 \text{ if and only if } u = 0.$$

Moreover, we have

$$\|\alpha u\| = |\alpha| \cdot \|u\|$$

where $|\alpha|$ is the absolute value of α in R.

We prove now an important inequality.

(15.3) LEMMA. *If* $\|u\| = \|v\| = 1$, *then* $|(u, v)| \leq 1$.

Proof. We have

$$(u - v, u - v) \geq 0.$$

This implies

$$(u, u) + (v, v) - 2(u, v) \geq 0$$

and since $(u, u) = (v, v) = 1$, we obtain

$$(u, v) \leq 1.$$

Similarly, from $(u + v, u + v) \geq 0$ we have

$$-(u, v) \leq 1.$$

Combining these inequalities we have $|(u, v)| \leq 1$.

(15.4) THEOREM (CAUCHY-SCHWARZ INEQUALITY). *For arbitrary vectors, $u, v \in V$ we have*

(15.5)
$$|(u, v)| \leq \|u\| \cdot \|v\|.$$

Proof. The result is trivial if either $\|u\|$ or $\|v\|$ is zero. Therefore, assume that $\|u\| \neq 0$, $\|v\| \neq 0$. Then

$$\frac{u}{\|u\|}, \quad \frac{v}{\|v\|}$$

are vectors of length 1, and by Lemma (15.3) we have

$$\left|\left(\frac{u}{\|u\|}, \frac{v}{\|v\|}\right)\right| \leq 1.$$

It follows that $|(u, v)| \leq \|u\| \cdot \|v\|$, and (15.5) is proved.

As a corollary we obtain the triangle inequality, which is a generalization of an inequality for real numbers,

$$|\alpha + \beta| \leq |\alpha| + |\beta|, \qquad \alpha, \beta \in R.$$

(15.6) COROLLARY (TRIANGLE INEQUALITY). *For all vectors $u, v \in V$ we have*

$$\|u + v\| \leq \|u\| + \|v\|.$$

Proof. By the Cauchy-Schwarz inequality and by the triangle inequality for R_1 we have

$$\begin{aligned}\|u + v\|^2 &= |(u + v, u + v)| = |(u, u) + (v, v) + 2(u, v)| \\ &\leq |(u, u)| + |(v, v)| + 2|(u, v)| \\ &\leq \|u\|^2 + \|v\|^2 + 2\|u\| \cdot \|v\| = (\|u\| + \|v\|)^2.\end{aligned}$$

Therefore,

$$(\|u\| + \|v\|)^2 - \|u + v\|^2 \geq 0,$$

and, upon factoring the difference of two squares, we have

$$(\|u\| + \|v\| - \|u + v\|)(\|u\| + \|v\| + \|u + v\|) \geq 0.$$

It follows that either both factors are ≥ 0 or both are ≤ 0. If both are ≥ 0, then

$$\|u\| + \|v\| - \|u + v\| \geq 0$$

as required. If both are ≤ 0, then in particular,

$$\|u\| + \|v\| + \|u + v\| \leq 0,$$

and since $\|u\|$, $\|v\|$ and $\|u + v\|$ are all ≥ 0, it follows that $\|u\| = \|v\| = 0$. Then $u = v = 0$ by Definition (15.1), and the corollary holds in this case also.

The triangle inequality expresses in a general vector space the familiar fact from high school geometry that the length of the third side of a triangle is less than or equal to the sum of the lengths of the other two sides. To make this interpretation, we think of the vectors u and v as directed line segments, so that the vector $u + v$ represents the third side of a triangle with sides u and v [see Figure (4.10)].

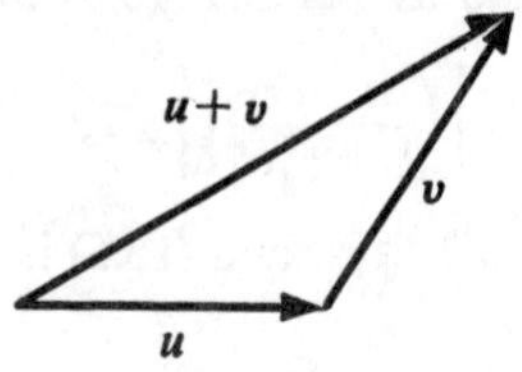

FIGURE 4.10

Another important concept is suggested by thinking about the geometry of the plane. The law of cosines in plane trigonometry states that the square of the length of one side of a triangle is equal to the sum of the squares of the lengths of the other two sides minus twice their product multiplied by the cosine of the angle between them. In order to see what form this statement should take in a general vector space, consider the accompanying Figure 4.11.

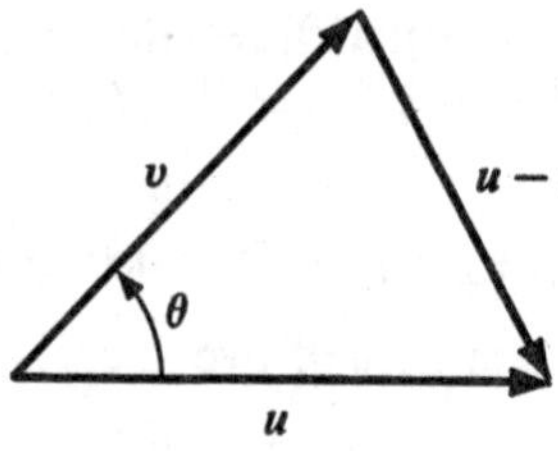

FIGURE 4.11

The law of cosines asserts that

$$\|u - v\|^2 = \|u\|^2 + \|v\|^2 - 2\|u\|\,\|v\| \cos \theta .$$

Using the fact that $\|u\|^2 = (u, u)$, etc., the formula becomes

$$(u - v, u - v) = (u, u) + (v, v) - 2\|u\|\,\|v\| \cos \theta$$

or, upon simplifying,

$$-2(u, v) = -2\|u\|\,\|v\| \cos \theta .$$

Therefore the law of cosines in the plane is equivalent to the statement that the cosine of the angle θ between two nonzero vectors u and v is given by

$$\cos \theta = \frac{(u, v)}{\|u\|\,\|v\|} .$$

This leads us to make the following definition about vectors in a general vector space with an inner product V.

(15.7) DEFINITION. Let $u, v \in V$. Define the *angle* θ between u and v as the angle, for $0 \leq \theta \leq \pi$, such that

$$\cos \theta = \frac{(u, v)}{\|u\| \, \|v\|}$$

[Note that $|\cos \theta| \leq 1$ by (15.5).] The vectors u and v are defined *orthogonal* if $(u, v) = 0$ or, in other words, if the cosine of the angle between them is zero. A basis $\{u_1, \ldots, u_n\}$ for a finite-dimensional space V with an inner product is called an *orthonormal basis* if:

1. $\|u_i\| = 1, \quad 1 \leq i \leq n.$
2. $(u_i, u_j) = 0, \quad i \neq j.$

An *orthonormal set of vectors* is any set of vectors $\{u_1, \ldots, u_k\}$ satisfying conditions (1) and (2).

(15.8) LEMMA. *Every orthonormal set of vectors is a linearly independent set.*

Proof. Let $\{u_1, \ldots, u_k\}$ be an orthonormal set, and suppose

$$\lambda_1 u_1 + \cdots + \lambda_k u_k = 0, \qquad \text{for } \lambda_i \in R.$$

Taking the inner product of both sides with u_1 and using the bilinear properties of the inner product, we obtain

$$\lambda_1(u_1, u_1) + \cdots + \lambda_k(u_k, u_1) = 0.$$

By conditions (1) and (2) we have $\lambda_1 = 0$. Taking the inner product with $u_2, \ldots, u_k$ in turn, we have $\lambda_2 = \cdots = \lambda_k = 0$, and we have proved that $\{u_1, \ldots, u_k\}$ are linearly independent.

Example A. (a) The simplest example of an orthonormal set of vectors is the familiar set of *unit vectors*

$$e_1 = \langle 1, 0, \ldots, 0 \rangle, \quad e_2 = \langle 0, 1, 0, \ldots, 0 \rangle, \quad \ldots, \quad e_n = \langle 0, \ldots, 0, 1 \rangle$$

in R_n, where the inner product is given by

$$(u, v) = \sum_{i=1}^{n} \alpha_i \beta_i,$$

for the vectors $u = \langle \alpha_1, \ldots, \alpha_n \rangle$, $v = \langle \beta_1, \ldots, \beta_n \rangle$. The unit vectors form an orthonormal basis of R_n with respect to the inner product given above.

(b) We shall prove that the functions

$$f_n(x) = \sin nx, \qquad n = 1, 2, \ldots$$

form an orthonormal set in the vector space $C(R)$ of continuous real-valued functions on R, with respect to the inner product

$$(f, g) = \frac{1}{\pi}\int_{-\pi}^{\pi} f(x)g(x)\,dx,$$

for continuous functions $f, g \in C(R)$. The statement that the functions $\{f_n\}$ form an orthonormal set is equivalent to proving the formulas

$$\frac{1}{\pi}\int_{-\pi}^{\pi} \sin nx \sin mx\,dx = \begin{cases} 1 & n = m \\ 0 & n \neq m. \end{cases}$$

First, for $n = m$ we have

$$\frac{1}{\pi}\int_{-\pi}^{\pi} (\sin nx)^2\,dx = \frac{1}{\pi}\int_{-\pi}^{\pi} \frac{1}{2}(1 - \cos 2nx)\,dx,$$

using the identity

$$\cos 2\theta = (\cos\theta)^2 - (\sin\theta)^2 = 1 - 2(\sin\theta)^2.$$

Then

$$\frac{1}{\pi}\int_{-\pi}^{\pi} \frac{1}{2}(1 - \cos 2nx)\,dx = 1 - \frac{1}{2n\pi}\sin 2nx\Big]_{-\pi}^{\pi} = 1.$$

For $n \neq m$, we have

$$\frac{1}{\pi}\int_{-\pi}^{\pi} \sin nx \sin mx\,dx = \frac{1}{2\pi}\int_{-\pi}^{\pi} [\cos(n - m)x - \cos(n + m)x]\,dx,$$

using the addition formulas for $\cos(A + B)$ and $\cos(A - B)$ to derive a formula for $\sin A \sin B$. The integral becomes

$$\frac{1}{2\pi}\left[\frac{1}{n - m}\sin(n - m)x\right]_{-\pi}^{\pi} - \left[\frac{1}{n + m}\sin(n + m)x\right]_{-\pi}^{\pi} = 0.$$

For applications of these formulas to differential equations and Fourier series, see the book by Boyce and DePrima, listed in the Bibliography.

The next theorem gives an inductive procedure for constructing an orthonormal basis from a given set of basis vectors.

(15.9) THEOREM (GRAM-SCHMIDT ORTHOGONALIZATION PROCESS). *Let V be a finite dimensional vector space with an inner product (u, v), and let $\{w_1, \ldots, w_n\}$ be a basis. Suppose $\{u_1, \ldots, u_r\}$ is an orthonormal basis for the subspace $S(w_1, \ldots, w_r)$. Define*

$$u_{r+1} = \frac{w}{\|w\|}$$

where

$$w = w_{r+1} - \sum_{i=1}^{r} (w_{r+1}, u_i)u_i.$$

Then $\{u_1, \ldots, u_{r+1}\}$ is an orthonormal basis for $S(w_1, \ldots, w_{r+1})$.

Proof. We have three statements to prove: first, that u_{r+1} has length 1; second, that $(u_{r+1}, u_i) = 0$ for $1 \le i \le r$; and third, that $S(w_1, \ldots, w_{r+1}) = S(u_1, \ldots, u_{r+1})$ (it follows from the last statement that $\{u_1, \ldots, u_{r+1}\}$ is a linearly independent set). Since $S(u_1, \ldots, u_r) = S(w_1, \ldots, w_r)$, it is clear that $w = w_{r+1} - \sum_1^r (w_{r+1}, u_i)u_i$ is different from zero. Then $u_{r+1} = w/\|w\|$ has length 1. To prove that $(u_{r+1}, u_j) = 0$ for $1 \le j \le r$ it is sufficient to prove that $(w, u_j) = 0$. We have

$$\begin{aligned}(w, u_j) &= \left(w_{r+1} - \sum_{i=1}^{r} (w_{r+1}, u_i)u_i, u_j\right) \\ &= (w_{r+1}, u_j) - \sum_{i=1}^{r} (w_{r+1}, u_i)(u_i, u_j) \\ &= (w_{r+1}, u_j) - (w_{r+1}, u_j) = 0.\end{aligned}$$

Finally, it is clear that $u_{r+1} \in S(w_1, \ldots, w_{r+1})$ and that $w_{r+1} \in S(u_1, \ldots, u_{r+1})$. Since $S(w_1, \ldots, w_r) = S(u_1, \ldots, u_r)$ by assumption, we have $S(u_1, \ldots, u_{r+1}) = S(w_1, \ldots, w_{r+1})$, as required. This completes the proof.

COROLLARY. *Every finite dimensional vector space V with an inner product has an orthonormal basis.*

Proof. Let $\{v_1, \ldots, v_n\}$ be a basis for V. The Gram-Schmidt process shows that, by mathematical induction, each subspace $S(v_1, v_2, \ldots, v_r)$ $r \le n$, of V has an orthonormal basis. In particular, $S(v_1, \ldots, v_n) = V$ has such a basis.

Example B. The Gram-Schmidt process can be looked at from the point of view of the problem of finding the distance from a point to a line, considered in Exercise 3 of Section 14.

Suppose, for simplicity, that L is a line through the origin in R_2, and let u_1 be a unit vector on L [see Figure (4.12)]. Let v be a vector not on L. The first step of the Gram-Schmidt process gives a vector

$$w = v - (v, u_1)u_1$$

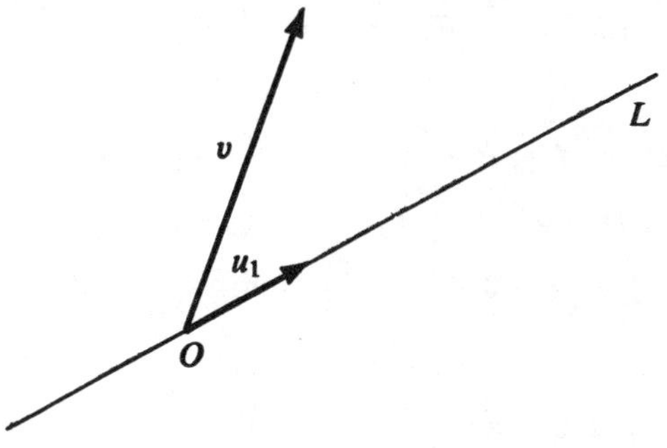

FIGURE 4.12

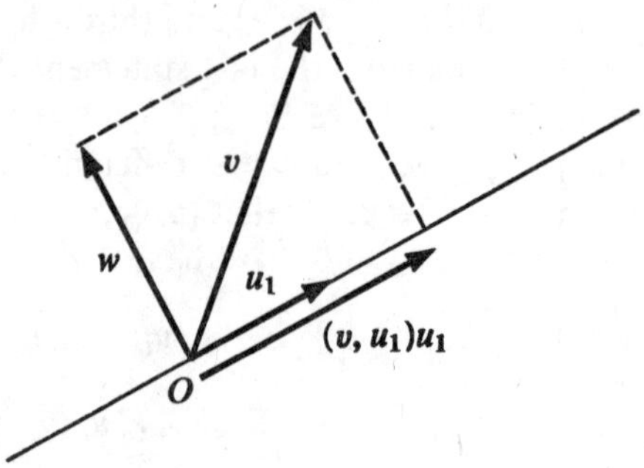

FIGURE 4.13

such that $(w, u_1) = 0$. Looked upon geometrically, this construction resolves the vector v into components along L and along a line perpendicular to L [see Figure (4.13)],

$$v = (v, u_1)u_1 + w = (v, u_1)u_1 + (v - (v, u_1)u_1).$$

The perpendicular distance from the terminal point of v to the line L is simply the length of the projection w,

$$\|w\| = \|v - (v, u_1)u_1\|.$$

In applying this formula, remember that u_1 is a unit vector to begin with.

Here is a numerical example. Find the perpendicular distance from the point (1, 5) to the line passing through the points (1, 1) and (−2, 0). A unit vector on the line is given by $u_1 = u/\|u\|$, where $u = \langle -3, -1 \rangle$. Then

$$u_1 = \frac{1}{\sqrt{10}}\langle -3, -1 \rangle.$$

To apply the projection method just described, we have in effect taken the point (1, 1) as the origin [see Figure (4.14)]. Then v is the vector given by the directed line segment from (1, 1) to (1, 5). Thus

$$v = \langle 0, 4 \rangle.$$

The projection of v perpendicular to the line is given by the vector

$$\begin{aligned} w &= v - (v, u_1)u_1 \\ &= \langle 0, 4 \rangle - \frac{1}{\sqrt{10}}(-4) \cdot \frac{1}{\sqrt{10}}\langle -3, -1 \rangle \\ &= \langle -\tfrac{6}{5}, \tfrac{18}{5} \rangle. \end{aligned}$$

The perpendicular distance is

$$\|w\| = \tfrac{1}{5}(36 + 324)^{1/2} = \tfrac{1}{5}\sqrt{360}.$$

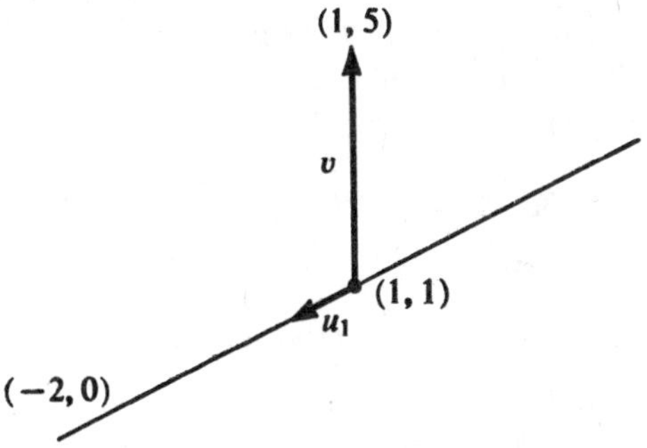

FIGURE 4.14

We turn now to the concept of length-preserving transformations on a general vector space with an inner product. Some examples of these transformations (rotations and reflections in R_2) have been discussed extensively in the preceding section.

(15.10) DEFINITION. Let V be a vector space with an inner product (u, v). A linear transformation $T \in L(V, V)$ is called an *orthogonal transformation*, provided that T preserves length, that is, that $\|T(u)\| = \|u\|$ for all $u \in V$.†

(15.11) THEOREM. *The following statements concerning a linear transformation $T \in L(V, V)$, where V is finite dimensional, are equivalent.*

1. *T is an orthogonal transformation.*
2. *$(T(u), T(v)) = (u, v)$ for all $u, v \in V$.*
3. *For some orthonormal basis $\{u_1, \ldots, u_n\}$ of V, the vectors $\{T(u_1), \ldots, T(u_n)\}$ also form an orthonormal set.*
4. *The matrix $\mathbf{A}$ of T with respect to an orthonormal basis satisfies the condition ${}^t\mathbf{A} \cdot \mathbf{A} = \mathbf{I}$ where ${}^t\mathbf{A}$ is the matrix obtained from $\mathbf{A}$ by interchanging rows and columns (called the* transpose *of* $\mathbf{A}$).

Proof. Statement (1) implies statement (2). We are given that $\|T(u)\| = \|u\|$ for *all* vectors $u \in V$. This implies that $(T(u), T(u)) = (u, u)$ for all vectors u. Replacing u by $u + v$ we obtain

$$(T(u + v), T(u + v)) = (u + v, u + v).$$

Now expand, using the bilinear and symmetric properties of the inner product, to obtain

$$(T(u), T(u)) + 2(T(u), T(v)) + (T(v), T(v)) = (u, u) + 2(u, v) + (v, v).$$

† In connection with this definition, see Exercise 12 at the end of this section.

Since $(T(u), T(u)) = (u, u)$ and $(T(v), T(v)) = (v, v)$, the last equation implies that $(T(u), T(v)) = (u, v)$ for all u and v.

Statement (2) implies statement (3). Let $\{u_1, \ldots, u_n\}$ be an orthonormal basis of V; then

$$(u_i, u_i) = 1, \qquad (u_i, u_j) = 0, \qquad i \neq j.$$

By statement (2) we have

$$(T(u_i), T(u_i)) = 1, \qquad (T(u_i), T(u_j)) = 0, \qquad i \neq j,$$

and $\{T(u_1), \ldots, T(u_n)\}$ is an orthonormal set.

Statement (3) implies statement (1). Suppose that for some orthonormal basis $\{u_1, \ldots, u_n\}$ of V the image vectors $\{T(u_1), \ldots, T(u_n)\}$ form an orthonormal set. Let

$$v = \xi_1 u_1 + \cdots + \xi_n u_n$$

be an arbitrary vector in V. Then

$$\|v\|^2 = (v, v) = (\xi_1 u_1 + \cdots + \xi_n u_n, \xi_1 u_1 + \cdots + \xi_n u_n) = \sum_1^n \xi_i^2$$

since $\{u_1, \ldots, u_n\}$ is an orthonormal set. Similarly, we have

$$\|T(v)\|^2 = (T(v), T(v)) = \left(\sum_1^n \xi_i T(u_i), \sum_1^n \xi_i T(u_i)\right) = \sum_1^n \xi_i^2.$$

Thus statement (1) is proved and we have shown the equivalence of the first three statements.

Finally, we prove that statements (3) and (4) are equivalent. Suppose that statement (3) holds and let $\{u_1, \ldots, u_n\}$ be an orthonormal basis for V. Let

$$T(u_i) = \sum_{j=1}^n \alpha_{ji} u_j.$$

Since $\{T(u_1), \ldots, T(u_n)\}$ is an orthonormal set, we have

$$(T(u_i), T(u_i)) = \left(\sum_{j=1}^n \alpha_{ji} u_j, \sum_{j=1}^n \alpha_{ji} u_j\right) = \sum_{j=1}^n \alpha_{ji}^2 = 1$$

and, if $i \neq j$,

$$(T(u_i), T(u_j)) = \left(\sum_{k=1}^n \alpha_{ki} u_k, \sum_{k=1}^n \alpha_{kj} u_k\right) = \sum_{k=1}^n \alpha_{ki}\alpha_{kj} = 0.$$

These equations imply that ${}^t\mathbf{A}\mathbf{A} = \mathbf{I}$, since the (i, k)th entry of ${}^t\mathbf{A}$ is α_{ki}. Conversely, ${}^t\mathbf{A}\mathbf{A} = \mathbf{I}$ implies that the equations above are satisfied and hence that $\{T(u_1), \ldots, T(u_n)\}$ is an orthonormal set. This completes the proof of the theorem.

(15.12) DEFINITION. A matrix $A \in M_n(R)$ is called an *orthogonal matrix* if ${}^t A \cdot A = I$; then A is simply the matrix of an orthogonal transformation with respect to an orthonormal basis of V.

EXERCISES

Throughout these exercises, V denotes a finite dimensional vector space over R with an inner product.

1. Find orthonormal bases, using the Gram-Schmidt process or otherwise, for the subspaces of R_4 generated by the following sets of vectors. It is understood that the usual inner product in R_4 is in use:

$$(\langle \alpha_1, \alpha_2, \alpha_3, \alpha_4 \rangle, \langle \beta_1, \beta_2, \beta_3, \beta_4 \rangle) = \sum_{i=1}^{4} \alpha_i \beta_i .$$

   a. $\langle 1, 1, 1, 0 \rangle, \langle -1, 1, 2, 1 \rangle$.
   b. $\langle 1, 1, 0, 0 \rangle, \langle 0, 1, 1, 0 \rangle, \langle 0, 0, 1, 1 \rangle$.
   c. $\langle -1, 1, 1, 1 \rangle, \langle 1, -1, 1, 1 \rangle, \langle 1, 1, -1, 1 \rangle$.
2. Find an orthonormal basis for the subspace of $C(R)$ generated by the functions $\{1, x, x^2\}$ with respect to the inner product $(f, g) = \int_0^1 f(t)g(t)\, dt$.
3. Use the Gram-Schmidt process as in Example B to find the perpendicular distance from the points to the corresponding lines in the following problems.
   a. Point $(0, 0)$ to the line through $(1, 1)$ and $(3, 0)$.
   b. Point $(-1, 0)$ to the line $y = x$.
   c. Point $(1, 1)$ to the line through $(-1, -1)$ and $(0, -2)$.
4. Use the methods of Example A to show that the functions $g_n(x) = \cos nx$, $n = 1, 2, \ldots$ form an orthonormal set in $C(R)$ with respect to the same inner product given in Example A,

$$(f, g) = \frac{1}{\pi} \int_{-\pi}^{\pi} f(x)g(x)\, dx .$$

5. Let $\{u_1, u_2, \ldots, u_n\}$ be an orthonormal basis for V.
   a. Show that if $v = \sum \xi_i u_i$, $w = \sum \eta_i u_i$, then

$$(v, w) = \sum \xi_i \eta_i .$$

   b. Show that every vector $v \in V$ can be expressed uniquely in the form

$$v = \sum_{i=1}^{n} (v, u_i) u_i .$$

6. Let u be a vector in R_n such that $\|u\| = 1$ (for the usual inner product). Prove that there exists an n-by-n orthogonal matrix whose first row is u.
7. Let $O(V)$ be the set of all orthogonal transformations on V. Prove $O(V)$ is a group with respect to the operation of multiplication.

8. Two vector spaces V and W with inner products (v_1, v_2) and $[w_1, w_2]$, respectively, are said to be *isometric* if there exists a one-to-one linear transformation T of V onto W such that $[Tv_1, Tv_2] = (v_1, v_2)$ for all $v_1, v_2 \in V$. Such a linear transformation T is called an *isometry*. Let V be a finite dimensional space with an inner product (u, v), and let $\{v_1, \ldots, v_n\}$ be an orthonormal basis. Prove that the mapping

$$T: \sum_{i=1}^{n} \xi_i v_i \to \langle \xi_1, \xi_2, \ldots, \xi_n \rangle$$

is an isometry of V onto R_n, where R_n is equipped with the usual inner product given in Example (*i*) at the beginning of the section.

9. Let V be a finite dimensional vector space with an inner product, and let W be a subspace of V. Let $W^\perp$ be the set of all vectors $v \in V$ such that $(v, w) = 0$ for all $w \in W$. Prove that $\dim(W) + \dim(W^\perp) = \dim V$.

10. Prove that, if W_1 and W_2 are subspaces of V such that $\dim W_1 = \dim W_2$, then there exists an orthogonal transformation T such that $T(W_1) = W_2$.

11. The vectors mentioned in this problem all belong to R_3, equipped with the usual inner product. A *plane* is a set P of vectors of the form $p + w$, $w \in W$, for some two-dimensional subspace W. From Chapter 2, we know that a set of vectors is a plane in R_3 if and only if it consists of all solutions of a linear equation

$$\alpha_1 x_1 + \alpha_2 x_2 + \alpha_3 x_3 + \alpha_4 = 0, \qquad \alpha_i \in R$$

where not all of $\alpha_1, \alpha_2, \alpha_3$ are zero. The two-dimensional subspace W associated with the plane is the set of solutions of the homogeneous equation

$$\alpha_1 x_1 + \alpha_2 x_2 + \alpha_3 x_3 = 0.$$

Note that if $n = \langle \alpha_1, \alpha_2, \alpha_3 \rangle$, then $(n, w) = 0$ for all $w \in W$, and (n, p) is a constant for all $p \in P$. The vector n is called a *normal* vector to the plane.

  a. Let n be a nonzero vector in R_3, and α a fixed real number. Show that the set of all vectors p such that

$$(p, n) = \alpha$$

  is a plane with a normal vector n, and two-dimensional subspace $S(n)^\perp$.

  b. Find a normal vector n and a basis for $S(n)^\perp$ for the plane

$$3x_1 - x_2 + x_3 - 1 = 0.$$

  c. Find the equation of the plane with normal vector $n = \langle 1, -1, 2 \rangle$, and containing $p = \langle -1, 1, 0 \rangle$. State the equation both in the form

$$\alpha_1 x_1 + \alpha_2 x_2 + \alpha_3 x_3 = \beta$$

  and

$$(n, p) = \alpha.$$

d. Show that a vector x lies on the plane with normal vector n, passing through p, if and only if

$$(x - p, n) = 0.$$

e. Let $a = \langle 2, 0, 1\rangle$, $b = \langle 1, 1, 0\rangle$. Find a vector $c \neq 0$ such that $(c, a) = (c, b) = 0$. [*Hint:* Show that c is a solution of a certain system of homogeneous equations.]

f. Find the equation of the plane passing through the points $\langle 2, 0, -1\rangle$, $\langle 1, 1, 1\rangle$, $\langle 0, 0, 1\rangle$. [*Hint:* Use the result of part (e) to find a normal vector.]

g. Let P be a plane, and let u be a vector not on P. Show that there is a unique vector p_0 on P such that for all $x \in P$,

$$(p_0 - x, p_0 - u) = 0.$$

[*Hint:* Suppose the equation of P is

$$(x - p, n) = 0,$$

where $p \in P$ and n is a normal vector. Then p_0 satisfies the equations

$$(p_0 - p, n) = 0$$
$$p_0 - u = \lambda n$$

for some $\lambda \in R$. Then it is necessary to solve for λ.]

h. Find the perpendicular distance from the vector $\langle 1, 1, 2\rangle$ to the plane $x_1 + x_2 - x_3 + 1 = 0$, using the result of part (g). Can you do the problem using the Gram-Schmidt process as in Example B?

12. Orthogonal transformations of a finite dimensional vector space V over R with an inner product are defined by reference to length, not orthogonality. Let T be a linear transformation which preserves orthogonality, in the sense that $(Tv, Tw) = 0$ whenever $(v, w) = 0$. Prove that T is a scalar multiple of an orthogonal transformation.

13. Let $v_1, \ldots, v_m$ be an orthonormal set of vectors in V, and let v be a vector such that $v \notin S(v_1, \ldots, v_m)$. Show that the vector $v' = v - \sum_{i=1}^{n} (v, v_i)v_i$ given by the Gram-Schmidt process has the shortest length among all vectors of the form $v - x$, $x \in S(v_1, \ldots, v_n)$.

Chapter 5

Determinants

In calculus, we can learn something about the behavior of a function $f: R \to R$ at a point x by computing the derivative $f'(x)$. For example, if f' is positive on an interval $[a, b]$, the function f is one-to-one on the interval, etc. The determinant plays a similar role in linear algebra. The determinant is a rule which assigns to each linear transformation $T: V \to V$ a number which tells something about the behavior of T (see Section 18). We approach the subject by thinking of the determinant as a function of the row vectors of a matrix of T with respect to some basis. If the matrix of T has real coefficients, the determinant is (up to a sign ± 1) the volume of the parallelopiped whose edges are the rows of the matrix.

16. DEFINITION OF DETERMINANTS

Let us begin with a study of the function $A(a_1, a_2)$ which assigns to each pair of vectors $a_1, a_2 \in R_2$ the area of the parallelogram with edges a_1 and a_2 (Figure 5.1). Instead of working out a formula for this function in

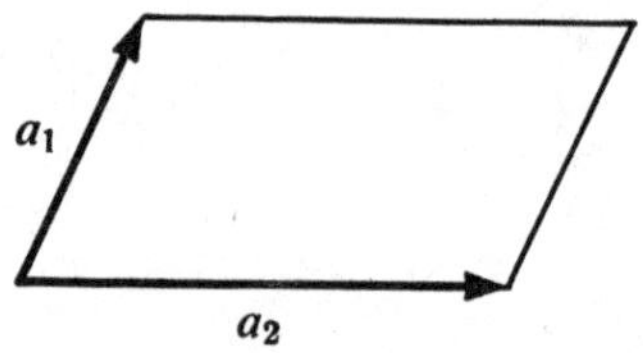

FIGURE 5.1

terms of the components of a_1 and a_2, let us see what are some of the general properties of the function A. We have, first of all,

(16.1) $$A(e_1, e_2) = 1, \quad \text{if } e_1 = \langle 1, 0\rangle, e_2 = \langle 0, 1\rangle.$$

Second, if we multiply one of the vectors by a positive real number λ, we multiply A by λ, since the area of a parallelogram is the product of the lengths of the base and height (see Figure 5.2) and the length of λa_1 is λ times the length of a_1.†

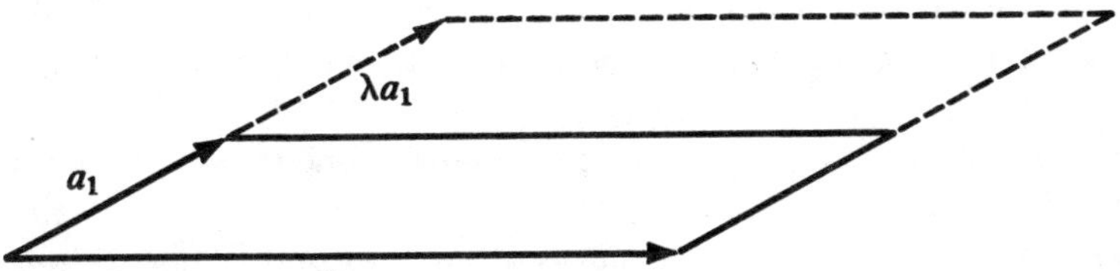

FIGURE 5.2

In terms of the function A we have:

(16.2) $$A(\lambda a_1, a_2) = A(a_1, \lambda a_2) = \lambda A(a_1, a_2), \quad \text{for } \lambda > 0.$$

Finally, the base and height of the parallelogram with edges a_1, a_2 are the same as those of the parallelogram with edges $a_1 + a_2$ and a_2 (see Figure 5.3), and hence we have:

(16.3) $$A(a_1 + a_2, a_2) = A(a_1, a_2) = A(a_1, a_2 + a_1).$$

Now stop! One would think that we have not yet described all the essential properties of the area function. We are going to prove, however, that there is one and only one function which assigns to each pair of

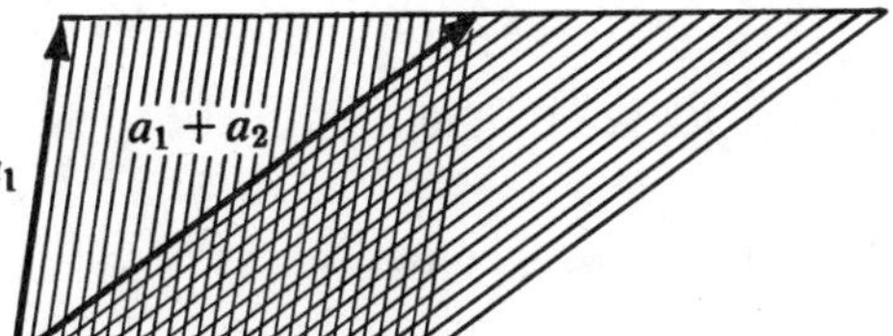

FIGURE 5.3

† As in Chapter 4, we define the length of $a = \langle \alpha_1, \alpha_2\rangle$ as $\sqrt{\alpha_1^2 + \alpha_2^2}$.

vectors in R_2 a nonnegative real number satisfying conditions (16.1), (16.2), and (16.3). Note also that the function A has a further property:

(16.4) *$A(a_1, a_2) \neq 0$ if and only if a_1 and a_2 are linearly independent.*

The statement (16.4) is perhaps the most important and interesting property of the area function. It states that two vectors are linearly dependent if and only if the area of the parallelogram determined by them is zero. In other words, the computation of a single number (the area) gives a test for linear dependence.

It is because of the obvious usefulness of the sort of test described in (16.4) that we shall define a function with properties similar to the area function in as general a setting as possible.

It will be convenient to begin by defining a function on sets of vectors from F_n which satisfy Axioms (16.1), (16.2), and (16.3) for arbitrary $\lambda \in F$. We shall derive consequences of these axioms in this section and postpone to the next section the task of proving that such a function really does exist. The reader will note that there is nothing logically wrong with this procedure; it is, for example, what we do in euclidean geometry, namely, to derive consequences of certain axioms before we have a construction of certain objects that satisfy the axioms. We return to the connection with areas and volumes in Section 19.

(16.5) DEFINITION. Let F be an arbitrary field. A *determinant* is a function which assigns to each n-tuple $\{a_1, \ldots, a_n\}$ of vectors in F_n an element of F, $D = D(a_1, \ldots, a_n)$ such that the following conditions are satisfied.

(*i*) $D(a_1, \ldots, a_{i-1}, a_i + a_j, a_{i+1}, \ldots, a_n) = D(a_1, \ldots, a_n)$, for $1 \leq i \leq n$ and $j \neq i$.

(*ii*) $D(a_1, \ldots, a_{i-1}, \lambda a_i, a_{i+1}, \ldots, a_n) = \lambda D(a_1, \ldots, a_n)$, for all $\lambda \in F$.

(*iii*) $D(e_1, \ldots, e_n) = 1$, if e_i is the ith unit vector

$$\langle 0, \ldots, 0, 1, 0, \ldots, 0 \rangle$$

with a 1 in the ith position and zeros elsewhere.

Now we shall derive some consequences of the definition.

(16.6) THEOREM. *Let D be a determinant function on F_n; then the following statements are valid.*

(a) *$D(a_1, \ldots, a_n)$ is multiplied by -1 if two of the vectors a_i and a_j are interchanged (where $i \neq j$).*

(b) *$D(a_1, \ldots, a_n) = 0$ if two of the vectors a_i and a_j are equal.*
(c) *D is unchanged if a_i is replaced by $a_i + \sum_{j \neq i} \lambda_j a_j$, for arbitrary $\lambda_j \in F$.*
(d) *$D(a_1, \ldots, a_n) = 0$, if $\{a_1, \ldots, a_n\}$ is linearly dependent.*
(e) *$D(a_1, \ldots, a_{i-1}, \lambda a_i + \mu a_i', a_{i+1}, \ldots, a_n)$*
$$= \lambda D(a_1, \ldots, a_i, \ldots, a_n) + \mu D(a_1, \ldots, a_i', \ldots, a_n)$$
for $1 \leq i \leq n$, for arbitrary field elements λ and μ, and for vectors a_i and $a_i' \in F_n$.

Proof. **(a)** We shall use the notation

$$D(\ldots, \underset{i}{a}, \ldots, \underset{j}{b}, \ldots)$$

to indicate that the ith argument of the function D is a and the jth is b, etc. Then we have, for arbitrary i and j,

$$\begin{aligned} D(\ldots, \underset{i}{a_i}, \ldots, \underset{j}{a_j}, \ldots) & \\ &= -D(\ldots, \underset{i}{-a_i}, \ldots, \underset{j}{a_j}, \ldots) && \text{by property (ii)} \\ &= -D(\ldots, \underset{i}{-a_i}, \ldots, \underset{j}{-a_i + a_j}, \ldots) && \text{by property (i)} \\ &= D(\ldots, \underset{i}{-a_i}, \ldots, \underset{j}{+a_i - a_j}, \ldots) && \text{by property (ii)} \\ &= D(\ldots, \underset{i}{-a_j}, \ldots, \underset{j}{a_i - a_j}, \ldots) && \text{by property (i) since } -a_i + (a_i - a_j) = -a_j \\ &= -D(\ldots, \underset{i}{-a_j}, \ldots, \underset{j}{-a_i + a_j}, \ldots) && \text{by property (ii)} \\ &= -D(\ldots, \underset{i}{-a_j}, \ldots, \underset{j}{-a_i}, \ldots) && \text{by properties (i) and (ii)} \\ &= -D(\ldots, \underset{i}{a_j}, \ldots, \underset{j}{a_i}, \ldots). \end{aligned}$$

(b) If $a_i = a_j$, $i \neq j$, then by statement **(a)** we have

$$\begin{aligned} D(a_1, \ldots, a_n) &= D(\ldots, \underset{i}{a_j}, \ldots, \underset{j}{a_i}, \ldots) \\ &= D(\ldots, 2a_i, \ldots, a_i, \ldots) \\ &= -2D(\ldots, -a_i, \ldots, a_i, \ldots) \\ &= -2D(\ldots, 0, \ldots, a_i, \ldots) = 0, \end{aligned}$$

and statement **(b)** is proved.

(c) Let $j \neq i$, and $\lambda \in F$. We may assume $\lambda \neq 0$. Then

$$\begin{aligned} D(\ldots, \underset{i}{a_i}, \ldots, \underset{j}{a_j}, \ldots) \\ &= \lambda^{-1} D(\ldots, \underset{i}{a_i}, \ldots, \underset{j}{\lambda a_j}, \ldots) && \text{by property (ii)} \\ &= \lambda^{-1} D(\ldots, \underset{i}{a_i + \lambda a_j}, \ldots, \underset{j}{\lambda a_j}, \ldots) && \text{by property (i)} \\ &= D(\ldots, a_i + \lambda a_j, \ldots, a_j, \ldots). \end{aligned}$$

Repeating this argument, we obtain statement (c).

(d) If $a_1, \ldots, a_n$ are linearly dependent, then some a_i can be expressed as a linear combination of the remaining vectors:

$$a_i = \sum_{j \neq i} \lambda_j a_j.$$

By proposition (c) we have

$$\begin{aligned} D(a_1, \ldots, a_n) &= D\Big(\ldots, \underset{i}{a_i - \sum_{j \neq i} \lambda_j a_j}, \ldots\Big) \\ &= D(\ldots, \underset{i}{0}, \ldots) \\ &= 0 D(\ldots, \underset{i}{0}, \ldots) && \text{by property (ii)} \\ &= 0. \end{aligned}$$

(e) Because of property (ii) it is sufficient to prove, for example, that

(16.7) $D(a_1 + a_1', a_2, \ldots, a_n) = D(a_1, a_2, \ldots, a_n) + D(a_1', a_2, \ldots, a_2).$

We may assume that $\{a_2, \ldots, a_n\}$ are linearly independent [otherwise, by statement (d) both sides of (16.7) are zero and there is nothing to prove]. By (7.4) the set $\{a_2, \ldots, a_n\}$ can be completed to a basis

$$\{\bar{a}_1, a_2, \ldots, a_n\}$$

of R_n. Then by statement (c) and Axiom (ii) we have:

(16.8) $D(\lambda_1 \bar{a}_1 + \sum_{i>1} \lambda_i a_i, a_2, \ldots, a_n) = \lambda_1 D(\bar{a}_1, a_2, \ldots, a_n)$, for all choices of $\lambda_1, \ldots, \lambda_n$.

Now let

$$\begin{aligned} a_1 &= \lambda_1 \bar{a}_1 + \lambda_2 a_2 + \cdots + \lambda_n a_n, \\ a_1' &= \mu_1 \bar{a}_1 + \mu_2 a_2 + \cdots + \mu_n a_n. \end{aligned}$$

Then

$$a_1 + a_1' = (\lambda_1 + \mu_1)\bar{a}_1 + (\lambda_2 + \mu_2) a_2 + \cdots + (\lambda_n + \mu_n) a_n.$$

By (16.8) we have

$$
\begin{aligned}
D(a_1 + a_1', a_2, \ldots, a_n) &= (\lambda_1 + \mu_1)D(\bar{a}_1, a_2, \ldots, a_n),\\
D(a_1, a_2, \ldots, a_n) &= \lambda_1 D(\bar{a}_1, a_2, \ldots, a_n),\\
D(a_1', a_2, \ldots, a_n) &= \mu_1 D(\bar{a}_1, a_2, \ldots, a_n).
\end{aligned}
$$

By the distributive law in R we obtain (16.7), and the proof of the theorem is completed.

Remark. Many authors use statements (e) and (a) instead of Axiom (i) in the definition of determinant. The point of our definition [using Axiom (i)] is that we are assuming much less and can still prove the fundamental rule (e) by using the fairly deep result (7.4) concerning sets of linearly independent vectors in F_n.

We shall now show that the properties of determinants which we have just derived give a quick and efficient method for computing the values of the determinant function in particular cases.

The idea is that the properties of $D(a_1, \ldots, a_n)$ stated in Definition (16.5) and in Theorem (16.6) are closely related to performing elementary row operations to the matrix whose rows are $\{a_1, \ldots, a_n\}$.

To make all this clear, we shall use the notation $D(\mathbf{A})$ for the determinant $D(a_1, \ldots, a_n)$ where $\{a_1, \ldots, a_n\}$ are the rows of the n-by-n matrix **A**. If

$$
\mathbf{A} = \begin{pmatrix} \alpha_{11} & \cdots & \alpha_{1n} \\ \alpha_{21} & \cdots & \alpha_{2n} \\ \cdots & \cdots & \cdots \\ \alpha_{n1} & \cdots & \alpha_{nn} \end{pmatrix}
$$

we shall often write

$$
D(\mathbf{A}) = \begin{vmatrix} \alpha_{11} & \cdots & \alpha_{1n} \\ \alpha_{21} & \cdots & \alpha_{2n} \\ \cdots & \cdots & \cdots \\ \alpha_{n1} & \cdots & \alpha_{nn} \end{vmatrix}.
$$

The last notation for determinants is the most convenient one to use for carrying out computations, but the reader is hereby warned that its use has the disadvantage of having almost the same notations for the very different objects of matrices and determinants.

The next theorem is the key to the practical calculation of determinants; it is stated in terms of the elementary row operations defined in Definition (6.9).

(16.9) THEOREM. *Let* **A** *be an n-by-n matrix with coefficients in F, having rows* $\{a_1, \ldots, a_n\}$.

a. *Let* **A**′ *be a matrix obtained from* **A** *by an elementary row operation of type I in Definition (6.9) (interchanging two rows). Then*

$$D(\mathbf{A}') = -D(\mathbf{A}).$$

b. *Let* **A**′ *be a matrix obtained from* **A** *by an elementary row operation of type II (replacing the row* a_i *by* $a_i + \lambda a_j$, *with* $\lambda \in F$, $i \neq j$). *Then*

$$D(\mathbf{A}') = D(\mathbf{A}).$$

c. *Let* **A**′ *be a matrix obtained from* **A** *by an elementary row operation of type III (replacing* a_i *by* μa_i, *for* $\mu \neq 0$ *in F). Then*

$$D(\mathbf{A}') = \mu D(\mathbf{A}).$$

Proof. Part (a) is simply a restatement of Theorem (16.6)(**a**). Part (b) is a restatement of Theorem (16.6)(**c**). Finally, part (c) is Axiom (ii) in Definition (16.5).

Example A. Compute the determinant

$$\begin{vmatrix} -1 & 0 & 1 & 1 \\ 2 & -1 & 0 & 2 \\ 1 & 2 & 1 & -1 \\ -1 & -1 & 1 & 0 \end{vmatrix}.$$

We know from Sections 6 and 12 that we can first apply elementary row operations to obtain a matrix **A**′ row equivalent to the given matrix

$$\mathbf{A} = \begin{pmatrix} -1 & 0 & 1 & 1 \\ 2 & -1 & 0 & 2 \\ 1 & 2 & 1 & -1 \\ -1 & -1 & 1 & 0 \end{pmatrix}$$

such that the nonzero rows of **A**′ are in echelon form. At this point we will know whether or not the rows of **A** are linearly independent. If the rows are linearly dependent then $D(\mathbf{A}) = 0$ by Theorem (16.6)(**d**). If the rows are linearly independent then the rows of **A**′ will all be different from zero, and from Section 12, Example C, we can apply further elementary row operations to reduce **A**′ to the identity matrix. Theorem (16.9) tells us how to keep track of the determinant at each step of the process, and the computation will be finished using the fact that the determinant of the identity matrix is 1, by Definition (16.5)(3).

Applying these remarks to our example we have

$$\begin{vmatrix} -1 & 0 & 1 & 1 \\ 2 & -1 & 0 & 2 \\ 1 & 2 & 1 & -1 \\ -1 & -1 & 1 & 0 \end{vmatrix} = \begin{vmatrix} -1 & 0 & 1 & 1 \\ 0 & -1 & 2 & 4 \\ 1 & 2 & 1 & -1 \\ -1 & -1 & 1 & 0 \end{vmatrix} \quad \text{(replacing } a_2 \text{ by } a_2 + 2a_1\text{)}$$

$$= \begin{vmatrix} -1 & 0 & 1 & 1 \\ 0 & -1 & 2 & 4 \\ 0 & 2 & 2 & 0 \\ 0 & -1 & 0 & -1 \end{vmatrix} \quad \begin{matrix}\text{(replacing } a_3 \text{ by } a_3 + a_1, \\ \text{and } a_4 \text{ by } a_4 - a_1\text{)}\end{matrix}$$

$$= \begin{vmatrix} -1 & 0 & 1 & 1 \\ 0 & -1 & 2 & 4 \\ 0 & 0 & 6 & 8 \\ 0 & 0 & 0 & -5 + \frac{8}{3} \end{vmatrix} \quad \begin{matrix}\text{(row operations of} \\ \text{type II applied to the} \\ \text{last two rows)}\end{matrix}$$

$$= \begin{vmatrix} -1 & 0 & 0 & 0 \\ 0 & -1 & 0 & 0 \\ 0 & 0 & 6 & 0 \\ 0 & 0 & 0 & -\frac{7}{3} \end{vmatrix} \quad \begin{matrix}\text{(more row operations of} \\ \text{type II as in Section 12)}\end{matrix}$$

$$= (-1)(-1)6(-\tfrac{7}{3}) \begin{vmatrix} 1 & 0 & 0 & 0 \\ 0 & 1 & 0 & 0 \\ 0 & 0 & 1 & 0 \\ 0 & 0 & 0 & 1 \end{vmatrix} \quad \begin{matrix}\text{(elementary row} \\ \text{operations of} \\ \text{type III)}\end{matrix}$$

$$= -\tfrac{42}{3} = -14.$$

We note that this whole procedure is easy to check for arithmetical errors.

EXERCISES

1. Compute the determinants of the following matrices.

a. $\begin{pmatrix} 1 & 1 \\ -1 & 1 \end{pmatrix}$ **b.** $\begin{pmatrix} 1 & 1 & 2 \\ 1 & 0 & 1 \\ -1 & 1 & 0 \end{pmatrix}$ **c.** $\begin{pmatrix} 1 & 1 & 0 & 1 \\ 1 & 0 & 1 & 1 \\ 1 & 1 & 0 & 1 \\ 0 & -1 & 1 & 1 \end{pmatrix}$,

d. The matrices in Exercise 7 of Section 12.

2. Let $\mathbf{A}$ be a matrix in triangular form (with zeros below the diagonal),

$$\mathbf{A} = \begin{pmatrix} \alpha_1 & & & * \\ 0 & \alpha_2 & & \\ \vdots & \ddots & \ddots & \\ 0 & \cdots & 0 & \alpha_n \end{pmatrix}$$

Prove that $D(\mathbf{A}) = \alpha_1\alpha_2 \cdots \alpha_n$.

3. Let $\mathbf{A}$ be a matrix in block-triangular form,

$$\mathbf{A} = \begin{pmatrix} \mathbf{A}_1 & & & * \\ 0 & \mathbf{A}_2 & & \\ \vdots & & \ddots & \\ 0 & \cdots & 0 & \mathbf{A}_r \end{pmatrix}$$

where the $\mathbf{A}_i$ are square matrices, possibly of different sizes. Prove that

$$D(\mathbf{A}) = D(\mathbf{A}_1)D(\mathbf{A}_2) \cdots D(\mathbf{A}_r).$$

4. Let $D^*(a_1, \ldots, a_n)$ be a function on vectors $a_1, \ldots, a_n$ in R_n to R such that, for all a_j, a_i', a_i'' in R_n,
   a. $D^*(e_1, \ldots, e_n) = 1$, where the e_i are the unit vectors.
   b. $D^*(a_1, \ldots, \lambda a_i, \ldots, a_n) = \lambda D^*(a_1, \ldots, a_n)$, for $\lambda \in R$.
   c. $D^*(a_1, \ldots, a_i' + a_i'', \ldots, a_n) = D^*(a_1, \ldots, a_i', \ldots, a_n) + D^*(a_1, \ldots, a_i'', \ldots, a_n)$.
   d. $D^*(a_1, \ldots, a_n) = 0$ if $a_i = a_j$ for $i \neq j$.

   Prove that D^* is a determinant function on R_n.

17. EXISTENCE AND UNIQUENESS OF DETERMINANTS

It is time to remove all doubts about whether a function D satisfying the conditions of Definition (16.5) exists or not. In this section we shall prove, first, that at most one such function can exist, and then we shall give a construction of such a function. In the course of the discussion we shall obtain some other useful properties of determinants.

(17.1) THEOREM. *Let D and D' be two functions satisfying conditions i, ii, and iii of Definition (16.5); then, for all $a_1, \ldots, a_n$ in F_n,*

$$D(a_1, \ldots, a_n) = D'(a_1, \ldots, a_n).$$

Proof. Consider the function Δ defined by

$$\Delta(a_1, \ldots, a_n) = D(a_1, \ldots, a_n) - D'(a_1, \ldots, a_n).$$

Then, because both the functions D and D' satisfy the conditions of (16.5) as well as of (16.6), Δ has the following properties.

(17.2) $\Delta(e_1, \ldots, e_n) = 0$.

(17.3) $\Delta(a_1, \ldots, a_n)$ changes sign if two of the vectors a_i and a_j are interchanged, and $\Delta(a_1, \ldots, a_n) = 0$, if $a_i = a_j$ for $i \neq j$.

(17.4) $\Delta(\ldots, \lambda a_i, \ldots) = \lambda\Delta(\ldots, a_i, \ldots)$, for $\lambda \in F$.

(17.5) $\Delta(\ldots, a_i + a_i', \ldots) = \Delta(\ldots, a_i, \ldots) + \Delta(\ldots, a_i', \ldots)$.

Now let $a_1, \ldots, a_n$ be arbitrary vectors in F_n. It is sufficient to prove that $\Delta(a_1, \ldots, a_n) = 0$, and we shall show that this is a consequence of the properties (17.2) to (17.5). Because $e_1, \ldots, e_n$ is a basis of F_n we can express

$$a_i = \lambda_{i1}e_1 + \cdots + \lambda_{in}e_n = \sum_{j=1}^{n} \lambda_{ij}e_j.$$

Using (17.4) and (17.5) applied to the first position, then to the second position, etc., we have

$$\begin{aligned}
\Delta(a_1, \ldots, a_n) &= \Delta\left(\sum_{j=1}^{n} \lambda_{1j}e_j, a_2, \ldots, a_n\right) \\
&= \sum_{j=1}^{n} \lambda_{1j}\Delta(e_j, a_2, \ldots, a_n) \\
&= \sum_{j_1=1}^{n} \lambda_{1j_1}\Delta\left(e_{j_1}, \sum_{j_2=1}^{n} \lambda_{2j_2}e_{j_2}, a_3, \ldots, a_n\right) \\
&= \sum_{j_1=1}^{n} \sum_{j_2=1}^{n} \lambda_{1j_1}\lambda_{2j_2}\Delta(e_{j_1}, e_{j_2}, a_3, \ldots, a_n) \\
&= \cdots = \sum_{j_1=1}^{n} \sum_{j_2=1}^{n} \cdots \sum_{j_n=1}^{n} \lambda_{1j_1}\lambda_{2j_2} \cdots \lambda_{nj_n}\Delta(e_{j_1}, \ldots, e_{j_n}) \\
&= \sum_{j_1,\ldots,j_n=1}^{n} \lambda_{1j_1} \cdots \lambda_{nj_n}\Delta(e_{j_1}, \ldots, e_{j_n})
\end{aligned}$$

where the last sum consists of n^n terms, and is obtained by letting $j_1, \ldots, j_n$ range independently between 1 and n inclusive. By (17.2) and (17.3) it follows easily by induction† that for all choices of $j_1, \ldots, j_n$, $\Delta(e_{j_1}, \ldots, e_{j_n}) = 0$. Therefore $\Delta(a_1, \ldots, a_n) = 0$, and the uniqueness theorem is proved.

Now we come to the proof of existence of determinants. There is no really simple proof of this fact in the general case. The proof we shall give will take time, care, and probably some exasperation before it is understood. It has the redeeming feature of proving much more than the statement that determinants exist; it also yields the important formula for the column expansion (17.7) of a determinant.

(17.6) THEOREM. *There exists a function $D(a_1, \ldots, a_n)$ satisfying the conditions of Definition (16.5).*

Proof. We use induction on n. For $n = 1$, the function $D(\alpha) = \alpha$, $\alpha \in F$, satisfies the requirements. Now suppose that D is a function on F_{n-1} that satisfies the conditions in Definition (16.5). Fix an index j, $1 \leq j \leq n$, and let the vectors $a_1, \ldots, a_n$ in F_n be given by

$$a_i = (\alpha_{i1}, \ldots, \alpha_{in}), \qquad \alpha_{ik} \in F, \quad 1 \leq i \leq n.$$

† What has to be proved by induction is that either $\Delta(e_{j_1}, \ldots, e_{j_n}) = 0$ (if two of the j's are equal) or $\Delta(e_{j_1}, \ldots, e_{j_n}) = \pm\Delta(e_1, \ldots, e_n)$ (if the j's are distinct).

Then *define*:

$$(17.7) \qquad D(a_1, \ldots, a_n) = (-1)^{1+j}\alpha_{1j}D_{1j} + \cdots + (-1)^{n+j}\alpha_{nj}D_{nj}$$

where, for $1 \le i \le n$, D_{ij} is the determinant of the vectors $a_1^{(i)}, \ldots, a_{n-1}^{(i)}$ in F_{n-1} obtained from the $n-1$ vectors $a_1, \ldots, a_{i-1}, a_{i+1}, \ldots, a_n$ by deleting the jth component in each case.

We shall prove that the function D defined by (17.7) satisfies the axioms for a determinant. By the uniqueness theorem (17.1) it will then follow that all the expansions (17.7) for different j are equal, which is an important result in its own right.

Let us look more closely at (17.7). It says that $D(a_1, \ldots, a_n)$ is obtained by taking the coefficients of the jth column of the matrix **A** with rows $a_1, \ldots, a_n$ and multiplying each of them by a power of (-1) times the determinant of certain vectors, which form the rows of a matrix obtained from **A**, by deleting the jth column and one of the rows.

First let $e_1, \ldots, e_n$ be the unit vectors in F_n; then the matrix **A** is given by

$$\mathbf{A} = \begin{pmatrix} 1 & & 0 \\ & 1 & \\ & & \ddots & \\ 0 & & & 1 \end{pmatrix}$$

where zeros fill all the vacant spaces. Then there is only one nonzero entry in the jth column, namely, $\alpha_{jj} = 1$. The matrix from whose rows D_{jj} is computed is the $(n-1)$-by-$(n-1)$ matrix

$$\begin{pmatrix} 1 & & 0 \\ & 1 & \\ & & \ddots & \\ 0 & & & 1 \end{pmatrix}.$$

Hence (17.7) becomes

$$D(e_1, \ldots, e_n) = (-1)^{j+j}\alpha_{jj}D_{jj} = 1$$

since $\alpha_{jj} = 1$, and since $D_{jj} = 1$ by the determinant axioms for F_{n-1}.

Next let us consider replacing a_i by λa_i for some $\lambda \in F$. Then the matrix $\mathbf{A}'$ whose rows are $a_1, \ldots, a_{i-1}, \lambda a_i, a_{i+1}, \ldots, a_n$ is

$$\mathbf{A}' = \begin{pmatrix} \alpha_{11} & \cdots & \alpha_{1j} & \cdots & \alpha_{1n} \\ \cdots & \cdots & \cdots & \cdots & \cdots \\ \alpha_{i-1,1} & \cdots & \alpha_{i-1,j} & \cdots & \alpha_{i-1,n} \\ \lambda\alpha_{i1} & \cdots & \lambda\alpha_{ij} & \cdots & \lambda\alpha_{in} \\ \cdots & \cdots & \cdots & \cdots & \cdots \\ \alpha_{n1} & \cdots & \alpha_{nj} & \cdots & \alpha_{nn} \end{pmatrix}.$$

Then (17.7) becomes

$$\text{(17.8)}\quad D(\ldots, \lambda a_i, \ldots) = (-1)^{1+j}\alpha_{1j}D'_{1j} + \cdots + (-1)^{i+j}\lambda\alpha_{ij}D'_{ij} + \cdots + (-1)^{n+j}\alpha_{nj}D'_{nj}$$

where D'_{ij} is defined for $\mathbf{A}'$ as D_{ij} is defined for $\mathbf{A}$. From this definition and the properties of D on F_{n-1} we have $D'_{kj} = \lambda D_{kj}$, for $k \neq i$, and $D'_{ij} = D_{ij}$. Then (17.8) yields the result that

$$D(\ldots, \lambda a_i, \ldots) = \lambda D(a_1, \ldots, a_n).$$

Finally, let $i \neq k$, and consider the determinant of the vectors

$$\underbrace{a_1, \ldots, a_i + a_k, \ldots, a_k, \ldots, a_n}_{i}.$$

Then the matrix $\mathbf{A}''$ whose rows are these vectors is

$$\mathbf{A}'' = \begin{array}{c} \\ \\ i \\ \\ k \\ \\ \\ \end{array}\begin{pmatrix} \alpha_{11} & \cdots & \alpha_{1n} \\ \cdots & \cdots & \cdots \\ \alpha_{i1} + \alpha_{k1} & \cdots & \alpha_{in} + \alpha_{kn} \\ \cdots & \cdots & \cdots \\ \alpha_{k1} & \cdots & \alpha_{kn} \\ \cdots & \cdots & \cdots \\ \alpha_{n1} & \cdots & \alpha_{nn} \end{pmatrix}.$$

Then (17.7) becomes

$$\begin{aligned} \text{(17.9)}\quad D'' &= D(\ldots, \underset{i}{a_i + a_k}, \ldots, \underset{k}{a_k}, \ldots, a_n) \\ &= (-1)^{1+j}\alpha_{1j}D''_{1j} + \cdots + (-1)^{i+j}(\alpha_{ij} + \alpha_{kj})D''_{ij} \\ &\quad + \cdots + (-1)^{k+j}\alpha_{kj}D''_{kj} + \cdots + (-1)^{n+j}\alpha_{nj}D''_{nj} \end{aligned}$$

where the D''_{ij} are defined as in (17.7) from the vectors which constitute the rows of $\mathbf{A}''$. From the induction hypothesis that

$$D(\ldots, \underset{i}{a'_i + a'_s}, \ldots) = D(a'_1, \ldots, a'_{n-1}), \qquad i \neq s \text{ in } F_{n-1}$$

we have now:

$$\text{(17.10)}\quad D''_{sj} = D_{sj}, \qquad \text{for } 1 \leq s \leq n \text{ and } s \neq i, k.$$

Inspection of $\mathbf{A}''$ yields also

$$D''_{ij} = D_{ij}.$$

But D''_{kj} is not so easy. The vectors contributing to D''_{kj} are obtained from $\mathbf{A}''$ by deleting the kth row and jth column. Since the ith row of $\mathbf{A}''$ is a sum $a_i + a_j$, we can apply statement (e) of (16.6) to express D''_{kj} as a sum of two determinants, of which the first is D_{kj} and the second

$\pm D_{ij}$, where the $\pm$ sign is determined by statement (a) of (16.6) and by an inductive argument is equal to $(-1)^{|k-i|+1}$. Thus we have

$$D''_{kj} = D_{kj} + (-1)^{|k-i|+1} D_{ij}.$$

We shall also need the facts that $(-1)^{|a|} = (-1)^a$ and $(-1)^{a+2b} = (-1)^a$ for all integers a, b. Substituting in (17.9), we obtain

$$\begin{aligned} D'' = (-1)^{1+j}\alpha_{1j}D_{1j} + \cdots &+ (-1)^{i+j}(\alpha_{ij} + \alpha_{kj})D_{ij} \\ &+ \cdots + (-1)^{k+j}\alpha_{kj}[D_{kj} + (-1)^{|k-i|+1}D_{ij}] \\ &+ \cdots + (-1)^{n+j}\alpha_{nj}D_{nj} \\ = D(a_1, \ldots, a_n) &+ [(-1)^{i+j} + (-1)^{k+j+|k-i|+1}]\,\alpha_{kj}D_{ij}. \end{aligned}$$

We are finished if we can show that the coefficient of $\alpha_{kj}D_{ij}$ is zero. We have

$$\begin{aligned} (-1)^{i+j} + (-1)^{k+j+|k-i|+1} &= (-1)^{i+j} + (-1)^{k+j}(-1)^{k-i}(-1) \\ &= (-1)^{i+j} + (-1)^{j-i+1} = (-1)^{i+j} + (-1)^{j+i+1} = 0. \end{aligned}$$

This completes the proof of the theorem.

In this section we have proved the existence of a determinant function $D(a_1, \ldots, a_n)$ of n vectors $a_1, \ldots, a_n$ in F_n. If these vectors are given by

$$\begin{aligned} a_1 &= \alpha_{11}e_1 + \cdots + \alpha_{1n}e_n \\ &\cdots\cdots\cdots\cdots\cdots\cdots\cdots\cdots, \\ a_n &= \alpha_{n1}e_1 + \cdots + \alpha_{nn}e_n \end{aligned}$$

that is, if we think of them as the rows of the matrix

$$\mathbf{A} = \begin{pmatrix} \alpha_{11} & \cdots & \alpha_{1n} \\ \cdots & \cdots & \cdots \\ \alpha_{n1} & \cdots & \alpha_{nn} \end{pmatrix},$$

then in the proof of Theorem (17.1) we have shown that

$$\textbf{(17.11)}\quad D(a_1, \ldots, a_n) = \sum_{j_1=1}^{n} \cdots \sum_{j_n=1}^{n} \alpha_{1j_1} \cdots \alpha_{nj_n} D(e_{j_1}, \ldots, e_{j_n})$$

where the sum is taken over the n^n possible choices of $(j_1, \ldots, j_n)$. Since $D(e_{j_1}, \ldots, e_{j_n}) = 0$ when two of the entries are the same, we can rewrite (17.11) in the form

$$\textbf{(17.12)}\quad D(a_1, \ldots, a_n) = \sum_{j_1, \ldots, j_n} \alpha_{1j_1} \cdots \alpha_{nj_n} D(e_{j_1}, \ldots, e_{j_n})$$

where it is understood that the sum is taken over the $n! = n(n-1) \times (n-2) \cdots 3 \cdot 2 \cdot 1$ possible choices of $\{j_1, \ldots, j_n\}$ in which all the j_i's are distinct. The formula (17.12) is called the *complete expansion* of the determinant. If we view the determinant $D(a_1, \ldots, a_n)$ as a function

$D(\mathbf{A})$ of the matrix whose rows are $a_1, \ldots, a_n$, then the complete expansion shows that, since the $D(e_{j_1}, \ldots, e_{j_n})$ are ± 1, the determinant is a sum (with coefficients ± 1) of products of the coefficients of the matrix **A**. If the matrix **A** has real coefficients, and is viewed as a point in the n^2-dimensional space, then (17.12) shows that $D(\mathbf{A})$ is a continuous function of **A**.

The idea of viewing the determinant $D(a_1, \ldots, a_n)$ as a function of a matrix **A** with rows $a_1, \ldots, a_n$ at once raises another problem. Let $c_1, \ldots, c_n$ be the columns of **A**. Then we can form $D(c_1, \ldots, c_n)$ and ask what is the relation of this function to $D(a_1, \ldots, a_n)$.

(17.13) THEOREM. *Let* **A** *be an n-by-n matrix with rows* $a_1, \ldots, a_n$ *and columns* $c_1, \ldots, c_n$*; then* $D(a_1, \ldots, a_n) = D(c_1, \ldots, c_n)$.

Proof. Let us use the complete expansion (17.12) and view (17.12) as defining a new function:

$$\textbf{(17.14)} \qquad D^*(c_1, \ldots, c_n) = \sum_{j_1, \ldots, j_n} \alpha_{1j_1} \cdots \alpha_{nj_n} D(e_{j_1}, \ldots, e_{j_n}) = D(a_1, \ldots, a_n).$$

We shall prove that $D^*(c_1, \ldots, c_n)$ satisfies the axioms for a determinant function; then Theorem (17.1) will imply that $D^*(c_1, \ldots, c_n) = D(c_1, \ldots, c_n)$.

First suppose that $c_1, \ldots, c_n$ are the unit vectors $e_1, \ldots, e_n$. Then the row vectors $a_1, \ldots, a_n$ are also the unit vectors and we have

$$D^*(e_1, \ldots, e_n) = D(e_1, \ldots, e_n) = 1.$$

From (17.14) it is clear, since each term in the sum has exactly one entry from a given column, that

$$D^*(\ldots, \underset{i}{\lambda c_i}, \ldots) = \lambda D^*(\ldots, c_i, \ldots).$$

Finally, let us consider

$$D^*(\ldots, \underset{i}{c_i + c_k}, \ldots), \qquad k \neq i.$$

That means that, for $1 \leq r \leq n$, α_{ri} is replaced by $\alpha_{ri} + \alpha_{rk}$. Making this substitution in (17.14), we can split up $D^*(\ldots, c_i + c_k, \ldots)$ as a sum, $D^*(\ldots, c_i + c_k, \ldots) = D^*(\ldots, \underset{i}{c_i}, \ldots) + D^*(\ldots, \underset{i}{c_k}, \ldots)$, and we shall be finished if we can show that $D^*(c_1, \ldots, c_n) = 0$ if two of the vectors, c_r and c_s, are equal, for $r \neq s$. In (17.14), consider a term

$$\alpha_{1j_1} \cdots \alpha_{kj_k} \cdots \alpha_{lj_l} \cdots \alpha_{nj_n} D(e_{j_1}, \ldots, e_{j_k}, \ldots, e_{j_l}, \ldots, e_{j_n})$$

such that $j_k = r, j_l = s$. There will also be a term in (17.14) of the form

$$\alpha_{1j_1} \cdots \alpha_{kj_l} \cdots \alpha_{lj_k} \cdots \alpha_{nj_n} D(e_{j_1}, \ldots, e_{j_l}, \ldots, e_{j_k}, \ldots, e_{j_n})$$

and the sum of these two terms will be zero, since $c_r = c_s$ and

$$D(\ldots, e_{j_k}, \ldots, e_{j_l}, \ldots) = -D(\ldots, e_{j_l}, \ldots, e_{j_k}, \ldots).$$

Thus each term of $D^*(c_1, \ldots, c_n)$ is canceled by another, and we have shown that $D^*(c_1, \ldots, c_n) = 0$ if $c_r = c_s$, for $r \neq s$. We have proved that D^* satisfies the axioms for a determinant function. By Theorem (17.1) we conclude that $D(a_1, \ldots, a_n) = D(c_1, \ldots, c_n)$, and Theorem (17.13) is proved.

We can now speak unambiguously of $D(\mathbf{A})$ for any n-by-n matrix $\mathbf{A}$ and know that $D(\mathbf{A})$ satisfies the axioms of a determinant function when viewed as a function either of rows or of columns. When

$$\mathbf{A} = \begin{pmatrix} \alpha_{11} & \cdots & \alpha_{1n} \\ \cdots & \cdots & \cdots \\ \alpha_{n1} & \cdots & \alpha_{nn} \end{pmatrix}$$

we shall frequently use the notation

$$D(\mathbf{A}) = \begin{vmatrix} \alpha_{11} & \cdots & \alpha_{1n} \\ \cdots & \cdots & \cdots \\ \alpha_{n1} & \cdots & \alpha_{nn} \end{vmatrix}.$$

Theorem (17.13) can be restated in the form

$$D(\mathbf{A}) = D({}^t\mathbf{A}), \tag{17.15}$$

where ${}^t\mathbf{A}$ is called the *transpose* of $\mathbf{A}$ and is obtained from $\mathbf{A}$ by interchanging rows and columns. Thus, if α_{ij} is the (i, j) entry of $\mathbf{A}$, α_{ji} is the (i, j) entry of ${}^t\mathbf{A}$.

EXERCISES

1. Prove that if $a_1 = \langle \xi, \eta \rangle$, $a_2 = \langle \lambda, \mu \rangle$ in R_2, then

$$D(a_1, a_2) = \xi\mu - \eta\lambda.$$

2. Prove that if $a = \langle \alpha_1, \alpha_2, \alpha_3 \rangle$, $b = \langle \beta_1, \beta_2, \beta_3 \rangle$, $c = \langle \gamma_1, \gamma_2, \gamma_3 \rangle$, then

$$D(a, b, c) = \alpha_1(\beta_2\gamma_3 - \gamma_2\beta_3) - \alpha_2(\beta_1\gamma_3 - \beta_3\gamma_1) + \alpha_3(\beta_1\gamma_2 - \beta_2\gamma_1).$$

18. THE MULTIPLICATION THEOREM FOR DETERMINANTS

We consider next what is perhaps the most important property of determinants and one that we shall use frequently in the later parts of the book. The definition we have given for determinants was chosen partly because it leads to a simple proof of this theorem.

A natural question to ask is the following. Suppose that $\mathbf{A} = (\alpha_{ij})$ and $\mathbf{B} = (\beta_{ij})$ are n-by-n matrices; then $\mathbf{AB}$ is an n-by-n matrix. Is there any relation between $D(\mathbf{AB})$ and $D(\mathbf{A})$ and $D(\mathbf{B})$? (For 2-by-2 matrices we have already settled this question in Exercise 4 of Section 14.)

The first step is the following preliminary result.†

(18.1) LEMMA. *Let $f(a_1, \ldots, a_n)$ be a function of n-tuples of vectors $a_i \in F_n$ to the field F which satisfies Axioms (i) and (ii) in the definition (16.5) of the determinant function; then for all $a_1, \ldots, a_n$ in F_n we have*

$$f(a_1, \ldots, a_n) = D(a_1, \ldots, a_n)f(e_1, \ldots, e_n)$$

where the e_i are the unit vectors and D is the determinant function on F_n.

Proof. If $f(e_1, \ldots, e_n) = 1$, then $f = D$ by Theorem (17.1), and the lemma is proved.

Now suppose $f(e_1, \ldots, e_n) \neq 1$ and consider the function

$$D'(a_1, \ldots, a_n) = \frac{D(a_1, \ldots, a_n) - f(a_1, \ldots, a_n)}{1 - f(e_1, \ldots, e_n)}. \qquad (18.2)$$

It is clear that D' satisfies the Axioms (i), (ii), and (iii) of (16.5). Hence, by Theorem (17.1), $D' = D$, and solving for $f(a_1, \ldots, a_n)$ in (18.2) we obtain the conclusion of the lemma.

Before giving the proof of the main theorem, we have to recall a fact about linear transformations on F_n. In Section 12, we showed that each n-by-n matrix $\mathbf{A}$ defined a linear transformation U of F_n into F_n, where the action of U on a vector $x = \langle \xi_1, \ldots, \xi_n \rangle$ is given by matrix multiplication of $\mathbf{A}$ with x regarded as a column vector:

$$x \to U(x) = \mathbf{A}\begin{pmatrix} \xi_1 \\ \vdots \\ \xi_n \end{pmatrix}.$$

In the course of the proof of the next theorem, a matrix $\mathbf{A}$ is used to define a linear transformation of F_n according to this definition.

We can now state the multiplication theorem:

(18.3) THEOREM. *Let* $\mathbf{A}$ *and* $\mathbf{B}$ *be n-by-n matrices; then* $D(\mathbf{AB}) = D(\mathbf{A})D(\mathbf{B})$.

† The idea of using this lemma as a key to the multiplication theorem comes from Schreier and Sperner (see the Bibliography).

Proof. Let $\mathbf{A} = (\alpha_{ij})$, $\mathbf{B} = (\beta_{ij})$ and let U be the linear transformation $U(a) = \mathbf{B}a$ where a is viewed as an n-by-1 column vector. Define a function f by

$$f(a_1, \ldots, a_n) = D[\,U(a_1), \ldots, U(a_n)\,], \qquad a_i \in F_n.$$

By Axioms (i) and (ii) of (16.5) for D it is clear that f satisfies Axioms (i) and (ii). By Lemma (18.1) we have $f(a_1, \ldots, a_n) = D(a_1, \ldots, a_n)f(e_1, \ldots, e_n)$ and, hence,

(18.4) $D[\,U(a_1), \ldots, U(a_n)\,] = D(a_1, \ldots, a_n)D[\,U(e_1), \ldots, U(e_n)\,].$

Now let $a_1, \ldots, a_n$ be the columns of the matrix $\mathbf{A}$. Then for $1 \le i \le n$ we compute $U(a_i)$.

Since $a_i = \langle \alpha_{1i}, \ldots, \alpha_{ni} \rangle$, $U(a_i)$ is the vector

$$\left\langle \sum_{k=1}^{n} \beta_{1k}\alpha_{ki}, \ldots, \sum_{k=1}^{n} \beta_{nk}\alpha_{ki} \right\rangle$$

which is the ith column of the matrix $\mathbf{BA}$. Similarly,

$$U(e_i) = \langle \beta_{1i}, \beta_{2i}, \ldots, \beta_{ni} \rangle$$

which is the ith column of the matrix $\mathbf{B}$. Then Equation (18.4) yields

$$D(\mathbf{BA}) = D(\mathbf{A})D(\mathbf{B})$$

and, since $D(\mathbf{A})D(\mathbf{B}) = D(\mathbf{B})D(\mathbf{A})$, the theorem is proved.

As a first application of the multiplication theorem, we prove:

(18.5) THEOREM. *Let* $\mathbf{A}$ *be an n-by-n matrix; then* $\mathbf{A}$ *has rank n if and only if* $D(\mathbf{A}) \neq 0$.

Proof. Part **(d)** of Theorem (16.6) shows that if $\mathbf{A}$ has rank less than n then $D(\mathbf{A}) = 0$. It remains to prove that if $\mathbf{A}$ has rank n then $D(\mathbf{A}) \neq 0$. (For another proof see Exercise 3 of Section 8.) Let $\{a_1, \ldots, a_n\}$ be the row vectors of $\mathbf{A}$. Since $\mathbf{A}$ has rank n, $\{a_1, \ldots, a_n\}$ is a basis for F_n; therefore for each i, $1 \le i \le n$, we can express the ith unit vector e_i as a linear combination of $\{a_1, \ldots, a_n\}$:

(18.6) $$e_i = \beta_{i1}a_1 + \cdots + \beta_{in}a_n.$$

Let $\mathbf{B}$ be the matrix (β_{ij}); then (18.6) implies that the matrix whose rows are $\{e_1, \ldots, e_n\}$ is the product matrix $\mathbf{BA}$. Since $D(e_1, \ldots, e_n) = 1$, we have by Theorem (18.3),

$$1 = D(\mathbf{BA}) = D(\mathbf{B})D(\mathbf{A})$$

and $D(\mathbf{A}) \neq 0$ as required.

We can now make the following important definition.

(18.7) DEFINITION. Let T be a linear transformation of a finite-dimensional vector space over an arbitrary field. Then the *determinant of* T, $D(T)$, is defined to be $D(\mathbf{A})$, when $\mathbf{A}$ is the matrix of T with respect to some basis of the vector space.

It is necessary to check that if $\mathbf{A}'$ is the matrix of T with respect to some other basis, then $D(\mathbf{A}) = D(\mathbf{A}')$. By Theorem (13.6) there exists an invertible matrix $\mathbf{X}$ such that

$$\mathbf{A}' = \mathbf{X}\mathbf{A}\mathbf{X}^{-1}.$$

Then by Theorem (18.3),

$$\begin{aligned} D(\mathbf{A}') &= D(\mathbf{X})D(\mathbf{A})D(\mathbf{X}^{-1}) \\ &= D(\mathbf{X})D(\mathbf{X}^{-1})D(\mathbf{A}) \\ &= D(\mathbf{X}\mathbf{X}^{-1})D(\mathbf{A}) \\ &= D(\mathbf{I})D(\mathbf{A}) \\ &= D(\mathbf{A}), \end{aligned}$$

where $\mathbf{I}$ is the n-by-n identity matrix.

We can now state the following improvement of Theorem (13.10).

(18.8) THEOREM. *Let $T \in L(V, V)$, where V is a finite-dimensional vector space over F. Then the following statements are equivalent.*

1. *T is invertible.*
2. *T is one-to-one.*
3. *T is onto.*
4. $D(T) \neq 0$.

Proof. The equivalence of (1), (2), and (3) was already proved. By Exercise 4 of Section 13,

$$\dim T(V) = \text{rank}\,(\mathbf{A}),$$

where $\mathbf{A}$ is the matrix of T with respect to some basis. If $n = \dim V$, then $T(V) = V$ if and only if the rank of $\mathbf{A}$ is n. By Theorem (18.5), the rank of $\mathbf{A}$ is n if and only if $D(\mathbf{A}) \neq 0$. Combining these statements, we see that (3) and (4) are equivalent, and the theorem is proved.

EXERCISES

1. Compute $D(\mathbf{AB})$ by multiplying out the matrices and also by Theorem (18.3) where

$$\mathbf{A} = \begin{pmatrix} 1 & 1 & -2 & 1 \\ 0 & -1 & 3 & 1 \\ 0 & 0 & 3 & 2 \\ 0 & 0 & 0 & -2 \end{pmatrix}, \qquad \mathbf{B} = \begin{pmatrix} 3 & 0 & 0 & 0 \\ -1 & 1 & 0 & 0 \\ 0 & 1 & -2 & 0 \\ 3 & 2 & 1 & 1 \end{pmatrix}.$$

2. Let $\mathbf{A}$ be an invertible n-by-n matrix. Show that

$$D(\mathbf{A}^{-1}) = D(\mathbf{A})^{-1}.$$

3. Compute the determinants of the elementary matrices $\mathbf{P}_{ij}$, $\mathbf{B}_{ij}(\lambda)$, $\mathbf{D}_i(\mu)$, defined in Example C of Section 12. Show that every invertible n-by-n matrix $\mathbf{A}$ is a product of elementary matrices, and hence that $D(\mathbf{A}) \neq 0$, thereby giving another proof of part of Theorem (18.8).
4. Fill in the details of the following steps in another proof of the important theorem (18.3). Notice that this approach avoids the use of Lemma (18.1).
   a. First suppose that $\mathbf{A}$ or $\mathbf{B}$ is not invertible. Prove that $\mathbf{AB}$ is not invertible, and hence that $D(\mathbf{AB}) = D(\mathbf{A})D(\mathbf{B})$ since both sides are equal to zero.
   b. Use Exercise 3 together with the basic properties of the determinant function from Section 16 to show that if $\mathbf{E}$ is an elementary matrix, then $D(\mathbf{EB}) = D(\mathbf{E})D(\mathbf{B})$ for all matrices $\mathbf{B}$.
   c. Suppose $\mathbf{A}$ is invertible. Factor $\mathbf{A}$ as a product of elementary matrices, $\mathbf{A} = \mathbf{E}_1 \cdots \mathbf{E}_s$. Using part (b), show that $D(\mathbf{A}) = D(\mathbf{E}_1) \cdots D(\mathbf{E}_s)$, and $D(\mathbf{AB}) = D(\mathbf{E}_1) \cdots D(\mathbf{E}_s)D(\mathbf{B}) = D(\mathbf{A})D(\mathbf{B})$.
5. Let T be an orthogonal transformation on a finite dimensional vector space V over the real numbers, with an inner product. Show that $D(T) = \pm 1$.

19. FURTHER PROPERTIES OF DETERMINANTS

In this section we take up a few of the many special topics one can study in the theory of determinants.

Row and Column Expansions, and Invertible matrices

Let $\mathbf{A} = (\alpha_{ij})$ be an n-by-n matrix. Define the (i, j) *cofactor* A_{ij} as $A_{ij} = (-1)^{i+j}D_{ij}$ where D_{ij} is the determinant of the $n - 1$ by $n - 1$ matrix, obtained by deleting the ith row and jth column of $\mathbf{A}$. Then the formulas (17.7) can be stated in the form

(19.1) $$D(\mathbf{A}) = \sum_{k=1}^{n} \alpha_{kj}A_{kj}, \qquad j = 1, 2, \ldots, n.$$

We shall refer to this formula as the *expansion of* $D(\mathbf{A})$ *along the* j*th column*. A related formula is

(19.2) $$\sum_{k=1}^{n} \alpha_{kj}A_{kl} = 0, \qquad j \neq l.$$

This is easily obtained from (19.1) as follows. Consider the matrix $\mathbf{A}'$ obtained from $\mathbf{A}$ by replacing the lth column of $\mathbf{A}$ by the jth column, for

$j \neq l$; then $\mathbf{A}'$ has two equal columns and hence $D(\mathbf{A}') = 0$, since the determinant function satisfies the conditions (*i*), (*ii*), and (*iii*) of (16.5) when considered as a function of either the row or the column vectors, according to the proof of Theorem (17.13). Taking the expansion of $D(\mathbf{A}')$ along the lth column, we obtain (19.2).

Let ${}^t\mathbf{A}$ be the transpose of $\mathbf{A}$; then the column expansions of $D({}^t\mathbf{A})$ yield the following *row expansions* of $D(\mathbf{A})$, since $D(\mathbf{A}) = D({}^t\mathbf{A})$ by (17.15).

(19.3) $$\sum_{k=1}^{n} \alpha_{jk}A_{jk} = D(\mathbf{A}), \qquad j = 1, 2, \ldots, n.$$

(19.4) $$\sum_{k=1}^{n} \alpha_{jk}A_{lk} = 0, \qquad j \neq l.$$

These formulas become especially interesting if we interpret them from the point of view of matrix multiplication. Let $\mathbf{A}^*$ be the matrix with A_{ji} in the (i, j) position, and let $\mathbf{I}$ be the matrix whose ith row is the ith unit vector e_i ($\mathbf{I}$ is the *identity matrix*). For any matrix $\mathbf{A} = (\alpha_{ij})$, $\lambda\mathbf{A}$ is the matrix whose (i, j) entry is $\lambda\alpha_{ij}$. Then the formulas (19.1) and (19.2) become

(19.5) $$\mathbf{A}^*\mathbf{A} = D(\mathbf{A})\mathbf{I}$$

while (19.3) and (19.4) become

(19.6) $$\mathbf{A}\mathbf{A}^* = D(\mathbf{A})\mathbf{I}.$$

We know that the matrix $\mathbf{I}$ plays the same role as 1 in the real number system:

$$\mathbf{AI} = \mathbf{IA} = \mathbf{A}$$

for all matrices $\mathbf{A}$. We consider again the problem of deciding if a matrix $\mathbf{A} \neq 0$ has a multiplicative inverse $\mathbf{A}^{-1}$ such that $\mathbf{A}^{-1}\mathbf{A} = \mathbf{A}\mathbf{A}^{-1} = \mathbf{I}$. Let us recall the following definition (see Section 12).

(19.7) DEFINITION. An n-by-n matrix $\mathbf{A}$ is *invertible* if there exists an n-by-n matrix $\mathbf{A}^{-1}$, called an *inverse* of $\mathbf{A}$, such that

$$\mathbf{A}\mathbf{A}^{-1} = \mathbf{A}^{-1}\mathbf{A} = \mathbf{I}.$$

(19.8) THEOREM. *An n-by-n matrix $\mathbf{A}$ is invertible if and only if $D(\mathbf{A}) \neq 0$. If $D(\mathbf{A}) \neq 0$, then an inverse $\mathbf{A}^{-1}$ is given by*

$$D(\mathbf{A})^{-1}\mathbf{A}^*$$

where $\mathbf{A}^*$ is the matrix whose (j, i) entry is $A_{ij} = (-1)^{i+j}D_{ij}$.

Proof. If $\mathbf{A}\mathbf{A}^{-1} = \mathbf{I}$, then $D(\mathbf{A}) \neq 0$ by the multiplication theorem. If

$D(\mathbf{A}) \neq 0$, then setting $\mathbf{A}^{-1} = D(\mathbf{A})^{-1}\mathbf{A}^*$ we have $\mathbf{A}^{-1}\mathbf{A} = \mathbf{A}\mathbf{A}^{-1} = \mathbf{I}$ by formulas (19.5) and (19.6). This completes the proof.

Example A. As an illustration of Theorem (19.8), we shall derive the following useful formula for $\mathbf{A}^{-1}$, in case $\mathbf{A}$ is a 2-by-2 matrix with determinant 1. Let

$$\mathbf{A} = \begin{pmatrix} \alpha & \beta \\ \gamma & \delta \end{pmatrix},$$

and let $$D(\mathbf{A}) = \alpha\delta - \beta\gamma = 1.$$

Then $A_{11} = +\delta$, $A_{12} = -\gamma$, $A_{21} = -\beta$, $A_{22} = +\alpha$, and by Theorem (19.8), since $D(\mathbf{A}) = 1$,

$$\mathbf{A}^* = \begin{pmatrix} A_{11} & A_{21} \\ A_{12} & A_{22} \end{pmatrix} = \begin{pmatrix} \delta & -\beta \\ -\gamma & \alpha \end{pmatrix} = \mathbf{A}^{-1}.$$

Determinants and Systems of Equations

We shall now combine our results to relate determinants to some of the questions studied in Chapter 2 concerning systems of linear equations and the rank of a matrix. The first relation gives an explicit formula for the solution of a system of n nonhomogeneous equations in n unknowns, one that is useful for theoretical purposes. It is less efficient than the methods developed in Chapter 2 for actually computing a solution of a particular system of equations, and cannot be applied to systems with a nonsquare coefficient matrix.

(19.9) THEOREM (CRAMER'S RULE). *A nonhomogeneous system of n linear equations in n unknowns*

(19.10) $$\begin{array}{c} \alpha_{11}x_1 + \cdots + \alpha_{1n}x_n = \beta_1 \\ \cdots\cdots\cdots\cdots\cdots\cdots \\ \alpha_{n1}x_1 + \cdots + \alpha_{nn}x_n = \beta_n \end{array}$$

has a unique solution if and only if the determinant of the coefficient matrix $D(\mathbf{A}) \neq 0$. If $D(\mathbf{A}) \neq 0$, the solution is given by

$$x_i = \frac{D(c_1, \ldots, c_{i-1}, b, c, \ldots, c_n)}{D(\mathbf{A})}, \qquad 1 \leq i \leq n$$

where $c_1, \ldots, c_n$ are the columns of $\mathbf{A}$ and $b = \langle \beta_1, \ldots, \beta_n \rangle$.

Proof. By (18.5), $D(\mathbf{A}) \neq 0$ if and only if the columns $c_1, \ldots, c_n$ are linearly independent, and thus the statement about the existence of a unique solution follows from the theorems in Sections 8 and 9 (why?). Finally, let $D(\mathbf{A}) \neq 0$, and let $x = \langle x_1, \ldots, x_n \rangle$ be a solution. Then

$$x_1c_1 + \cdots + x_nc_n = b,$$

and we have

$$\begin{aligned} D(c_1, \ldots, c_{i-1}, b, c_{i+1}, \ldots, c_n) &= D\Big(c_1, \ldots, c_{i-1}, \sum_{j=1}^{n} x_jc_j, c_{i+1}, \ldots, c_n\Big) \\ &= \sum_{j=1}^{n} x_jD(c_1, \ldots, c_{i-1}, c_j, c_{i+1}, \ldots, c_n) \\ &= x_iD(c_1, \ldots, c_n) = x_iD(\mathbf{A}), \end{aligned}$$

proving the theorem.

One important consequence of these formulas is that when $D(\mathbf{A}) \neq 0$ the solution $\langle x_1, \ldots, x_n \rangle$ of a system (19.10) with real coefficients depends continuously on the coefficient matrix $\mathbf{A}$.

In Sections 8 and 9 we defined the rank of an m-by-n matrix and proved that the row rank and column rank are the same. In (18.5) we showed that for an n-by-n matrix $\mathbf{A}$, the rank is n if and only if $D(\mathbf{A}) \neq 0$. By using this result it is possible to prove a connection between determinants and rank for arbitrary matrices. We first define an *r-rowed minor determinant* of $\mathbf{A}$ as the determinant of an r-by-r matrix obtained from $\mathbf{A}$ by deleting rows and columns. For example, the two-rowed minors of

$$\begin{pmatrix} 1 & -1 & 0 \\ 2 & 3 & 1 \end{pmatrix}$$

are

$$\begin{vmatrix} 1 & -1 \\ 2 & 3 \end{vmatrix}, \quad \begin{vmatrix} 1 & 0 \\ 2 & 1 \end{vmatrix}, \quad \begin{vmatrix} -1 & 0 \\ 3 & 1 \end{vmatrix}.$$

(19.11) THEOREM. *The rank of an m-by-n matrix $\mathbf{A}$ is s if and only if there exists a nonzero s-rowed minor and all $(s + k)$-rowed minors are zero for $k = 1, 2, \ldots$.*

Proof. Let us call the number s defined in the statement of the theorem det rank $\mathbf{A}$. We prove first that rank $\mathbf{A} = r$ implies det rank $\mathbf{A} \geq r$. From Section 9, rank $\mathbf{A} = r$ implies that there exists an r-by-r matrix of rank r obtained by deleting rows and columns from $\mathbf{A}$. Then (18.5) implies det rank $\mathbf{A} \geq r$.

Conversely, det rank $\mathbf{A} = s$ implies that there exist s linearly independent rows of $\mathbf{A}$; hence rank $\mathbf{A} \geq s$. Combining the inequalities, we have det rank $\mathbf{A}$ = rank $\mathbf{A}$, and the theorem is proved.

Determinants and Volumes

We conclude the chapter with the n-dimensional interpretation of the determinant as a volume function. A *volume function* V in R_n is a function which assigns to each n-tuple of vectors $\{a_1, \ldots, a_n\}$ in R_n a real number $V(a_1, \ldots, a_n)$ such that

$$V(a_1, \ldots, a_n) \geq 0,$$

$$V(\ldots, a_i + a_k, \ldots) = V(a_1, \ldots, a_n), \qquad i \neq k,$$

$$V(\ldots, \lambda a_i, \ldots) = |\lambda| V(a_1, \ldots, a_n), \qquad \lambda \in R,$$

$$V(e_1, \ldots, e_n) = 1, \qquad e_i = \text{unit vectors}.$$

Such a function can be interpreted as the volume of the n-dimensional parallelopiped, with edges $a_1, \ldots, a_n$, which consists of all vectors $x = \sum \lambda_i a_i$, for $0 \leq \lambda_i \leq 1$. The connection between volume functions and determinants is given in the following theorem.

(19.12) THEOREM. *There is one and only one volume function $V(a_1, \ldots, a_n)$ on R_n, which is given by*

$$V(a_1, \ldots, a_n) = |D(a_1, \ldots, a_n)|.$$

Proof. Clearly, $|D(a_1, \ldots, a_n)|$ is a volume function. Now let V be a volume function, and define

$$V^*(a_1, \ldots, a_n) = \begin{cases} \dfrac{V(a_1, \ldots, a_n) D(a_1, \ldots, a_n)}{|D(a_1, \ldots, a_n)|}, & D \neq 0 \\ 0, & \text{if } D(a_1, \ldots, a_n) = 0. \end{cases}$$

Then one verifies easily that V^* satisfies the axioms for a determinant function and hence that

$$V^*(a_1, \ldots, a_n) = D(a_1, \ldots, a_n)$$

for all $a_1, \ldots, a_n$ in R_n. It then follows from the definition of V^* that $V(a_1, \ldots, a_n) = |D(a_1, \ldots, a_n)|$, and the theorem is proved.

An inductive definition of the k-dimensional volume of a k-dimensional parallelopiped in R_n, based on the fact that the area of a parallelogram is the product of the base and the height, is given in Section 3, Chapter 10, of Birkoff and MacLane (see the Bibliography). They show, by an interesting argument, that the k-dimensional volume in R_n is given by a certain determinant.

Because of the interpretation of the absolute value of the determinant function as a volume, the following result is almost intuitively obvious, if we believe that n-dimensional geometry is not too different from the geometry in R_2 and R_3. The proof is not so easy, however, and requires both the multiplication theorem and the theory of orthogonal transformations from Section 15.

(19.13) THEOREM (HADAMARD'S INEQUALITY). *Let $\{a_1, \ldots, a_n\}$ be vectors in R_n. Then*

$$|D(a_1, \ldots, a_n)| \leq \|a_1\| \cdots \|a_n\|,$$

where $\|a_i\|$ is the length of the vector $a_i = \langle \alpha_{i1}, \ldots, \alpha_{in} \rangle$ in R_n, given by $\|a_i\| = (\alpha_{i1}^2 + \cdots + \alpha_{in}^2)^{1/2}$.

Remark. The theorem says that the volume of a parallelopiped with edges $a_1, \ldots, a_n$ is less than or equal to the product of the lengths of the edges.

Proof. We may assume that all the vectors $a_i \neq 0$; otherwise both sides of the inequality are zero and there is nothing to prove. We now show that it is sufficient to prove the theorem in case $\|a_i\| = 1$, for $1 \leq i \leq n$. Suppose we have the theorem in this case. Then in general,

$$\begin{aligned} |D(a_1, \ldots, a_n)| &= \|a_1\| \cdots \|a_n\| D\left(\frac{a_1}{\|a_1\|}, \ldots, \frac{a_n}{\|a_n\|}\right) \\ &\leq \|a_1\| \cdots \|a_n\|, \end{aligned}$$

since $a_i/\|a_i\|$ has length 1 for $1 \leq i \leq n$.

Now assume $u_1, \ldots, u_n$ are vectors of length 1; we have to prove that $|D(u_1, \ldots, u_n)| \leq 1$. If T is an arbitrary orthogonal transformation of R_n, then formula (18.4) in the proof of Theorem (18.3) shows that

$$|D(T(u_1), \ldots, T(u_n))| = |D(T)|\, |D(u_1, \ldots, u_n)|.$$

Since T is an orthogonal transformation we have $|D(T)| = 1$ (see Exercise 5, Section 18), so that

$$|D(T(u_1), \ldots, T(u_n))| = |D(u_1, \ldots, u_n)|.$$

We may assume that $\{u_1, \ldots, u_n\}$ is a linearly independent set, otherwise $|D(u_1, \ldots, u_n)| = 0$, and the inequality holds in an obvious way. By Exercise 10 of Section 15, there exists an orthogonal transformation T of R_n such that $T(u_i) \in S(e_2, \ldots, e_n)$ for $i = 2, \ldots, n$, where $\{e_1, \ldots, e_n\}$ are the usual unit vectors in R_n. Then the matrix whose rows are $\{T(u_1), \ldots, T(u_n)\}$ has the form

$$\mathbf{X} = \begin{pmatrix} \lambda_{11} & \lambda_{12} & \cdots & \lambda_{1n} \\ 0 & \lambda_{22} & \cdots & \lambda_{2n} \\ \cdots & \cdots & \cdots & \cdots \\ 0 & \lambda_{n2} & \cdots & \lambda_{nn} \end{pmatrix}.$$

Moreover, since $\|T(u_i)\| = 1$ for $1 \le i \le n$, we know that the sum of the squares of the elements in each row is 1. Expanding $D(\mathbf{X})$ along the first column we have

$$|D(\mathbf{X})| = |\lambda_{11}| \begin{vmatrix} \lambda_{22} & \cdots & \lambda_{2n} \\ \cdots & \cdots & \cdots \\ \lambda_{n2} & \cdots & \lambda_{nn} \end{vmatrix}.$$

We may assume as an induction hypothesis that the determinant on the right-hand side has absolute value less than or equal to one, because the rows of the corresponding matrix are vectors of length one in R_{n-1}, because of the form of the matrix $\mathbf{X}$. Since $|\lambda_{11}| \le 1$, we have

$$|D(u_1, \ldots, u_n)| = |D(\mathbf{X})| \le 1,$$

and the theorem is proved.

Even and Odd Permutations

In this part of the section, we show how the multiplication theorem for determinants can be used to derive some important facts about permutations.

(19.14) DEFINITION. Let X be the set consisting of the positive integers $\{1, 2, \ldots, n\}$. A *permutation* of X is a one-to-one mapping σ of X onto X. The set of all permutations of X will be denoted by $P(X)$. A permutation σ will often be described using the notation

$$\sigma = \begin{pmatrix} 1 & 2 & \cdots & n \\ j_1 & j_2 & \cdots & j_n \end{pmatrix},$$

where $\sigma(1) = j_1$, $\sigma(2) = j_2, \ldots, \sigma(n) = j_n$.

For example,

$$\begin{pmatrix} 1 & 2 & 3 \\ 3 & 1 & 2 \end{pmatrix}$$

stands for the permutation σ such that $\sigma(1) = 3$, $\sigma(2) = 1$, $\sigma(3) = 2$.

(19.15) THEOREM. *The set of permutations $P(X)$ forms a group under the operation $\sigma\tau$, defined by*

$$\sigma\tau(x) = \sigma(\tau(x)), \qquad x \in X, \qquad \sigma, \tau \in P(X).$$

Proof. We recall the definition of a group from Section 11 [Definition (11.10)]. What has to be proved is that the following statements hold:

a. $\sigma\tau \in P(X)$ for all $\sigma, \tau \in P(X)$.
b. $(\rho\sigma)\tau = \rho(\sigma\tau)$ for $\rho, \sigma, \tau \in P(X)$ (associative law).

c. There exists an element $1 \in P(X)$ such that $1\sigma = \sigma 1 = \sigma$ for all $\sigma \in P(X)$.

d. For each $\sigma \in P(X)$ there exists an element $\tau \in P(X)$ such that $\sigma\tau = \tau\sigma = 1$.

The proofs of these facts are almost identical to the proofs that the set of invertible linear transformations on a vector space V form a group. We shall leave the details of checking (*a*) to (*d*) to the reader, with the discussion in Section 11 as a guide.

Multiplication of permutations can be carried out effectively using the notation for permutations which we have introduced. For example, the permutations of $X = \{1, 2, 3\}$ are

$$1 = \begin{pmatrix} 1 & 2 & 3 \\ 1 & 2 & 3 \end{pmatrix}, \quad \sigma_1 = \begin{pmatrix} 1 & 2 & 3 \\ 2 & 1 & 3 \end{pmatrix}, \quad \sigma_2 = \begin{pmatrix} 1 & 2 & 3 \\ 1 & 3 & 2 \end{pmatrix};$$

$$\sigma_3 = \begin{pmatrix} 1 & 2 & 3 \\ 3 & 2 & 1 \end{pmatrix}, \quad \sigma_4 = \begin{pmatrix} 1 & 2 & 3 \\ 3 & 1 & 2 \end{pmatrix}, \quad \sigma_5 = \begin{pmatrix} 1 & 2 & 3 \\ 2 & 3 & 1 \end{pmatrix}.$$

Using the definition $\sigma\tau(x) = \sigma(\tau(x))$ one can show that the *multiplication table* of the group $P(X)$ is given by

	1	σ_1	σ_2	σ_3	σ_4	σ_5
1	1	σ_1	σ_2	σ_3	σ_4	σ_5
σ_1	σ_1	1	σ_5	σ_4		
σ_2						
σ_3				etc.		
σ_4						
σ_5						

where the entries in the table are computed as follows:

$$\sigma_1\sigma_2 = \begin{pmatrix} 1 & 2 & 3 \\ 2 & 1 & 3 \end{pmatrix}\begin{pmatrix} 1 & 2 & 3 \\ 1 & 3 & 2 \end{pmatrix} = \begin{pmatrix} 1 & 2 & 3 \\ 2 & 3 & 1 \end{pmatrix} = \sigma_5$$

$$\sigma_1\sigma_3 = \begin{pmatrix} 1 & 2 & 3 \\ 2 & 1 & 3 \end{pmatrix}\begin{pmatrix} 1 & 2 & 3 \\ 3 & 2 & 1 \end{pmatrix} = \begin{pmatrix} 1 & 2 & 3 \\ 3 & 1 & 2 \end{pmatrix} = \sigma_4.$$

Now let V be a vector space over R with a basis of n vectors $\{v_1, \ldots, v_n\}$. For each permutation $\sigma \in P(X)$, we define a linear transformation $T_\sigma \in L(V, V)$ by the rule

$$T_\sigma(v_i) = v_{\sigma(i)}, \qquad i = 1, 2, \ldots, n.$$

Then for all $\sigma, \tau \in P(X)$,

$$T_\sigma T_\tau = T_{\sigma\tau}$$

since

$$\begin{aligned}(T_\sigma T_\tau)v_i &= T_\sigma(v_{\tau(i)}) = T_{\sigma(\tau(i))} = T_{\sigma\tau(i)} \\ &= T_{\sigma\tau}v_i, \qquad i = 1, 2, \ldots, n.\end{aligned}$$

(19.16) DEFINITION. The *signature* $\epsilon(\sigma)$ of a permutation σ is defined by the formula $\epsilon(\sigma) = D(T_\sigma)$, where T_σ is the linear transformation defined above.

(19.17) THEOREM. $\epsilon(\sigma) = \pm 1$ *for all* $\sigma \in P(X)$. *Moreover,*

$$\epsilon(\sigma\tau) = \epsilon(\sigma)\epsilon(\tau)$$

for all $\sigma, \tau \in P(X)$.

Proof. Let $\mathbf{A}_\sigma$ be the matrix of T_σ with respect to the basis $\{v_1, \ldots, v_n\}$. Then we have

$$\epsilon(\sigma) = D(T_\sigma) = D(\mathbf{A}_\sigma),$$

and it follows from the definition of T_σ that the columns of $\mathbf{A}_\sigma$ are simply a rearrangement of the columns of the identity matrix $\mathbf{I}$. Therefore, since interchanging two columns of $\mathbf{A}_\sigma$ changes the sign of the determinant, we have

$$D(\mathbf{A}_\sigma) = \pm D(\mathbf{I}) = \pm 1.$$

The second statement, that $\epsilon(\sigma\tau) = \epsilon(\sigma)\epsilon(\tau)$, follows from Theorem (18.3), since

$$\begin{aligned}\epsilon(\sigma\tau) &= D(T_{\sigma\tau}) = D(\mathbf{A}_{\sigma\tau}) = D(\mathbf{A}_\sigma \mathbf{A}_\tau) \\ &= D(\mathbf{A}_\sigma)D(\mathbf{A}_\tau) = \epsilon(\sigma)\epsilon(\tau).\end{aligned}$$

This completes the proof of the theorem.

(19.18) DEFINITION. A transposition $\sigma = (ij)$ is a permutation such that $\sigma(i) = j$ and $\sigma(j) = i$ for some pair of distinct integers i and j, and is such that $\sigma(x) = x$ for all x different from i or j.

(19.19) THEOREM. *Every permutation* $\sigma \in P(X)$ *is a product of transpositions.*

Proof. Let $X = \{1, 2, \ldots, n\}$. Assume as an induction hypothesis that every permutation of $\{1, 2, \ldots, n-1\}$ is a product of transpositions. The verification that the first cases of the induction hypothesis are true are easy and will be omitted. Let $\sigma \in P(X)$. If $\sigma(n) = n$, then σ can be regarded as a permutation of $\{1, 2, \ldots, n-1\}$; and by the induction hypothesis, we can assume σ is a product of transpositions. Now suppose $\sigma(n) = i \neq n$. Then

$$[\,(in)\sigma\,](n) = (in)\sigma(n) = (in)(i) = n.$$

By the induction hypothesis, $(in)\sigma$ is a product of transpositions. Since

$$(in)\,[\,(in)\sigma\,] = \sigma,$$

it follows that σ is also a product of transpositions.

For example, if $X = \{1, 2, 3\}$, the transpositions are

$$\sigma_2 = \begin{pmatrix} 1 & 2 & 3 \\ 2 & 1 & 3 \end{pmatrix}, \qquad \sigma_3 = \begin{pmatrix} 1 & 2 & 3 \\ 1 & 3 & 2 \end{pmatrix}, \qquad \sigma_4 = \begin{pmatrix} 1 & 2 & 3 \\ 3 & 2 & 1 \end{pmatrix},$$

and it is easily checked that σ_5 and σ_6 are products of two transpositions.

(19.20) DEFINITION. A permutation $\sigma \in P(X)$ is *even* if $\epsilon(\sigma) = 1$ and *odd* if $\epsilon(\sigma) = -1$.

(19.21) THEOREM. *A permutation σ is even if and only if σ can be factored in at least one way as a product of an even number of transpositions. If σ is a transposition, then σ is odd. Even and odd permutations multiply in the following way:*

$$\begin{aligned} (\textit{even})(\textit{even}) &= \textit{even} \\ (\textit{even})(\textit{odd}) &= \textit{odd} \\ (\textit{odd})(\textit{odd}) &= \textit{even}. \end{aligned}$$

Proof. We first verify that if σ is a transposition $\sigma = (ij)$, then $\epsilon(\sigma) = -1$. This follows from the observation that if $\sigma = (ij)$, then the matrix $\mathbf{A}_\sigma$ of T_σ is obtained from the identity matrix by interchanging the ith and jth columns.

Next, suppose σ is factored as a product of transpositions, according to Theorem (19.19),

$$\sigma = \tau_1\tau_2 \ldots \tau_k,$$

where each τ_i is a transposition. By Theorem (19.17)

$$\epsilon(\sigma) = \epsilon(\tau_1)\epsilon(\tau_2) \cdots \epsilon(\tau_k) = (-1)^k.$$

Therefore σ is even (that is, $\epsilon(\sigma) = +1$) if and only if k is even.

The statements about how permutations multiply is clear from the formula $\epsilon(\sigma\tau) = \epsilon(\sigma)\epsilon(\tau)$, and the theorem is proved.

It may appear that the introduction of the signature $\epsilon(\sigma)$ is irrelevant to the concepts of even and odd permutations. The difficulty is that if we simply define an even permutation σ to be a product of an even number of transpositions, then it is not easy to show directly that σ cannot also be expressed (in a different way) as a product of an odd number of transpositions. The theory of the signature $\epsilon(\sigma)$ takes care of this problem.

Finally, we can use the theory of the permutation group $P(X)$ to give a new formula for the complete expansion of a determinant, given in Section 17.

(19.22) THEOREM. *Let* $A = (\alpha_{ij})$ *be an n-by-n matrix with coefficients in a field F. Then*

$$D(A) = \sum_{\sigma \in P(X)} \epsilon(\sigma)\alpha_{1\sigma(1)}\alpha_{2\sigma(2)} \cdots \alpha_{n\sigma(n)}.$$

Proof. The result is immediate from formula (17.12) and the definition of $\epsilon(\sigma)$.

The formula for $D(A)$ given above is useful because it can be used to define the determinant of a matrix with elements in a commutative ring (see the book by Jacobson, listed in the Bibliography).

EXERCISES

1. List the methods you know for finding the inverse of a matrix. Test the following matrices to see whether or not they are invertible; if they are invertible, find an inverse.

$$\begin{pmatrix} 2 & -1 \\ -2 & 1 \end{pmatrix}, \quad \begin{pmatrix} 2 & 1 \\ 1 & 1 \end{pmatrix}, \quad \begin{pmatrix} 3 & 1 & 0 \\ 1 & 2 & 1 \\ 0 & -1 & 2 \end{pmatrix}.$$

2. Find a solution of the system

$$\begin{aligned} 3x_1 + x_2 &= 1 \\ x_1 + 2x_2 + x_3 &= 2 \\ - x_2 + 2x_3 &= -1 \end{aligned}$$

both by Cramer's rule and by the methods of Chapter 2.

3. Find the ranks of the following matrices using determinants:

$$\begin{pmatrix} 1 & 2 & 3 & 4 \\ -1 & 2 & 1 & 0 \end{pmatrix}, \quad \begin{pmatrix} -1 & 0 & 1 & 2 \\ 1 & 1 & 3 & 0 \\ -1 & 2 & 4 & 1 \end{pmatrix}.$$

4. Prove that the equation of the line through the distinct vectors $\langle \alpha, \beta \rangle$, $\langle \gamma, \delta \rangle$ in R_2 is given by

$$\begin{vmatrix} x_1 & x_2 & 1 \\ \alpha & \beta & 1 \\ \gamma & \delta & 1 \end{vmatrix} = 0.$$

5. Prove that the following is the equation of the hyperplane in R_3 containing the non-collinear vectors $\langle \alpha_1, \alpha_2, \alpha_3 \rangle$, $\langle \beta_1, \beta_2, \beta_3 \rangle$, $\langle \gamma_1, \gamma_2, \gamma_3 \rangle$.

$$\begin{vmatrix} x_1 & x_2 & x_3 & 1 \\ \alpha_1 & \alpha_2 & \alpha_3 & 1 \\ \beta_1 & \beta_2 & \beta_3 & 1 \\ \gamma_1 & \gamma_2 & \gamma_3 & 1 \end{vmatrix} = 0.$$

6. Show that the linear transformation $T: a \to T(a)$ of $R_2 \to R_2$, which takes $a = \langle x_1, x_2 \rangle$ onto $T(a) = \langle y_1, y_2 \rangle$ where

$$\begin{aligned} y_1 &= 3x_1 - x_2 \\ y_2 &= x_1 + 2x_2, \end{aligned}$$

carries the square consisting of all points P such that $\overrightarrow{OP} = \lambda e_1 + \mu e_2$, for $0 \leq \lambda$ and $\mu \leq 1$, onto a parallelogram. Show that the area of this parallelogram is the absolute value of the determinant of the matrix of the transformation T,

$$\begin{pmatrix} 3 & -1 \\ 1 & 2 \end{pmatrix}.$$

7. Show that the area of the triangle in the plane with the vertices (α_1, α_2), (β_1, β_2), (γ_1, γ_2) is given by the absolute value of

$$\frac{1}{2}\begin{vmatrix} \alpha_1 & \alpha_2 & 1 \\ \beta_1 & \beta_2 & 1 \\ \gamma_1 & \gamma_2 & 1 \end{vmatrix}.$$

8. Show that the volume of the tetrahedron with vertices $(\alpha_1, \alpha_2, \alpha_3)$, $(\beta_1, \beta_2, \beta_3)$, $(\gamma_1, \gamma_2, \gamma_3)$, $(\delta_1, \delta_2, \delta_3)$ is given by the absolute value of

$$\frac{1}{6}\begin{vmatrix} \alpha_1 & \alpha_2 & \alpha_3 & 1 \\ \beta_1 & \beta_2 & \beta_3 & 1 \\ \gamma_1 & \gamma_2 & \gamma_3 & 1 \\ \delta_1 & \delta_2 & \delta_3 & 1 \end{vmatrix}.$$

9. Let $p_1, \ldots, p_n$ be vectors in R_n and let

$$p_i = \langle \alpha_{i1}, \ldots, \alpha_{in} \rangle, \qquad 1 \leq i \leq n.$$

Prove that $p_1, \ldots, p_n$ lie on a linear manifold of dimension $< n - 1$ if and only if

$$\begin{vmatrix} x_1 & \cdots & x_n & 1 \\ \alpha_{11} & \cdots & \alpha_{1n} & 1 \\ \cdots & \cdots & \cdots & \cdots \\ \alpha_{n1} & \cdots & \alpha_{nn} & 1 \end{vmatrix} = 0$$

for all $x_1, \ldots, x_n$ in R. Prove that, if $p_1, \ldots, p_n$ do not lie on a linear manifold of dimension of $< n - 1$, then $p_1, \ldots, p_n$ lie on a unique hyperplane whose equation is given by the above formula.

10. Prove the following formula for the *van der Monde determinant*:

$$\begin{vmatrix} 1 & \xi_1 & \xi_1^2 & \cdots & \xi_1^{n-1} \\ 1 & \xi_2 & \xi_2^2 & \cdots & \xi_2^{n-1} \\ \cdots & \cdots & \cdots & \cdots & \cdots \\ 1 & \xi_n & \xi_n^2 & \cdots & \xi_n^{n-1} \end{vmatrix} = \pm \prod_{i<j} (\xi_i - \xi_j).$$

(*Hint:* Let $c_1, c_2, \ldots, c_n$ be the columns of the van der Monde matrix. Show that

$$D(c_1, \ldots, c_n) = D(c_1, c_2 - \xi_1 c_1, \ldots, c_{n-1} - \xi_1 c_{n-2}, c_n - \xi_1 c_{n-1})$$

$$= \begin{vmatrix} 1 & 0 & 0 & \cdots & 0 \\ 1 & \xi_2 - \xi_1 & \xi_2^2 - \xi_1\xi_2 & \cdots & \xi_2^{n-1} - \xi_1\xi_2^{n-2} \\ \cdots & \cdots & \cdots & \cdots & \cdots \\ 1 & \xi_n - \xi_1 & \xi_n^2 - \xi_1\xi_n & \cdots & \xi_n^{n-1} - \xi_1\xi_n^{n-2} \end{vmatrix}.$$

Then take the row expansion along the first row, factor out appropriate factors from the result, and use induction.)

11. Show that if $\{a_1, \ldots, a_n\}$ are nonzero and mutually orthogonal vectors in R_n, then $|D(a_1, \ldots, a_n)| = \|a_1\| \, \|a_2\| \cdots \|a_n\|$.

Chapter 6

Polynomials and Complex Numbers

A prerequisite for understanding the deeper theorems about linear transformations is a knowledge of factorization of polynomials as products of prime polynomials. This topic is developed from its beginning in this chapter, along with the facts about complex numbers that will be needed in later chapters.

20. POLYNOMIALS

Everyone is familiar with the concept of a polynomial $\alpha_0 + \alpha_1 x + \alpha_2 x^2 + \cdots + \alpha_n x^n$ where the α_i are real numbers. However, there are some questions which should be answered: Is a polynomial a function or, if not, what is it? What is x? Is it a variable, or an indeterminate, or a number?

In this section we give one approach to polynomials which answers these questions, and we prove the basic theorem on prime factorization of polynomials in preparation for the theory of linear transformations to come in the next chapter.

(20.1) DEFINITION. Let F be an arbitrary field. A *polynomial* with coefficients in F is by definition a sequence

$$f = \{\alpha_0, \alpha_1, \alpha_2, \ldots\}, \qquad \alpha_i \in F,$$

such that for some positive integer M depending on f, $\alpha_M = \alpha_{M+1} = \cdots = 0$. We make the following definitions. If

$$g = \{\beta_0, \beta_1, \beta_2, \ldots\},$$

then $f = g$ if and only if $\alpha_0 = \beta_0$, $\alpha_1 = \beta_1$, $\alpha_2 = \beta_2, \ldots$. We define addition as for vectors:

$$f + g = \{\alpha_0 + \beta_0, \alpha_1 + \beta_1, \ldots\}$$

and multiplication by the rule

$$fg = \{\gamma_0, \gamma_1, \gamma_2, \ldots\}$$

where, for each k,

$$\gamma_k = \sum_{i+j=k} \alpha_i\beta_j = \alpha_0\beta_k + \alpha_1\beta_{k-1} + \cdots + \alpha_k\beta_0.$$

It is clear that both $f + g$ and fg are polynomials; that is, $\alpha_M + \beta_M$ and γ_M are zero for sufficiently large M.

For example, let F be the field of real numbers, and let

$$f = \{0, 1, 0, 0, 0, 0, \ldots\}$$
$$g = \{1, 1, -1, 0, 0, 0, \ldots\}.$$

Then

$$f + g = \{1, 2, -1, 0, 0, 0, \ldots\}$$

and

$$\begin{aligned} f \cdot g = \{&0 \cdot 1, 0 \cdot 1 + 1 \cdot 1, 0 \cdot -1, +1 \cdot 1 + 0 \cdot 1, 0 \cdot 0 + 1 \cdot -1 \\ &+ 0 \cdot 1 + 0 \cdot 1, 0 \cdot 0 + 1 \cdot 0 + 0 \cdot 0 + 0 \cdot -1 + 0 \cdot 0, 0, 0, \ldots\} \\ = \{&0, 1, 1, -1, 0, 0, \ldots\}. \end{aligned}$$

Computations using these definitions are cumbersome, and after the next theorem we shall derive a more familiar and convenient expression for polynomials.

(20.2) THEOREM. *The polynomials with coefficients in F satisfy all the axioms for a commutative ring [Definition (11.8)].*

Proof. Since addition of polynomials is defined as vector addition, it is clear that all the axioms concerning addition alone are satisfied.

The commutative law for multiplication is clear by (20.1). For the associative law, let

$$h = \{\delta_0, \delta_1, \delta_2, \ldots\}.$$

Then the kth coefficient of $(fg)h$ is

$$\sum_{r+s=k} \left(\sum_{i+j=r} \alpha_i\beta_j\right)\delta_s = \sum_{i+j+s=k} (\alpha_i\beta_j)\delta_s$$

while the kth coefficient of $f(gh)$ is

$$\sum_{i+t=k} \alpha_i \left(\sum_{j+s=t} \beta_j \delta_s \right) = \sum_{i+j+s=k} \alpha_i(\beta_j \delta_s)$$

and both expressions are equal because of the associative law in F. Finally, we check the distributive law for multiplication. The kth coefficient of $f(g + h)$ is

$$\sum_{i+j=k} \alpha_i(\beta_j + \delta_j)$$

while the kth coefficient of $fg + fh$ is

$$\sum_{i+j=k} \alpha_i \beta_j + \sum_{i+j=k} \alpha_i \delta_j$$

and these expressions are equal because of the distributive law in F. This completes the proof.

It is clear that the mapping

$$\alpha \rightarrow \{\alpha, 0, 0, \ldots\} = \alpha'$$

is a one-to-one mapping of F into the polynomials such that

$$(\alpha + \beta)' = \alpha' + \beta', \qquad (\alpha\beta)' = \alpha'\beta'.$$

If we let F' be the set of all polynomials α' obtained in this way, then F' is a field that is *isomorphic* with F, and we shall identify the elements of F with the polynomials that correspond to them; that is, we shall write

$$\alpha = \{\alpha, 0, 0, \ldots\}.$$

Now let x be the polynomial defined by the sequence

$$x = \{0, 1, 0, \ldots\}.$$

Then

$$x^2 = \{0, 0, 1, 0, \ldots\}$$

and (remembering that we index the coefficients starting from 0), in general,

$$x^i = \{0, 0, \ldots, \underset{i}{1}, \ldots\}.$$

Moreover, it is easily checked that we have

$$\alpha x^i = \{\alpha, 0, \ldots\}\{0, \ldots, \underset{i}{1}, \ldots\} = \{0, \ldots, \underset{i}{\alpha}, \ldots\} \qquad \alpha \in F, \quad i = 1, 2, \ldots.$$

Therefore an arbitrary polynomial

$$f = \{\alpha_0, \alpha_1, \alpha_2, \ldots\}$$

can be expressed uniquely in the form

(20.3) $$f = \alpha_0 + \alpha_1 x + \alpha_2 x^2 + \cdots + \alpha_r x^r.$$

We shall use the notation $F[x]$ for the set of polynomials with coefficients in F.

Example A. Compute

$$(3 - x + x^2)(2 + 2x + x^2 - x^3).$$

We see that x^5 is the highest power of x appearing in the product with a nonzero coefficient. Then we compute the product by leaving spaces for the coefficients of $x^0, x^1, \ldots, x^5$ and filling them in by inspection. Thus the product is

$$(3 \cdot 2) + (3 \cdot 2 - 2 \cdot 1)x + (3 - 2 + 2)x^2 + (-3 - 1 + 2)x^3 + (1 + 1)x^4 - x^5 = 6 + 4x + 3x^2 - 2x^3 + 2x^4 - x^5.$$

(20.4) DEFINITION. Let $f = \alpha_0 + \alpha_1 x + \alpha_2 x^2 + \cdots$ be a polynomial in $F[x]$. We say that the *degree* of f is r, and write $\deg f = r$, if $\alpha_r \neq 0$ and $\alpha_{r+1} = \alpha_{r+2} = \cdots = 0$. We say that the polynomial $0 = 0 + 0x + 0x^2 + \cdots$ does not have a degree.

(20.5) THEOREM. *Let f, g be nonzero polynomials in $F[x]$; then:*

$$\deg(f + g) \leq \max\{\deg f, \deg g\}, \qquad \text{if } f + g \neq 0,$$
$$\deg fg = \deg f + \deg g.$$

Proof. Let

$$f = \alpha_0 + \alpha_1 x + \cdots + \alpha_r x^r, \qquad \alpha_r \neq 0,$$
$$g = \beta_0 + \beta_1 x + \cdots + \beta_s x^s, \qquad \beta_s \neq 0.$$

Then $$f + g = (\alpha_0 + \beta_0) + (\alpha_1 + \beta_1)x + \cdots + (\alpha_t + \beta_t)x^t,$$

where $t = \max\{r, s\}$, proving the first statement. For the second statement of the theorem, we observe that fg has no nonzero terms $\gamma_i x^i$ for $i > r + s$ and that the coefficient of x^{r+s} is exactly $\alpha_r\beta_s$, which is nonzero since the product of two nonzero elements of a field is different from zero (why?). This completes the proof.

(20.6) COROLLARY. *If f, $g \in F[x]$, then $fg = 0$ implies that $f = 0$ or $g = 0$. If $fg = hg$, and if $g \neq 0$, then $f = h$.*

Proof. If both $f \neq 0$ and $g \neq 0$ then, by (20.5), $\deg fg \geq 0$ and hence $fg \neq 0$. For the proof of the second part let $fg = hg$. Then $(f - h)g = 0$ and, since $g \neq 0$, $f - h = 0$ by the first part of the proof.

(20.7) COROLLARY. *Let $f \neq 0$ be a polynomial in $F[x]$; then there exists a polynomial $g \in F[x]$ such that $fg = 1$ if and only if* $\deg f = 0$. *Consequently, $F[x]$ is definitely not a field.*

Proof. If $\deg f = 0$, then $f = \alpha \in F$ and we have $\alpha\alpha^{-1} = 1$ by the axioms for a field. Conversely, if $fg = 1$ for some $g \in F[x]$, then by (20.5) we have

$$\deg f + \deg g = \deg 1 = 0.$$

Since $\deg f$ is a nonnegative integer, this equation implies that $\deg f = 0$, and (20.7) is proved.

(20.8) THEOREM (DIVISION PROCESS). *Let f, g be polynomials in $F[x]$ such that $g \neq 0$; then there exist the uniquely determined polynomials Q, R called the quotient and the remainder, respectively, such that*

$$f = Qg + R$$

where either $R = 0$ or $\deg R < \deg g$.

Proof. If $f = 0$, then we may take $Q = R = 0$. Now let

$$\text{(20.9)} \qquad \begin{aligned} f &= \alpha_0 + \alpha_1 x + \cdots + \alpha_r x^r, & \alpha_r \neq 0,\ r \geq 0,\\ g &= \beta_0 + \beta_1 x + \cdots + \beta_s x^s, & \beta_s \neq 0,\ s \geq 0. \end{aligned}$$

We use induction on r to prove the existence of Q and R. First let $r = 0$. If $s > 0$, then

$$f = 0 \cdot g + f$$

satisfies our requirements while, if $s = 0$, then by (20.7) we have $gg^{-1} = 1$ and can write

$$f = (fg^{-1})g + 0.$$

Now assume that $r > 0$ and that the existence of Q and R has been proved for polynomials of degree $\leq r - 1$. Consider f and g as in (20.9). If $s > r$, then $f = 0 \cdot g + f$ satisfies the requirements and there is nothing to prove. Finally, let $s \leq r$. Then, using the distributive law in $F[x]$, we obtain

$$\alpha_r\beta_s^{-1}x^{r-s}g = \beta_0' + \beta_1' x + \cdots + \alpha_r x^r + 0x^{r+1} + \cdots$$

with coefficients $\beta_i' \in F$ where $\beta_r' = \alpha_r$. The polynomial

$$\text{(20.10)} \qquad f_1 = f - \alpha_r\beta_s^{-1}x^{r-s}g$$

had degree $\leq r - 1$, since the coefficients of x^r are canceled out. By the induction hypothesis there exist polynomials Q and R with either $R = 0$ or $\deg R < \deg g$ such that

$$f_1 = Qg + R.$$

Substituting (20.10) in this formula, we obtain

$$f = (Q + \alpha_r \beta_s^{-1} x^{r-s})g + R,$$

and the existence part of the theorem is proved.

For the uniqueness, let

$$f = Qg + R = Q'g + R'$$

where both R and R' satisfy the requirements of the theorem. We show first that $R = R'$. Otherwise, $R - R' \neq 0$ and we have

$$R - R' = (Q' - Q)g,$$

where $\deg(R - R') < \deg g$ by (20.5), while $(Q' - Q)g = \deg(Q' - Q) + \deg g \geq \deg g$. This is a contradiction, and so we must have $R = R'$. Then we obtain

$$(Q - Q')g = 0, \qquad g \neq 0$$

and by (20.6) we have $Q - Q' = 0$. This completes the proof of the theorem.

Example B. To illustrate the division process, let

$$f = 3x^3 + x^2 - x + 1,$$
$$g = x^2 + 2x - 1.$$

Then
$$f - 3xg = -5x^2 + 2x + 1,$$
$$-5x^2 + 2x + 1 + 5g = 12x - 4.$$

Since $\deg(2x - 4) < \deg g$, the division is finished and we have

$$f = (3x - 5)g + 12x - 4.$$

Then
$$Q = 3x - 5, \qquad R = 12x - 4.$$

Now we come to the important concept of polynomial function.

(20.11) Definition. Let $f = \sum \alpha_i x^i \in F[x]$ and let $\xi \in F$. We define an element $f(\xi) \in F$ by

$$f(\xi) = \sum \alpha_i \xi^i$$

and call $f(\xi)$ the *value of the polynomial f when ξ is substituted for x*. For a fixed polynomial $f \in F[x]$ the *polynomial function* $f(x)$ is the rule which assigns to each $\xi \in F$ the element $f(\xi) \in F$.† A *zero* of a polynomial

† The notation $f(x)$ is sometimes used for a polynomial f; this notation suppresses the distinction between polynomials (which are sequences) and polynomial functions. The need for the distinction arises from the fact that for finite fields F, two different polynomials in $F[x]$ may correspond to the same polynomial function. For example, let F be the field of two elements [see Exercise 1(c) of Section 4]. Then $x^2 - x$ and 0 are distinct polynomials which define the same polynomial function.

f is an element $\xi \in F$ such that $f(\xi) = 0$; a zero of f will also be called a *solution* or *root* of the polynomial equation $f(x) = 0$.

The next two results govern the connection between finding the zeros of a polynomial and factoring the polynomial. First we require an important lemma.

(20.12) LEMMA. *Let $f,\ g \in F[x]$ and let $\xi \in F$; then $(f \pm g)(\xi) = f(\xi) \pm g(\xi)$ and $(fg)(\xi) = f(\xi)g(\xi)$.*

Proof. Let $f = \alpha_0 + \alpha_1 x + \cdots + \alpha_r x^r$ and $g = \beta_0 + \beta_1 x + \cdots + \beta_s x^s$. Then

$$\begin{aligned}(f+g)(\xi) &= (\alpha_0 + \beta_0) + (\alpha_1 + \beta_1)\xi + (\alpha_2 + \beta_2)\xi^2 + \cdots \\ &= (\alpha_0 + \alpha_1\xi + \alpha_2\xi^2 + \cdots) + (\beta_0 + \beta_1\xi + \beta_2\xi^2 + \cdots) \\ &= f(\xi) + g(\xi).\end{aligned}$$

Similarly, $(f - g)(\xi) = f(\xi) - g(\xi)$. Next we have

$$\begin{aligned}(fg)(\xi) &= \alpha_0\beta_0 + (\alpha_0\beta_1 + \alpha_1\beta_0)\xi + (\alpha_0\beta_2 + \alpha_1\beta_1 + \alpha_2\beta_0)\xi^2 + \cdots \\ &= (\alpha_0 + \alpha_1\xi + \alpha_2\xi^2 + \cdots)(\beta_0 + \beta_1\xi + \beta_2\xi^2 + \cdots) \\ &= f(\xi)g(\xi).\end{aligned}$$

This completes the proof of the lemma.

(20.13) REMAINDER THEOREM. *Let $f \in F[x]$ and let $\xi \in F$; then the remainder obtained upon dividing f by $x - \xi$ is equal to $f(\xi)$:*

$$f = Q(x - \xi) + f(\xi).$$

Proof. By the division process we have

$$f = Q(x - \xi) + r$$

where r is either 0 or an element of F. Substituting ξ for x, we obtain by (20.12) the desired result: $r = f(\xi)$.

(20.14) FACTOR THEOREM. *Let $f \in F[x]$ and let $\xi \in F$; then*

$$f = (x - \xi)Q$$

for some $Q \in F[x]$ if and only if $f(\xi) = 0$.

The proof is immediate by the remainder theorem.

Now let A be a commutative ring (actually, we have in mind the particular ring $A = F[x]$). If $r, s \in A$, then we say that $r \mid s$ (read "*r divides s*" or "*r is a factor of s*" or "*s is a multiple of r*") if $s = rt$ for some $t \in A$. An element u of A is called a *unit* if $u \mid 1$. An element is, clearly, a unit if and only if it is a factor of every element of A and hence is uninteresting from the point of view of factorization. Since every nonzero element of a field is a unit, it is of no interest to study questions of factorization in a field. An element $p \in A$ is called a *prime*† when p is neither 0 nor a unit and when $p = ab$ implies that either a or b is a unit. Two distinct elements are *relatively prime* if their only common divisors are units. An element $d \in A$ is called a *greatest common divisor* of $r_1, r_2, \ldots, r_k \in A$ if $d \mid r_i$, for $1 \leq i \leq k$, and if d' is such that $d' \mid r_i$, $1 \leq i \leq k$, then $d' \mid d$.

We are now going to study these concepts for the ring of polynomials $F[x]$. An almost identical discussion holds for the ring of integers Z, and the reader will find it worthwhile to write out this application.

We note first that the set of units in $F[x]$ coincides with the "constant" polynomials $\alpha \in F$ where $\alpha \neq 0$, that is, the polynomials of degree zero.

(20.15) THEOREM. *Let $f_1, \ldots, f_k$ be arbitrary nonzero polynomials in $F[x]$; then:*

1. *The elements $f_1, \ldots, f_k$ possess at least one greatest common divisor d.*
2. *The greatest common divisor d is uniquely determined up to a unit factor and can be expressed in the form*
$$d = h_1 f_1 + \cdots + h_k f_k$$
for some polynomials $\{h_i\}$ in $F[x]$.

Proof. Consider the set S of all polynomials of the form

$$\sum_{i=0}^{k} g_i f_i,$$

where the $\{g_i\}$ are arbitrary polynomials. Then S contains the set of polynomials $\{f_1, \ldots, f_k\}$ and has the property that, if $p \in S$ and $h \in F[x]$, then $ph \in S$. Since the degrees of nonzero elements of S are in $N \cup \{0\}$, by the well-ordering principle (2.5B) we can find a nonzero polynomial

$$d = h_1 f_1 + \cdots + h_k f_k \in S$$

such that $\deg d \leq \deg d'$ for all nonzero $d' \in S$.

We prove first that $d \mid f_i$ for $1 \leq i \leq k$. By the division process we have, for $1 \leq i \leq k$,

$$f_i = dq_i + r_i$$

where either $r_i = 0$, or $\deg r_i < \deg d$ and

$$r_i = f_i - dq_i \in S.$$

† The primes in $F[x]$, where F is a field, are sometimes called *irreducible polynomials.*

Because of the choice of d as a polynomial of least degree in S we have $r_i = 0$ and, hence, d divides each f_i for $1 \le i \le k$.

Now let d' be another common divisor of $f_1, \ldots, f_k$. Then there are polynomials g_i such that $f_i = d'g_i$, $1 \le i \le k$, and

$$d = \sum h_i f_i = \sum h_i d' g_i = d'(\sum h_i g_i).$$

Therefore $d' \mid d$, and d is a greatest common divisor of $\{f_1, \ldots, f_k\}$.

Finally, let e be another greatest common divisor of $\{f_1, \ldots, f_k\}$. Then $d \mid e$ and $e \mid d$. Therefore there exist polynomials u and v such that $e = du$, $d = ev$. Then $e = euv$ and $e(1 - uv) = 0$. By (20.6) we have $1 - uv = 0$, so that u and v are units. This completes the proof of the theorem.

(20.16) Corollary. *Let $r_1, \ldots, r_k$ be elements of $F[x]$ that have no common factors other than units; then there exist elements $x_1, \ldots, x_k$ in $F[x]$ such that*

$$x_1 r_1 + \cdots + x_k r_k = 1.$$

(20.17) Corollary. *Let p be a prime in $F[x]$ and let $p \mid ab$; then either $p \mid a$ or $p \mid b$.*

Proof. Suppose p does not divide a. Then a and p are relatively prime, and by (20.15) we have

$$au + pv = 1$$

for some $u, v \in F[x]$. Then

$$abu + pvb = b$$

and, since $p \mid ab$, we have $p \mid b$ by the distributive law in $F[x]$.

(20.18) Unique Factorization Theorem. *Let $a \neq 0$ be an element of $F[x]$; then either a is a unit or*

$$a = p_1 \cdots p_s, \qquad s \ge 1,$$

where $p_1, \ldots, p_s$ are primes. Moreover,

(20.19) $$p_1 \cdots p_s = q_1 \cdots q_t,$$

where the $\{p_i\}$ and $\{q_j\}$ are primes, implies that $s = t$, and for a suitable indexing of the p's and q's we have

$$p_1 = \epsilon_1 q_1, \ldots, p_t = \epsilon_t q_t$$

where the ϵ_i are units.

Proof. The existence of at least one factorization of a into primes is clear by induction on $\deg a$.

For the uniqueness assertion, we use induction on s, the result being clear if $s = 1$. Given (20.19) we apply (20.17) to conclude that p_1 divides some q_j, and we may assume that $j = 1$. Then $q_1 = p_1\epsilon_1$ for some unit ϵ_1. Then (20.19) becomes

$$p_1 \cdots p_s = \epsilon_1 p_1 q_2 \cdots q_t.$$

By the cancellation law we have

$$p_2 \cdots p_s = \epsilon_1 q_2 \cdots q_t = q_2' q_3 \cdots q_t$$

where $q_2' = \epsilon_1 q_2$. The result now follows, by the induction hypothesis.

We have followed the approach to the unique factorization theorem via the theory of the greatest common divisor because the greatest common divisor will be used in Chapter 7. It is interesting that the uniqueness of factorization can be proved by using nothing but the well-ordering principle of sets of natural numbers and the simplest facts concerning degrees of polynomials. Neither the division process nor the theory of the greatest common divisor is needed.

The following proof was discovered in 1960 by Charles Giffen while he was an undergraduate at the University of Wisconsin.

Suppose the uniqueness of factorization is false in $F[x]$. Then by the well-ordering principle there will be a polynomial of least degree

$$f = p_1 \cdots p_r = q_1 \cdots q_s, \tag{20.20}$$

where the $\{p_i\}$ and $\{q_j\}$ are primes, which has two essentially different factorizations. We may assume that $r > 1$ and $s > 1$ and that no p_i coincides with a q_j, for otherwise we could cancel p_i and q_j and obtain a polynomial of lower degree than that of f with two essentially different factorizations. We may assume also that each p_i and q_j has leading coefficient (that is, the coefficient of the highest power of x) equal to 1. By interchanging the p's and q's, if necessary, we may arrange matters so that $\deg p_r \le \deg q_s$. Then for a suitable power x^t of x the coefficient of $x^{\deg q_s}$ in the polynomial $q_s - x^t p_r$ will cancel and we will have $0 \le \deg(q_s - x^t p_r) < \deg q_s$. Now form the polynomial

$$f_1 = f - q_1 \cdots q_{s-1}(x^t p_r).$$

By (20.20) we have

$$f_1 = q_1 \cdots q_{s-1}(q_s - x^t p_r) \tag{20.21}$$

and so, by what has been said, $f_1 \neq 0$ and $\deg f_1 < \deg f$. But from the form of f_1 we see that $p_r \mid f_1$ and, since prime factorization is unique for polynomials of degree $<\deg f$, we conclude from (20.21) that $p_r \mid (q_s - x^t p_r)$, since p_r is distinct from all the primes $q_1, \ldots, q_{s-1}$. Then

$$q_s - x^t p_r = hp_r$$

and

$$q_s = p_r(h + x^t)$$

which is a contradiction. This completes the proof of unique factorization.†

We conclude this section with the observation familiar to us from high school algebra that, although $F[x]$ is not a field, $F[x]$ can be embedded in a field, and in exactly the way that the integers can be embedded in the field of rational numbers. Some of the details will be omitted.

Consider all pairs (f, g), for f and $g \in F[x]$, where $g \neq 0$. Define two such pairs (f, g) and (h, k) as *equivalent* if $fk = gh$; in this case we write $(f, g) \sim (h, k)$. Then the relation $\sim$ has the properties:

1. $(f, g) \sim (f, g)$.
2. $(f, g) \sim (h, k)$ implies $(h, k) \sim (f, g)$.
3. $(f, g) \sim (h, k)$, $(h, k) \sim (p, q)$ imply $(f, g) \sim (p, q)$.

[For the proof of property (3) the cancellation law (20.6) is required.] Now define a fraction f/g, with $g \neq 0$, to be the set of all pairs (h, k), $k \neq 0$, such that $(h, k) \sim (f, g)$. Then we can state:

4. Every pair (f, g) belongs to one and only one fraction f/g.
5. Two fractions f/g and r/s coincide if and only if $fs = gr$.

Now we define:

6. $f/g + r/s = (fs + gr)/gs$.
7. $(f/g)(r/s) = fr/gs$.

It can be proved first of all that the operations of addition and multiplication of fractions are defined independently of the representatives of the fractions. In other words, one has to show that if $f/g = f_1/g_1$ and $r/s = r_1/s_1$ then

$$\frac{fs + gr}{gs} = \frac{f_1 s_1 + g_1 r_1}{g_1 s_1}$$

and that a similar statement holds for multiplication.

---

† Note that the same argument establishes the uniqueness of factorization in the ring of integers Z. In more detail, if uniqueness of factorization does not hold in Z, there exists a smallest positive integer $m = p_1 \cdots p_r = q_1 \cdots q_s$ which has two essentially different factorizations. Then we may assume that no p_i coincides with a q_j and that r and s are greater than 1. We may also assume that $p_r < q_s$ and form $m_1 = m - q_1 \cdots q_{s-1} p_r$. Then $m_1 < m$, and $p_r \mid m_1$. Since $m_1 = q_1 \cdots q_{s-1}(q_s - p_r)$, it follows that $p_r \mid (q_s - p_r)$, which is a contradiction. This argument first came to the attention of the author in Courant and Robbins (see the Bibliography).

Now we shall state a result. The proof offers no difficulties, and will be omitted.

(20.22) THEOREM. *The set of fractions f/g, $g \neq 0$, with respect to the operations of addition and multiplication previously defined, forms a field $F(x)$. The mapping $f \to fg/g = \varphi(f)$, where $f \in F[x]$, is a one-to-one mapping of $F[x]$ into $F(x)$ such that $\varphi(f + h) = \varphi(f) + \varphi(h)$ and $\varphi(fh) = \varphi(f)\varphi(h)$ for $f, h \in F[x]$.*

The field $F(x)$ we have constructed is called the *field of rational functions* in one variable with coefficients in F; it is also called the *quotient field* of the polynomial ring $F[x]$. If we identify the polynomial $f \in F[x]$ with the rational function $\varphi(f) = fg/g$, $g \neq 0$, then we may say that the field $F(x)$ contains the polynomial ring $F[x]$.

Example C. Find the greatest common divisor of the polynomials

$$f = x^2 - x - 2, \qquad g = x^3 + 1$$

in $R[x]$. Referring to Theorem (20.15) we see that although the existence of the greatest common divisor d is guaranteed by the theorem, the proof gives practically no idea about how to find d. We show how to use the prime factorization of f and g in $R[x]$ to find d. In Exercise 8, another method for computing the greatest common divisor will be given, which also exhibits polynomials a and b such that $d = af + bg$.

(20.23) THEOREM. *Let f, g be nonzero polynomials in $F[x]$, and let*

$$f = p_1^{a_1} \cdots p_r^{a_r}, \qquad g = p_1^{b_1} \cdots p_r^{b_r},$$

where $p_1, \ldots, p_r$ are primes and the exponents $a_i, b_i \geq 0$ (by allowing the exponents to be zero, we can express f and g in terms of the same set of primes $\{p_1, \ldots, p_r\}$). Then the greatest common divisor d of f and g is given (up to a unit factor) by

$$d = p_1^{c_1} \cdots p_r^{c_r},$$

where for each i, c_i is the smaller of the exponents a_i and b_i.

Proof. It is clear that $d \mid f$ and $d \mid g$. Now let $d' \mid f$ and $d' \mid g$. By the uniqueness of factorization, we conclude that

$$d' = \epsilon p_1^{d_1} \cdots p_r^{d_r}, \qquad \epsilon \text{ a unit},$$

where $d_i \leq a_i$ and $d_i \leq b_i$ for $1 \leq i \leq r$. It follows that $d' \mid d$ and the theorem is proved.

To apply the theorem to our example, we have to factor f and g into their prime factors in $R[x]$. We have

$$f = (x - 2)(x + 1),$$
$$g = (x + 1)(x^2 - x + 1).$$

The polynomials of degree one are all clearly primes in $R[x]$. The polynomial $x^2 - x + 1$ has no real roots, since $1 - 4 = -3 < 0$, and therefore, by the factor theorem (20.14), $x^2 - x + 1$ is a prime. We can now apply Theorem (20.23) to conclude that $x + 1$ is the greatest common divisor of f and g.

EXERCISES

1. Use the method of proof of Theorem (20.8) to find the quotient Q and the remainder R such that
$$f = Qg + R$$
where
$$f = 2x^4 - x^3 + x - 1,$$
$$g = 3x^3 - x^2 + 3.$$
2. Prove that a polynomial $f \in F[x]$ has at most $\deg f$ distinct zeros in F, where F is any field.
3. Let $f = ax^2 + bx + c$, for a, b, c real numbers and $a \neq 0$. Prove that f is a prime in $R[x]$ if and only if $b^2 - 4ac < 0$. Prove that if $b^2 - 4ac = D \geq 0$ then
$$f = a\left(x - \frac{-b + \sqrt{D}}{2a}\right)\left(x - \frac{-b - \sqrt{D}}{2a}\right).$$
4. Let F be any field and let $f \in F[x]$ be a polynomial of degree ≤ 3. Prove that f is a prime in $F[x]$ if and only if f either has degree 1 or has no zeros in F. Is the same result valid if $\deg f > 3$?
5. Prove that, if a rational number m/n, for m and n relatively prime integers, is a root of the polynomial equation
$$a_0x^r + a_1x^{r-1} + \cdots + a_r = 0$$
where the $a_i \in Z$, then $n \mid a_0$ and $m \mid a_r$.† Use this result to list the possible rational roots of the equations
$$2x^3 - 6x^2 + 9 = 0,$$
$$x^3 - 8x^2 + 12 = 0.$$
6. Prove that if m is a positive integer which is not a square in Z then $\sqrt{m}$ is irrational (use Exercise 5).
7. Factor the following polynomials into their prime factors in $Q[x]$ and $R[x]$.
   a. $2x^3 - x^2 + x + 1$.
   b. $3x^3 + 2x^2 - 4x + 1$.
   c. $x^6 + 1$.
   d. $x^4 + 16$.

† This argument uses the fact that the law of unique factorization holds for the integers Z, as we pointed out in the footnote to Giffen's proof of unique factorization in $F[x]$.

8. Let $a, b \in F[x]$ for $a, b \neq 0$. Apply the division process to obtain

$$\begin{aligned} a &= bq_0 + r_0 & & \\ b &= r_0q_1 + r_1, & \deg r_1 &< \deg r_0 \\ r_0 &= r_1q_2 + r_2, & \deg r_2 &< \deg r_1 \\ &\cdots & & \\ r_i &= r_{i+1}q_{i+2} + r_{i+2}, & \deg r_{i+2} &< \deg r_{i+1}. \end{aligned}$$

Show that for some i_0, $r_{i_0} \neq 0$ and $r_{i_0+1} = 0$. Prove that $r_{i_0} = (a, b)$.

9. Find the greatest common divisor of the following pairs of polynomials in the ring $R[x]$.
   a. $4x^3 + 2x^2 - 2x - 1$, $2x^3 - x^2 + x + 1$.
   b. $x^3 - x + 1$, $2x^4 + x^2 + x - 5$.

10. Let F be the field of two elements defined in Exercise 4 of Section 2. Factor the following polynomials into primes in $F[x]$: $x^2 + x + 1$, $x^3 + 1$, $x^4 + x^2 + 1$, $x^4 + 1$.

11. Find the greatest common divisor of $x^5 + x^4 + x^3 + x^2 + x + 1$ and $x^3 + x^2 + x + 1$ in $F[x]$, where F is the field of two elements as in Exercise 10.

21. COMPLEX NUMBERS

The field of real numbers of R has the drawback that not every quadratic equation with real coefficients has a solution in R. This fact was circumvented by mathematicians of the eighteenth and nineteenth centuries by assuming that the equation $x^2 + 1 = 0$ had a solution i, and they investigated the properties of the new system of "imaginary" numbers obtained by considering the real numbers together with the new number i. Although today we do not regard the complex numbers as any more imaginary than real numbers, it was clear that mathematicians such as Euler used the "imaginary" number i with some hesitation, since it was not constructed in a clear way from the real numbers.

Whatever the properties of the new number system, the eighteenth- and nineteenth-century mathematicians insisted upon making the new numbers obey the same rules of algebra as the real numbers. In particular, they reasoned, the new number system had to contain all such expressions as

$$\alpha + \beta i + \gamma i^2 + \cdots$$

where $\alpha, \beta, \gamma, \ldots$ were real numbers. Since $i^2 = -1$, $i^3 = -i$, etc., any such expression could be simplified to an expression like

$$\alpha + \beta i, \qquad \alpha, \beta \in R.$$

The rules of combination of these numbers were easily found to be

$$(\alpha + \beta i) + (\gamma + \delta i) = (\alpha + \gamma) + (\beta + \delta)i,$$
$$(\alpha + \beta i)(\gamma + \delta i) = \alpha\gamma + \beta\delta i^2 + \alpha\delta i + \beta i\gamma,$$
$$= (\alpha\gamma - \beta\delta) + (\alpha\delta + \beta\gamma)i.$$

This was the situation when the Irish mathematician W. R. Hamilton became interested in complex numbers in the 1840's. He realized first that the complex numbers would not seem quite so imaginary if there were a rigorous way of constructing them from the real numbers, and we shall give his construction.

(21.1) DEFINITION. The system of *complex numbers* C is the two-dimensional vector space R_2 over the real numbers R together with two operations, called addition and multiplication, addition being the vector addition defined in R_2 and multiplication being defined by the rule

$$\langle \alpha, \beta \rangle \langle \gamma, \delta \rangle = \langle \alpha\gamma - \beta\delta, \alpha\delta + \beta\gamma \rangle.$$

(21.2) THEOREM. *The complex numbers form a field.*

Proof. The axioms for a field [see Definition (2.1)] which involve only addition are satisfied because addition in C is vector addition of pairs of real numbers. We have to check the associative and distributive laws, the existence of 1, and the existence of multiplicative inverses. We leave checking the associative and commutative laws for multiplication and the distributive laws to the reader. The element $1 = \langle 1, 0 \rangle$ has the property that $1z = z1 = z$ for all $z \in C$. Finally, let $z = \langle \alpha, \beta \rangle \neq 0$. We wish to show that there exists an element $w = \langle x, y \rangle$ such that $zw = 1$. This means that we have to solve the following equations for x and y.

$$\alpha x - \beta y = 1$$
$$\beta x + \alpha y = 0.$$

The coefficient matrix is

$$\begin{pmatrix} \alpha & -\beta \\ \beta & \alpha \end{pmatrix},$$

whose determinant is $\alpha^2 + \beta^2$. Since α and β are real numbers, and are not both equal to zero, we have $\alpha^2 + \beta^2 > 0$, and the equations can be solved. Thus z^{-1} exists, and C forms a field.

The mapping $\alpha \to \langle \alpha, 0 \rangle = \alpha'$ of $R \to C$ is a one-to-one mapping such that

$$(\alpha + \beta)' = \alpha' + \beta', \qquad (\alpha\beta)' = \alpha'\beta'.$$

In this sense we may say that R is contained in C, and we shall write $\alpha = \langle \alpha, 0 \rangle$.

There is no longer anything mysterious about the equation $x^2 + 1 = 0$. Remembering that $1 = \langle 1, 0 \rangle$, we see that $\pm\langle 0, 1 \rangle$ are the solutions of the equation, so that if we define i by

$$i = \langle 0, 1 \rangle,$$

then $i^2 = -1$. For all $\beta = \langle \beta, 0 \rangle \in R$ we have $\beta i = \langle 0, \beta \rangle$, and hence every complex number $z = \langle \alpha, \beta \rangle$ can be expressed as

$$z = \langle \alpha, \beta \rangle = \langle \alpha, 0 \rangle + \langle 0, \beta \rangle = \alpha \cdot 1 + \beta i.$$

We shall call α the *real part* of z and β the *imaginary part*. Moreover, $\alpha + \beta i = \gamma + \delta i$ if and only if $\alpha = \gamma$ and $\beta = \delta$.

Another point that Hamilton emphasized was that not only i but every complex number $z = \alpha + \beta i$ is a root of a quadratic equation with real coefficients. To find the equation, we compare

$$z^2 = (\alpha^2 - \beta^2) + (2\alpha\beta)i$$

with $z = \alpha + \beta i$. We find that

$$z^2 - 2\alpha z = -\alpha^2 - \beta^2$$

so that the equation satisfied by $z = \alpha + \beta i$ is

(21.3) $$z^2 - 2\alpha z + (\alpha^2 + \beta^2) = 0.$$

We know that there is another root of this equation, and we find that it is given by

$$\bar{z} = \alpha - \beta i.$$

Since the constant term of a quadratic polynomial $z^2 + Az + B$ is easily seen by the factor theorem to be the product of the zeros, we see at once that

$$z\bar{z} = \alpha^2 + \beta^2.$$

If we let $|z| = \sqrt{\alpha^2 + \beta^2}$, then $|z|$ is the length of the vector $\langle \alpha, \beta \rangle$, and is called the *absolute value* of z. We have shown that

(21.4) $$z\bar{z} = |z|^2.$$

From this formula we obtain a simple formula for z^{-1}, if $z \neq 0$, namely,

(21.5) $$z^{-1} = \frac{\bar{z}}{|z|^2}.$$

The complex number $\bar{z}$ is called the *conjugate* of z; it is the other root of the quadratic equation with real coefficients satisfied by z. The operation of taking conjugates has the properties

$$\overline{z_1 + z_2} = \bar{z}_1 + \bar{z}_2, \qquad \overline{z_1 z_2} = \bar{z}_1 \bar{z}_2 .$$

Thus $z \to \bar{z}$ is an isomorphism of C onto C, and we say it is an *automorphism* of C. Using this automorphism, we can derive the formula $|z_1 z_2| = |z_1|\,|z_2|$, since

$$|z_1 z_2|^2 = z_1 z_2 \overline{z_1 z_2} = z_1 z_2 \bar{z}_1 \bar{z}_2 = (z_1 \bar{z}_1)(z_2 \bar{z}_2) = |z_1|^2 |z_2|^2 .$$

If $z_1 = \alpha + \beta i$ and $z_2 = \gamma + \delta i$, then this formula gives the remarkable identity for real numbers,

$$(\alpha\gamma - \beta\delta)^2 + (\alpha\delta + \beta\gamma)^2 = (\alpha^2 + \beta^2)(\gamma^2 + \delta^2)$$

which asserts that a product of two sums of two squares can be expressed as a sum of two squares.†

We come next to the important *polar representation* of complex numbers. Let $z = \langle \alpha, \beta \rangle$; then letting $\rho = \sqrt{\alpha^2 + \beta^2} = |z|$, we can write

$$\alpha = \rho \cos \theta, \qquad \beta = \rho \sin \theta,$$

where θ is the angle determined by the rays joining the origin to the points $(1, 0)$ and (α, β). Thus,‡

$$z = \alpha + \beta i = \rho(\cos \theta + i \sin \theta) = |z|(\cos \theta + i \sin \theta)$$

where we note that

$$|z| = |\rho|, \qquad |\cos \theta + i \sin \theta| = 1 .$$

If $w = |w|(\cos \varphi + i \sin \varphi)$, then we obtain

$$\begin{aligned} zw &= |z|\,|w|(\cos \theta + i \sin \theta)(\cos \varphi + i \sin \varphi) \\ &= |zw|\,[\,(\cos \theta \cos \varphi - \sin \theta \sin \varphi) + i(\sin \theta \cos \varphi + \cos \theta \sin \varphi)\,] . \end{aligned}$$

From the addition theorems for the sine and cosine functions, this formula becomes

(21.6) $$zw = |z|\,|w|\,[\,\cos(\theta + \varphi) + i \sin(\theta + \varphi)\,]$$

which says in geometrical terms that to multiply two complex numbers we must multiply their absolute values and add the angles they make with the "real" axis.

---

† For a discussion of this formula and analogous formulas for sums of four and eight squares, see Section 35.

‡ We write $i \sin \theta$ instead of the more natural $(\sin \theta)i$ in order to keep the number of parentheses down to a minimum.

An important application of (21.6) is the following theorem.

(21.7) DE MOIVRE'S THEOREM. *For all positive integers n,*

$$(\cos \theta + i \sin \theta)^n = \cos n\theta + i \sin n\theta.$$

De Moivre's theorem has several important applications. If for a fixed n we expand $(\cos \theta + i \sin \theta)^n$ by using the binomial formula and compare real and imaginary parts on both sides of the equation in (21.7), we then obtain formulas expressing $\cos n\theta$ and $\sin n\theta$ as polynomials in $\sin \theta$ and $\cos \theta$.

Another important application is the construction of the *roots of unity.*

(21.8) THEOREM. *For each positive integer n, the equation $x^n = 1$ has exactly n distinct complex roots, $z_1, z_2, \ldots, z_n$, which are given by*

$$z_1 = \cos \frac{2\pi}{n} + i \sin \frac{2\pi}{n}, \ldots, z_k = z_1^k = \cos \frac{2\pi k}{n} + i \sin \frac{2\pi k}{n}, \qquad k = 1, \ldots, n.$$

The proof is left to the reader.

We come finally to what is perhaps the most important property of the field of complex numbers from the point of view of algebra.

(21.9) DEFINITION. A field F is said to be *algebraically closed* if every polynomial $f \in F[x]$ of positive degree has at least one zero in F.

The next theorem is sometimes called "The Fundamental Theorem of Algebra," and although modern algebra no longer extolls it in quite such glowing terms, it is nevertheless a basic result concerning the complex field.

(21.10) THEOREM. *The field of complex numbers is algebraically closed.*

Many proofs of this theorem have been found, all of which rest on the completeness axiom for the real numbers or on some topological property of the real numbers which is equivalent to the completeness axiom. The reader will find a proof very much in the spirit of this course in Schreier and Sperner's book, and other proofs may be found in Birkhoff and MacLane's book (see the Bibliography for both), or in any book on functions of a complex variable.

(21.11) THEOREM. *Let F be an algebraically closed field. Then every prime polynomial in $F[x]$ has (up to a unit factor) the form $x - a$, $a \in F$. Every polynomial $f \in F[x]$ can be factored in the form*

$$\prod_{i=1}^{n} (x - a_i), \qquad a_i \in F.$$

Proof. Let F be algebraically closed and let $p \in F[x]$ be a prime polynomial. By Definition (21.9) there is an element $a \in F$ such that $p(a) = 0$. By the factor theorem (20.14), $x - a$ is a factor of p. Since p is prime, p is a constant multiple of $x - a$, and the first statement is proved. The second statement is immediate from the first.

The disadvantage of this theorem is that, although it asserts the existence of the zeros of a polynomial, it gives no information about how the zeros depend on the coefficients of the polynomial f. This problem belongs to the subject of the Galois theory (Van der Waerden, Vol. I, Chapter 5; see the Bibliography).

We conclude this section with an application of Theorem (21.11) to polynomials with real coefficients.

(21.12) THEOREM. *Let $f = \alpha_0 + \alpha_1 x + \cdots + \alpha_n x^n \in R[x]$. If $u \in C$ is a zero of f, then $\bar{u}$ is also a zero of f; if $u \neq \bar{u}$, then*

$$(x - u)(x - \bar{u}) = x^2 - (u + \bar{u})x + u\bar{u}$$

is a factor of f.

Proof. If u is a zero of f, then

$$\alpha_0 + \alpha_1 u + \cdots + \alpha_n u^n = 0.$$

Taking the conjugate of the left side and using the fact that $u \to \bar{u}$ is an automorphism of C such that $\bar{\alpha} = \alpha$ for $\alpha \in R$, we obtain

$$\alpha_0 + \alpha_1 \bar{u} + \cdots + \alpha_n \bar{u}^n = 0,$$

which is the first assertion of the theorem. The second is immediate by the factor theorem.

An important consequence of the last theorem is its corollary:

(21.13) COROLLARY. *Every prime polynomial in $R[x]$ has the form (up to a unit factor)*

$$x - \alpha, \qquad \text{or} \qquad x^2 + \alpha x + \beta, \qquad \alpha^2 - 4\beta < 0.$$

Proof. Let $f \in R[x]$ be a prime polynomial; then f has a zero $u \in C$. If $u = \alpha \in R$, then $f = \xi(x - \alpha)$ for some $\xi \in R$. If $u \notin R$, then $\bar{u} \neq u$ and, by (21.12),

$$(x - u)(x - \bar{u}) = x^2 - (u + \bar{u})x + u\bar{u}$$

is a factor of f. Since $u + \bar{u}$ and $u\bar{u}$ belong to R, it follows that f is (up to a unit factor) $x^2 + \alpha x + \beta$. The condition $\alpha^2 - 4\beta < 0$ follows from the fact that $(u + \bar{u})^2 - 4\,u\bar{u} = (u - \bar{u})^2 = 4\delta^2 i^2$, if $u = \gamma + \delta i$.

EXERCISES

1. Express in the form $\alpha + \beta i$:

$$(3 + i)(-2 + 4i), \qquad \frac{1}{3 + 2i}, \qquad \frac{2 + i}{2 - i}.$$

2. Derive formulas for $\cos 3\theta$ and $\sin 3\theta$ in terms of $\cos\theta$ and $\sin\theta$.
3. Find all solutions of the equation $x^5 = 2$.
4. Let $a_0 + a_1x + \cdots + a_{r-1}x^{r-1} + x^r = (x - u_1)(x - u_2)\cdots(x - u_r)$ be a polynomial in $C[x]$ with leading coefficient $a_r = 1$ and with zeros $u_1, \ldots, u_r$ in C. Prove that $a_0 = \pm u_1u_2, \ldots u_r$ and $a_{r-1} = -(u_1 + u_2 + \cdots + u_r)$.
5. Prove that the field of complex numbers C is isomorphic† with the set of all 2-by-2 matrices with real coefficients of the form

$$\begin{pmatrix} \alpha & -\beta \\ \beta & \alpha \end{pmatrix}, \qquad \alpha, \beta \in R,$$

where the operations are addition and multiplication‡ of matrices.

6. Prove that the complex numbers of absolute value 1 form a group with respect to the operation of multiplication.
7. Prove that the mapping

$$\begin{pmatrix} \cos\theta & -\sin\theta \\ \sin\theta & \cos\theta \end{pmatrix} \to \cos\theta + i\sin\theta$$

is an isomorphism between the group of rotations in the plane and the multiplicative group of complex numbers of absolute value 1.

8. Let $f(x) = \alpha_0 + \alpha_1x + \cdots + \alpha_rx^r$ be a polynomial of degree r with coefficients $\alpha_i \in Q$, the field of rational numbers, and let $u \in C$ be a zero of f. Let $Q[u]$ be the set of complex numbers of the form

$$z = \beta_0 + \beta_1u + \cdots + \beta_{r-1}u^{r-1}, \qquad \beta_i \in Q.$$

Prove that if $z, w \in Q[u]$ then $z \pm w$ and $zw \in Q[u]$. Prove that $Q[u]$ is a field if f is a prime polynomial in $Q[x]$. [*Hint:* In case $f(x)$ is a prime, the main difficulty is proving that if $z = \beta_0 + \beta_1u + \cdots +$

---

† Two fields F and F' are said to be isomorphic if there exists a one-to-one mapping $\alpha \to \alpha'$ of F onto F' such that $(\alpha + \beta)' = \alpha' + \beta'$ and $(\alpha\beta)' = \alpha'\beta'$ for all $\alpha, \beta \in F$.

‡ For the definitions and properties of addition and multiplication of matrices, see Section 12.

$\beta_{r-1}u^{r-1} \neq 0$ then there exists $w \in Q[u]$ such that $zw = 1$. Since $z \neq 0$, the polynomial

$$g(x) = \beta_0 + \beta_1 x + \cdots + \beta_{r-1}x^{r-1} \neq 0$$

in $Q[x]$. Since $\deg f(x) = r$, it follows that $f(x)$ and $g(x)$ are relatively prime. Therefore there exist polynomials $a(x)$ and $b(x)$ such that

$$a(x)g(x) + b(x)f(x) = 1.$$

Upon substituting u for x, we obtain

$$a(u)g(u) = 1$$

and, since $g(u) = z$, we have produced an inverse for z.]

9. Recall from Section 2 that a field F is called an *ordered field* if there exists a subset P of F (called the set of positive elements) such that (a) sums and products of elements in P are in P, and (b) for each element α in F, one and only one of the following possibilities holds: $\alpha \in P$, $\alpha = 0$, $-\alpha \in P$. Prove that the field of complex numbers is not an ordered field.

Chapter 7

The Theory of a Single Linear Transformation

The main topic of this chapter is an introduction to the theory of a single linear transformation on a vector space. The goal is to find a basis of the vector space such that the matrix of the linear transformation with respect to this basis is as simple as possible. Since the end result is still somewhat complicated, mathematicians have given several different versions of what their favorite (or most useful) form of the matrix should be. We shall give several of these theorems in this chapter.

22. BASIC CONCEPTS

In this section, we introduce some of the basic ideas needed to investigate the structure of a single linear transformation. These are the minimal polynomial of a linear transformation and the concepts of characteristic roots and characteristic vectors.

Throughout the discussion, F denotes an arbitrary field, and V a finite dimensional vector space over F. In Section 11 we saw that $L(V, V)$ is a vector space over F. Theorem (13.3) asserts that if we select a basis $\{v_1, \ldots, v_n\}$ of V then the mapping which assigns to $T \in L(V, V)$ its matrix with respect to the basis $\{v_1, \ldots, v_n\}$ is an isomorphism of the vector space $L(V, V)$ onto the vector space $M_n(F)$ of all n-by-n matrices, viewed as a vector space of n^2-tuples. From Section 5, $M_n(F)$ is an n^2-dimensional vector space. If $\{\mathbf{A}_1, \ldots, \mathbf{A}_{n^2}\}$ is a basis for $M_n(F)$ over F, then the linear transformations $T_1, \ldots, T_{n^2}$, whose matrices with respect to $\{v_1, \ldots, v_n\}$

are $\mathbf{A}_1, \ldots, \mathbf{A}_{n^2}$, respectively, form a basis of $L(V, V)$ over F. In particular, the matrices which have a 1 in one position and zeros elsewhere form a basis of $M_n(F)$; therefore the linear transformations $T_{ij} \in L(V, V)$ defined by

$$T_{ij}(v_j) = v_i, \qquad T_{ij}(v_k) = 0, \qquad k \neq j,$$

form a basis of $L(V, V)$ over F.

For the rest of this section let T be a fixed linear transformation of V. Since $L(V, V)$ has dimension n^2 over F the $n^2 + 1$ powers of T,

$$1, \; T, \; T^2, \; \ldots, \; T^{n2}$$

are linearly dependent. That means that there exist elements of F, $\xi_0, \xi_1, \ldots, \xi_{n^2}$, not all zero, such that

$$\xi_0 1 + \xi_1 T + \xi_2 T^2 + \cdots + \xi_{n^2} T^{n^2} = 0.$$

In other words, there exists a nonzero polynomial

$$f(x) = \xi_0 + \xi_1 x + \cdots + \xi_{n^2} x^{n^2} \in F[x]$$

such that $f(T) = 0$.

As we shall see, the study of these polynomial equations is the key to most of the deeper properties of the transformation T.

Let us make the idea of substituting a linear transformation in a polynomial absolutely precise.

(22.1) DEFINITION. Let $f(x) = \lambda_0 + \lambda_1 x + \cdots + \lambda_r x^r \in F[x]$ and let $T \in L(V, V)$; then $f(T)$ denotes the linear transformation

$$f(T) = \lambda_0 \cdot 1 + \lambda_1 T + \cdots + \lambda_r T^r$$

where 1 is the identity transformation on V. Similarly, we may define $f(\mathbf{A})$ where $\mathbf{A}$ is an n-by-n matrix over F, with 1 replaced by the identity matrix $\mathbf{I}$.

(22.2) LEMMA. *Let $T \in L(V, V)$ and let $f, g \in F[x]$; then:*

a. $f(T)T = Tf(T)$.
b. $(f \pm g)(T) = f(T) \pm g(T)$.
c. $(fg)(T) = f(T)g(T)$.

Of course, the same lemma holds for matrices. The proof is similar to the proof of (20.12), and will be omitted.

(22.3) THEOREM. *Let $T \in L(V, V)$; then $1, T, T^2, \ldots, T^{n^2}$ are linearly dependent in $L(V, V)$. Therefore there exists a uniquely determined integer $r \leq n^2$ such that*

$$1, T, T^2, \ldots, T^{r-1} \qquad \textit{are linearly independent},$$
$$1, T, T^2, \ldots, T^{r-1}, T^r \qquad \textit{are linearly dependent}.$$

Then we have

$$T^r = \xi_0 1 + \xi_1 T + \cdots + \xi_{r-1} T^{r-1}, \qquad \xi_i \in F.$$

Let $m(x) = x^r - \xi_{r-1}x^{r-1} - \cdots - \xi_0 \cdot 1 \in F[x]$. *Then* $m(x)$ *has the following properties*:

1. $m(x) \neq 0$ *in* $F[x]$ *and* $m(T) = 0$.
2. *If* $f(x)$ *is any polynomial in* $F[x]$ *such that* $f(T) = 0$, *then* $m(x) \mid f(x)$ *in* $F[x]$.

Proof. The existence of the polynomial $m(x)$ and the statement (1) concerning it follow from the introductory remarks in this section. Now let $f(x)$ be any polynomial in $F[x]$ such that $f(T) = 0$. Because $1, T, \ldots, T^{r-1}$ are linearly independent, there does not exist a polynomial $R(x) \neq 0$ of degree $<r$ such that $R(T) = 0$. Now apply the division process to $f(x)$ and $m(x)$, and obtain

$$f(x) = m(x)Q(x) + R(x)$$

where either $R(x) = 0$ or $\deg R(x) < r = \deg m(x)$. By Lemma (22.2) we have

$$R(T) = (f - mQ)(T) = f(T) - m(T)Q(T) = 0$$

and by the preceding remark we have $R(x) = 0$ in $F[x]$. This proves that $m(x) \mid f(x)$, and the theorem is proved.

(22.4) Definition. Let $T \in L(V, V)$. The polynomial $m(x) \in F[x]$ defined in Theorem (22.3) is called a *minimal polynomial* of T; $m(x)$ is characterized as the nonzero polynomial of least degree such that $m(T) = 0$, and it is uniquely determined up to a constant factor.

The remarks about the uniqueness of $m(x)$ are clear by part (2) of Theorem (22.3). To see this, let $m(x)$ and $m'(x)$ be two nonzero polynomials of degree r such that $m(T) = m'(T) = 0$. Then by the proof of part (2) of Theorem (22.3) we have $m(x) \mid m'(x)$ and $m'(x) \mid m(x)$. It follows from the discussion in Section 20 that $m(x)$ and $m'(x)$ differ by a unit factor in $F[x]$ and, since the units in $F[x]$ are simply the constant polynomials, the uniqueness of $m(x)$ is proved.

We remark that Theorem (22.3) also holds for any matrix $\mathbf{A} \in M_n(F)$. If $T \in L(V, V)$ has the matrix $\mathbf{A}$ with respect to the basis $\{v_1, \ldots, v_n\}$ of V, it follows from Theorem (13.3) that T and $\mathbf{A}$ have the same minimal polynomial.

A thorough understanding of the definition and properties of the minimal polynomial will be absolutely essential in the rest of this chapter.

Example A. It is one thing to know the existence of the minimal polynomial and quite another thing to compute it in a particular case. For our first example, we consider a 2-by-2 matrix

$$\mathbf{A} = \begin{pmatrix} \alpha & \beta \\ \gamma & \delta \end{pmatrix}, \qquad \alpha, \beta, \gamma, \delta \in F.$$

We have

$$\mathbf{A}^2 = \begin{pmatrix} \alpha^2 + \beta\gamma & \alpha\beta + \beta\delta \\ \gamma\alpha + \gamma\delta & \gamma\beta + \delta^2 \end{pmatrix} = \begin{pmatrix} \alpha^2 + \beta\gamma & \beta(\alpha + \delta) \\ \gamma(\alpha + \delta) & \gamma\beta + \delta^2 \end{pmatrix}.$$

According to Theorem (22.3), we should expect to have to compute $\mathbf{A}^3$, $\mathbf{A}^4$, and then find a relation of linear dependence among $\{\mathbf{I}, \mathbf{A}, \mathbf{A}^2, \mathbf{A}^3, \mathbf{A}^4\}$. But something surprising happens. From the formulas for $\mathbf{A}$ and $\mathbf{A}^2$ we have

$$\mathbf{A}^2 - (\alpha + \delta)\mathbf{A} = \begin{pmatrix} \beta\gamma - \alpha\delta & 0 \\ 0 & \beta\gamma - \alpha\delta \end{pmatrix}.$$

Therefore,

$$\text{(22.5)} \qquad \mathbf{A}^2 - (\alpha + \delta)\mathbf{A} - (\beta\gamma - \alpha\delta)I = 0.$$

We have shown that $\mathbf{A}$ satisfies the equation

$$F(x) = x^2 - (\alpha + \delta)x + (\alpha\delta - \beta\gamma) = 0.$$

By Theorem (22.3), we have

$$m(x) \mid F(x),$$

if $m(x)$ is the minimal polynomial of $\mathbf{A}$. This means that $\deg m(x) = 1$ or 2. If $\deg m(x) = 1$, then $\mathbf{A}$ satisfies an equation

$$\mathbf{A} - \alpha\mathbf{I} = 0$$

and $\mathbf{A}$ is simply a scalar multiple of the identity matrix. Our computations show that in all other cases, the minimal polynomial is the polynomial $F(x)$ given above.

As an illustration, we list a few minimal polynomials of 2-by-2 matrices with entries in R.

Matrix	*Minimal Polynomial*
$\begin{pmatrix} 3 & 0 \\ 0 & 3 \end{pmatrix}$	$x - 3$
$\begin{pmatrix} 3 & 0 \\ 0 & -3 \end{pmatrix}$	$x^2 - 9$
$\begin{pmatrix} 1 & 2 \\ -1 & 3 \end{pmatrix}$	$x^2 - 4x + 5$

Example B. We now turn our attention to 3-by-3 matrices. For larger matrices, we must try to compute the powers of **A** in a transparent way (or have a better method!) as the following computation will show. Let

$$\mathbf{A} = \begin{pmatrix} \alpha & \beta & \gamma \\ \delta & \epsilon & \zeta \\ \eta & \theta & \xi \end{pmatrix}.$$

Then

$$\mathbf{A}^2 = \begin{pmatrix} \alpha^2 + \beta\delta + \gamma\eta & \alpha\beta + \beta\epsilon + \gamma\theta & \alpha\gamma + \beta\zeta + \gamma\xi \\ \alpha\delta + \delta\epsilon + \zeta\eta & \zeta\beta + \epsilon^2 + \xi\theta & \delta\gamma + \epsilon\zeta + \xi\zeta \\ \alpha\eta + \theta\delta + \xi\eta & \eta\beta + \epsilon\theta + \xi\theta & \eta\gamma + \theta\zeta + \xi^2 \end{pmatrix};$$

$$\mathbf{A}^3 = \begin{pmatrix} \alpha(\alpha^2 + \beta\delta + \gamma\eta) + \beta(\alpha\delta + \delta\epsilon + \zeta\eta) + \gamma(\alpha\eta + \theta\delta + \xi\eta) & \text{etc.} \\ \delta(\alpha^2 + \beta\delta + \gamma\eta) + \epsilon(\alpha\delta + \delta\epsilon + \zeta\eta) + \zeta(\alpha\eta + \theta\delta + \xi\eta) & \text{etc.} \\ \eta(\alpha^2 + \beta\delta + \gamma\eta) + \theta(\alpha\delta + \delta\epsilon + \zeta\eta) + \xi(\alpha\eta + \theta\delta + \xi\eta) & \text{etc.} \end{pmatrix}$$

A direct but lengthy calculation shows that

$$\text{(22.6)} \quad \mathbf{A}^3 - (\alpha + \epsilon + \xi)\mathbf{A}^2 + (\epsilon\alpha + \xi\alpha + \xi\epsilon - \delta\beta - \gamma\eta - \zeta\theta)\mathbf{A} - D(\mathbf{A}) = 0.$$

This result will serve to compute the minimal polynomial of an arbitrary 3-by-3 matrix. Either the polynomial given by (22.6) is the minimal polynomial, or **A** satisfies a polynomial equation of degree one or two. Notice that as in the case of 2-by-2 matrices, the degree of the minimal polynomial in this case is less than or equal to the number of rows (or columns) of **A**. We give a few examples.

Matrix	*Minimal Polynomial*
$\begin{pmatrix} 2 & 1 & 0 \\ 1 & 0 & 1 \\ -1 & 1 & 0 \end{pmatrix}$	$x^3 - 2x^2 - 2x + 3$
$\begin{pmatrix} 0 & -1 & 0 \\ 0 & 0 & 0 \\ 0 & 0 & 2 \end{pmatrix}$	$x^3 - 2x^2$
$\begin{pmatrix} 0 & 0 & -1 \\ 0 & 0 & 0 \\ 0 & 0 & 0 \end{pmatrix}$	x^2

When using formulas such as (22.5) and (22.6) to find the minimal polynomial of **A**, it must be checked that **A** satisfies no polynomial equation of lower degree. In the case of the third matrix above, this check shows that x^2 is actually the minimal polynomial.

The calculations leading to (22.6) should convince the reader that we have reached the limit of the experimental method, except in special cases. One of the objectives of this chapter will be to gain some theoretical insight toward what to expect in general.

The first idea that might be expected to give additional information is to study not only polynomials $f(x)$ such that $f(T)$ sends the whole vector space V into zero, but also polynomials in T which send individual vectors to zero. The simplest case is that of a polynomial $x - \alpha$. For a given $\alpha \in F$, we then might ask for vectors $v \in V$ such that $(T - \alpha)v = 0$. This problem leads to the following important definition.

(22.7) DEFINITION. Let $T \in L(V, V)$. An element $\alpha \in F$ is called a *characteristic root* (or *eigenvalue*, or *proper value*) of T if there exists a vector $v \neq 0$ in V such that $T(v) = \alpha v$. A nonzero vector v such that $T(v) = \alpha v$ is called a *characteristic vector* (*eigenvector* or *proper vector*) belonging to the characteristic root α.

There may be many characteristic vectors belonging to a given characteristic root. For example, the identity linear transformation 1 on V has $1 \in F$ as its only characteristic root, but every nonzero vector in V is a characteristic vector belonging to this characteristic root.

We can also define characteristic roots and characteristic vectors for matrices.

(22.7)′ Let $\mathbf{A}$ be an n-by-n matrix with entries in F. An element $\alpha \in F$ is called a *characteristic root* of $\mathbf{A}$ if there exists some nonzero column vector $\mathbf{x}$ in F_n such that $\mathbf{Ax} = \alpha\mathbf{x}$; such a vector $\mathbf{x}$ is called a *characteristic vector* belonging to α.

The correspondence between linear transformations and matrices shows that α is a characteristic root of $T \in L(V, V)$ if and only if α is a characteristic root of the matrix of T with respect to any basis of the vector space.

Example C. We shall give an interpretation of characteristic roots in calculus. Let V be the subspace of the vector space $\mathscr{F}(R)$ of all functions $f: R \to R$, consisting of differentiable functions. Then the operation of differentiation d† defines a linear transformation

$$d: V \to \mathscr{F}(R).$$

What does it mean for a function $\mathscr{F} \in V$ to be a characteristic vector of d? It means that $f \neq 0$, and for some $\alpha \in R$, that

$$df = \alpha f.$$

Do we know any such functions? The exponential functions $\{e^{\alpha t}\}$ are all characteristic vectors for d, and the theory of linear differential

† In this section, we shall use d to denote the operation of differentiation, and D, as usual, will be used for determinants.

equations of the first order shows that these are the only characteristic vectors of d.

We conclude this section with two more general theorems, which throw some light on the preceding examples.

(22.8) THEOREM. *Let $v_1, v_2, \ldots, v_r$ be characteristic vectors belonging to distinct characteristic roots $\alpha_1, \ldots, \alpha_r$ of $T \in L(V, V)$. Then $\{v_1, \ldots, v_r\}$ are linearly independent.*

Proof. We use induction on r. The result is clear if $r = 1$, and so we assume an induction hypothesis that any set of fewer than r of the $\{v_i\}$ is a linearly independent set. Suppose

$$\eta_1 v_1 + \eta_2 v_2 + \cdots + \eta_r v_r = 0, \qquad \eta_i \in F. \tag{22.9}$$

We wish to prove that all $\eta_i = 0$. Suppose some $\eta_i \neq 0$; we prove that this contradicts the induction hypothesis. We may assume all $\eta_i \neq 0$, otherwise we have already contradicted the induction hypothesis. Applying T to (22.9), we obtain

$$\eta_1 \alpha_1 v_1 + \eta_2 \alpha_2 v_2 + \cdots + \eta_r \alpha_r v_r = 0.$$

Multiplying (22.9) by α_1 and subtracting, we obtain

$$\eta_1(\alpha_1 - \alpha_1)v_1 + \eta_2(\alpha_2 - \alpha_1)v_2 + \cdots + \eta_r(\alpha_r - \alpha_1)v_r = 0.$$

The term involving v_1 drops out. Since the $\{\alpha_i\}$ are distinct, the coefficients of $\{v_2, \ldots, v_r\}$ are different from zero, and we have contradicted the induction hypothesis. This completes the proof of the theorem.

Example D. Test the functions in $\mathscr{F}(R)$, $\{e^{\alpha_1 t}, e^{\alpha_2 t}, \ldots, e^{\alpha_s t}\}$, with distinct α_i, for linear dependence. This is not too easy to do directly. But consider the vector space generated by the functions,

$$V = S(e^{\alpha_1 t}, \ldots, e^{\alpha_s t}).$$

Then the derivative $d \in L(V, V)$(why?). Moreover, the functions $e^{\alpha_1 t}, \ldots, e^{\alpha_s t}$ are characteristic vectors of d belonging to distinct characteristic roots, by Example C. Therefore, the functions are linearly independent by Theorem (22.8).

The next result gives an important application of determinants to the problem of finding characteristic roots of linear transformations.

(22.10) THEOREM. *Let T be a linear transformation on a finite dimensional vector space over F, and let $\alpha \in F$. Then α is a characteristic root of T if and only if the determinant $D(T - \alpha 1) = 0$, where 1 is the identity transformation on V.*

Proof. First suppose α is a characteristic root of T. Then from the definition, there exists a nonzero vector $v \in V$ such that

$$Tv = \alpha v.$$

Then
$$(T - \alpha 1)v = 0,$$

and since $v \neq 0$, $T - \alpha 1$ is not one-to-one. Therefore $D(T - \alpha 1) = 0$ by Theorem (18.8).

Conversely, suppose $D(T - \alpha 1) = 0$. By Theorem (18.8), $T - \alpha 1$ is not one-to-one. Therefore there exist vectors v_1 and v_2, with $v_1 \neq v_2$, such that

$$(T - \alpha 1)(v_1) = (T - \alpha 1)(v_2).$$

Then, letting $v = v_1 - v_2$, we have $v \neq 0$ and

$$(T - \alpha 1)(v) = 0,$$

proving that α is a characteristic root of T.

Example E. Find a characteristic root α and a characteristic vector corresponding to it, for the linear transformation $T: \mathbf{x} \rightarrow \mathbf{Ax}$ of R_2, where

$$\mathbf{x} = \begin{pmatrix} x_1 \\ x_2 \end{pmatrix}, \quad \mathbf{A} = \begin{pmatrix} -3 & -2 \\ 2 & 2 \end{pmatrix}.$$

First of all, we check that $\mathbf{A}$ is the matrix of T with respect to the basis

$$\mathbf{e}_1 = \begin{pmatrix} 1 \\ 0 \end{pmatrix}, \quad \mathbf{e}_2 = \begin{pmatrix} 0 \\ 1 \end{pmatrix}$$

of R_2. Upon computing $\mathbf{Ae}_1$ and $\mathbf{Ae}_2$, we see that

$$T(\mathbf{e}_1) = \begin{pmatrix} -3 \\ 2 \end{pmatrix} = -3\mathbf{e}_1 + 2\mathbf{e}_2,$$

$$T(\mathbf{e}_2) = \begin{pmatrix} -2 \\ 2 \end{pmatrix} = -2\mathbf{e}_1 + 2\mathbf{e}_2.$$

Therefore $\mathbf{A}$ is the matrix of T with respect to the basis $\{\mathbf{e}_1, \mathbf{e}_2\}$. By Theorem (22.10), α is a characteristic root of T if and only if

$$D(T - \alpha 1) = D(\mathbf{A} - \alpha I) = 0.$$

We have

$$D(\mathbf{A} - \alpha I) = \begin{vmatrix} -3 - \alpha & -2 \\ 2 & 2 - \alpha \end{vmatrix} = (-3 - \alpha)(2 - \alpha) + 4$$
$$= \alpha^2 + \alpha - 2 = (\alpha + 2)(\alpha - 1).$$

Therefore -2 and 1 are characteristic roots.

We shall now find a characteristic vector for the characteristic root -2. This means we have to solve the equation

(22.11) $$T\mathbf{x} = -2\mathbf{x}.$$

We have

$$T\mathbf{x} = \begin{pmatrix} -3 & -2 \\ 2 & 2 \end{pmatrix} \begin{pmatrix} x_1 \\ x_2 \end{pmatrix} = \begin{pmatrix} -3x_1 - 2x_2 \\ 2x_1 + 2x_2 \end{pmatrix}.$$

Therefore a vector x will satisfy (22.11) if and only if

$$-3x_1 - 2x_2 = -2x_1,$$
$$2x_1 + 2x_2 = -2x_2$$

or

$$-x_1 - 2x_2 = 0,$$
$$2x_1 + 4x_2 = 0.$$

A nontrivial solution is $\langle 2, -1\rangle$. We conclude that

$$\mathbf{x} = \begin{pmatrix} 2 \\ -1 \end{pmatrix}$$

is a characteristic vector for the characteristic root -2 of T.

EXERCISES

1. Let $T \in L(V, V)$ and $f, g \in F[x]$. Prove that $f(T)g(T) = g(T)f(T)$.
2. Let V and W be vector spaces over F of dimensions m and n, respectively. Find a basis for $L(V, W)$.
3. Find the minimal polynomials of

$$\begin{pmatrix} 2 & 0 \\ 3 & -1 \end{pmatrix}, \quad \begin{pmatrix} 0 & 1 & 0 \\ 0 & 0 & 1 \\ 1 & 0 & 0 \end{pmatrix}, \quad \begin{pmatrix} -1 & 1 \\ 1 & 0 \end{pmatrix}, \quad \begin{pmatrix} 0 & 1 & 3 \\ 0 & 0 & 2 \\ 0 & 0 & 0 \end{pmatrix}.$$

4. a. Let $T \in L(V, V)$, and let $\{v_1, \ldots, v_n\}$ be a basis of V consisting of characteristic vectors for T belonging to characteristic roots $\xi_1, \ldots, \xi_n$, respectively. Then $Tv_i = \xi_i v_i$, $i = 1, \ldots, n$. Prove that $f(T) = 0$, where

$$f(x) = (x - \xi_1)(x - \xi_2) \cdots (x - \xi_n).$$

   b. Prove that the minimal polynomial of T is $\Pi(x - \xi_j)$, where the ξ_j are the distinct characteristic roots of T.

   c. Show that the matrices

$$\begin{pmatrix} -1 & 0 & 0 & 0 \\ 0 & -1 & 0 & 0 \\ 0 & 0 & 2 & 0 \\ 0 & 0 & 0 & -1 \end{pmatrix} \quad \text{and} \quad \begin{pmatrix} 2 & 0 & 0 & 0 \\ 0 & 2 & 0 & 0 \\ 0 & 0 & 2 & 0 \\ 0 & 0 & 0 & -1 \end{pmatrix}$$

   have the same minimal polynomials.

5. Prove that if $T \in L(V, V)$ then T is invertible if and only if the constant term of the minimal polynomial of T is different from zero. Describe

how to compute T^{-1} from the minimal polynomial. In particular, show that T^{-1} can always be expressed as a polynomial $f(T)$ in T.

6. a. Let T be an invertible linear transformation on V. Show that α is a characteristic root of T if and only if $\alpha \neq 0$ and α^{-1} is a characteristic root of T^{-1}.
   b. Let $T \in L(V, V)$ be arbitrary. Prove that for any polynomial $f \in F[x]$, if α is a characteristic root of T, then $f(\alpha)$ is a characteristic root of $f(T)$.
7. Show that $v \neq 0$ in V is a characteristic vector of T (belonging to some characteristic root) if and only if $T(W) \subset W$, where $W = S(v)$.
8. Prove that similar matrices have the same characteristic roots.
9. Find the characteristic roots, and characteristic vectors corresponding to them, for each of the following matrices (with entries from R).

$$\text{a. } \begin{pmatrix} 4 & 5 \\ -1 & -2 \end{pmatrix} \qquad \text{b. } \begin{pmatrix} -3 & -2 \\ 2 & 1 \end{pmatrix} \qquad \text{c. } \begin{pmatrix} 1 & 0 \\ 1 & -2 \end{pmatrix}.$$

10. Let $T \in L(V, V)$ be a linear transformation such that $T^m = 0$ for some $m > 0$. Prove that all the characteristic roots of T are zero.
11. Show that a linear transformation $T \in L(V, V)$ has at most $n = \dim V$ distinct characteristic roots.
12. Show that if $T \in L(V, V)$ has the maximum number $n = \dim V$ distinct characteristic roots, then there exists a basis of V consisting of characteristic vectors. What will be the matrix of T with respect to such a basis?

23. INVARIANT SUBSPACES

Let $T \in L(V, V)$. A nonzero vector $v \in V$ is a characteristic vector of T if and only if the one-dimensional subspace $S = S(v)$ is invariant relative to T in the sense that $T(s) \in S$ for all $s \in S$ (see Section 22). The search for invariant subspaces is the key to the deeper properties of a single linear transformation.

For a concrete example, let T be the linear transformation of R_2 such that for some basis $\{v_1, v_2\}$ of R_2,

$$T(v_1) = v_2,$$
$$T(v_2) = v_1.$$

The geometrical behavior of T becomes clear when we find the characteristic vectors of T. These are $w_1 = v_1 + v_2$ and $w_2 = v_1 - v_2$. The matrix of T with respect to the basis $\{w_1, w_2\}$ is

$$\mathrm{A} = \begin{pmatrix} 1 & 0 \\ 0 & -1 \end{pmatrix}.$$

We can see now that T is a reflection with respect to the line through the origin in the direction of the vector w_1; it sends each vector in $S(w_1)$

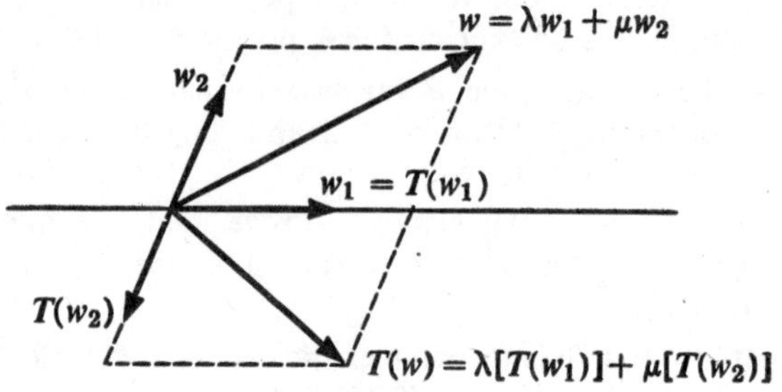

FIGURE 7.1

onto itself and sends w_2 onto its mirror image $-w_2$ with respect to the line $S(w_1)$. Figure 7.1 shows how the image $T(w)$ of an arbitrary vector w can be described geometrically.

The concept illustrated here is the simplest case of the following basic idea.

(23.1) DEFINITION. Let $T \in L(V, V)$. A subspace W of V is called an *invariant subspace relative* to T (or simply a *T-invariant subspace* or *T-subspace*) if $T(w) \in W$ for all $w \in W$.

We recall that a generator $v \neq 0$ of a one-dimensional T-invariant subspace is called a *characteristic vector* of T. If $Tv = \xi v$ for $\xi \in F$, then ξ is called a *characteristic root* of T and v is said to *belong to* the *characteristic root* ξ.

The next result shows one way to construct T-invariant subspaces.

(23.2) LEMMA. *Let $T \in L(V, V)$ and let $f(x) \in F[x]$; then the set of all vectors $v \in V$ such that $f(T)(v) = 0$ [that is, the null space of $f(T)$] is a T-invariant subspace—notation, $n[f(T)]$.*

Proof. Since $f(T) \in L(V, V)$, the null space $n[f(T)]$ is a subspace of V. We have to prove that if $w \in n[f(T)]$ then $T(w) \in n[f(T)]$. We have

$$f(T)[T(w)] = [f(T)T](w) = [Tf(T)](w) = T[f(T)(w)] = 0,$$

since $f(T)T = Tf(T)$ in $L(V, V)$, and the lemma is proved.

We observe in our example of the reflection that the basis vector w_1 generates the null space of the transformation $T - 1$ and w_2 generates the null space of the transformation $T + 1$. The polynomials $x + 1$ and $x - 1$ are exactly the prime factors of the minimal polynomial $x^2 - 1$ of T. The chief result of this section is a far-reaching generalization of this example.

(23.3) DEFINITION. Let $V_1, \ldots, V_s$ be subspaces of V. The space V is said to be the *direct sum* of $\{V_1, \ldots, V_s\}$ (notation, $V = V_1 \oplus \cdots \oplus V_s$) if, first, every vector $v \in V$ can be expressed as a sum,

$$v = v_1 + \cdots + v_s, \qquad v_i \in V_i, \quad 1 \le i \le s, \tag{23.4}$$

and if, second, the expressions (23.4) are unique, in the sense that if

$$v_1 + \cdots + v_s = v_1' + \cdots + v_s', \qquad v_i, v_i' \in V_i, \quad 1 \le i \le s$$

then

$$v_i = v_i', \qquad 1 \le i \le s.$$

(23.5) LEMMA. *Let $V_1, \ldots, V_s$ be subspaces of V; then V is the direct sum $V_1 \oplus \cdots \oplus V_s$ if and only if:*

a. *$V = V_1 + \cdots + V_s$, that is, every vector $v \in V$ can be expressed in at least one way as a sum*

$$v = v_1 + \cdots + v_s, \qquad v_i \in V_i, \quad 1 \le i \le s.$$

b. *If $v_i \in V_i$, for $1 \le i \le s$, are vectors such that*

$$v_1 + \cdots + v_s = 0,$$

then $v_1 = v_2 = \cdots = v_s = 0$.

Proof. If V is the direct sum $V_1 \oplus \cdots \oplus V_s$, then part (a) is satisfied. If

$$v_1 + \cdots + v_s = 0, \qquad v_i \in V_i,$$

then we have

$$v_1 + \cdots + v_s = 0 + \cdots + 0, \qquad v_i, 0 \in V_i, \quad 1 \le i \le s,$$

and by the second part of Definition (23.3) we have $v_1 = \cdots = v_s = 0$, as required.

Now suppose that conditions (a) and (b) of the lemma are satisfied. To prove that $V = V_1 \oplus \cdots \oplus V_s$ it is sufficient to prove the uniqueness assertion of Definition (23.3). If

$$v_1 + \cdots + v_s = v_1' + \cdots + v_s', \qquad v_i, v_i' \in V_i, \quad 1 \le i \le s,$$

then we can rewrite this equation as

$$(v_1 - v_1') + \cdots + (v_s - v_s') = 0, \qquad v_i - v_i' \in V_i, \quad 1 \le i \le s.$$

By condition (b) we have $v_i - v_i' = 0$ for $1 \le i \le s$ and hence $v_i = v_i'$ for all i. This completes the proof of the lemma.

The next result gives a useful criterion for a vector space to be expressed as a direct sum.

(23.6) LEMMA. *Let V be a vector space over F, and suppose there exist nonzero linear transformations $\{E_1, \ldots, E_s\}$ in $L(V, V)$ such that the following conditions are satisfied.*

a. $1 = E_1 + \cdots + E_s$,
b. $E_iE_j = E_jE_i = 0 \quad$ if $i \neq j,\ 1 \leq i,\ j \leq s$.

Then we have $E_i^2 = E_i$, $1 \leq i \leq s$. Moreover, V is the direct sum $V = E_1V \oplus \cdots \oplus E_sV$, and each subspace E_iV is different from zero.

Proof. From (a) and (b) we have

$$E_i = E_i \cdot 1 = E_i(E_1 + \cdots + E_s) = E_i^2 + \sum_{j \neq i} E_iE_j = E_i^2,$$

proving the first statement. For the second statement, we note that E_iV is a nonzero subspace for $1 \leq i \leq s$, since E_i is a nonzero linear transformation. Let $v \in V$; then

$$v = 1v = E_1v + \cdots + E_sv,$$

proving that $V = E_1V + \cdots + E_sV$. Now suppose $v_1 \in E_1V, \ldots, v_s \in E_sV$, and $v_1 + \cdots + v_s = 0$. Then

(23.7) $$E_i(v_1 + \cdots + v_s) = 0, \qquad 1 \leq i \leq s.$$

Moreover, $E_iv_j = 0$ if $i \neq j$ because $v_j \in E_jV$, and $E_iE_j = 0$. Finally $E_iv_i = v_i$, since $v_i = E_iv$ for some $v \in V$, and $E_iv_i = E_i^2v = E_iv = v_i$, using the fact that $E_i^2 = E_i$, $1 \leq i \leq s$. The equation (23.7) implies that $v_1 = \cdots = v_s = 0$, and the lemma follows from Lemma (23.5).

(23.8) DEFINITION. A linear transformation $E \in L(V, V)$ satisfying $E^2 = E \neq 0$ is called an *idempotent* linear transformation.

Finally, we are ready to state our main theorem. It can be stated in the following intuitive way. Let $T \in L(V, V)$ and let $m(x)$ be the minimal polynomial of T. By the theorems of Section 20, $m(x)$ can be factored into primes in $F[x]$, say

$$m(x) = p_1(x)^{e_1} \ldots p_s(x)^{e_s}$$

where the $\{p_i\}$ are distinct primes and the e_i are positive integers. By Lemma (23.2) the null spaces

$$n(p_i(T)^{e_i}), \qquad 1 \leq i \leq s,$$

are T-subspaces of V. The theorem asserts simply that V is their direct sum. As we shall see, although it is by no means the best theorem that can be proved in this direction, this theorem already goes a long way toward solving the problem of finding a basis of V such that the matrix of T with respect to this basis is as simple as possible. The result is often called the *primary decomposition theorem*.

(23.9) THEOREM. *Let $T \in L(V, V)$, $I \neq 0$, and let*

$$m(x) = p_1(x)^{e_1} \cdots p_s(x)^{e_s}$$

be the minimal polynomial of T, factored into powers of distinct primes $p_i(x) \in F[x]$. Then there exist polynomials $\{f_1(x), \ldots, f_s(x)\}$ in $F[x]$ such that linear transformation $E_i = f_i(T)$, $1 \leq i \leq s$, satisfy

$$E_i \neq 0, \quad 1 \leq i \leq s,$$
$$1 = E_1 + \cdots + E_s,$$
$$E_iE_j = E_jE_i = 0, \quad i \neq j,$$

and $$E_iV = n(p_i(T)^{e_i}), \quad \textit{the null space of } p_i(T)^{e_i}, 1 \leq i \leq s.$$

The subspaces $n(p_i(T)^{e_i})$, $1 \leq i \leq s$, are T-invariant subspaces, and we have

$$V = n(p_1(T)^{e_1}) \oplus \cdots \oplus n(p_s(T)^{e_s}).$$

Proof. Let

$$q_i(x) = \frac{m(x)}{p_i(x)^{e_i}}, \qquad 1 \leq i \leq s.$$

Then the $\{q_i(x)\}$ are polynomials in $F[x]$ with no common prime factors. Hence by Corollary (20.16) there exist polynomials $a_i(x)$, $1 \leq i \leq s$, such that

$$1 = q_1(x)a_1(x) + \cdots + q_s(x)a_s(x).$$

Substituting T, we have by Lemma (22.2) the result that

$$1 = q_1(T)a_1(T) + \cdots + q_s(T)a_s(T).$$

Now let $f_i(x) = q_i(x)a_i(x)$, $1 \leq i \leq s$, and let $E_i = f_i(T)$. We have already proved that

$$1 = E_1 + \cdots + E_s.$$

Next, we see that if $i \neq j$,

$$E_iE_j = q_i(T)a_i(T)q_j(T)a_j(T) = 0$$

since $m(x) \mid q_i(x)q_j(x)$ if $i \neq j$, and hence $q_i(T)q_j(T) = 0$.

We have, for all $v \in V$,

$$p_i(T)^{e_i}E_iv = p_i(T)^{e_i}q_i(T)a_i(T)v = m(T)a_i(T)v$$
$$= 0,$$

proving that $E_iV \subset n(p_i(T)^{e_i})$. Moreover, $V = E_1V + \cdots + E_sV$ from what has been proved. If $E_iV = 0$ for some i, then $V = \sum_{j \neq i} E_jV$, and $q_i(T)V = \sum_{j \neq i} q_i(T)E_jV = 0$, since $m(x) \mid q_i(x)q_j(x)$, so that

$$q_i(T)E_j = q_i(T)q_j(T)a_j(T) = 0$$

Then $q_i(T) = 0$, contradicting the assumption that $m(x)$ is the minimal polynomial of T. Therefore $E_iV \neq 0$ for $1 \leq i \leq s$. By Lemma (23.6), we have $V = E_1V \oplus \cdots \oplus E_sV$. Moreover, each subspace E_iV is T-invariant since

$$TE_iV = Tf_i(T)V = f_i(T)TV \subset E_iV, \qquad 1 \leq i \leq s.$$

The only statement remaining to be proved is that $E_iV = n(p_i(T)^{e_i})$. We have already shown that $E_iV \subset n(p_i(T)^{e_i})$. Now let $v \in n(p_i(T)^{e_i})$, and express v in the form $v = E_1v_1 + \cdots + E_sv_s$ for some $v_i \in V$. Then

$$p_i(T)^{e_i}v = p_i(T)^{e_i}E_1v_1 + \cdots + p_i(T)^{e_i}E_sv_s = 0,$$

and since V is the direct sum of the T-invariant subspaces E_iV, we have

$$p_i(T)^{e_i}E_1v_1 = \cdots = p_i(T)^{e_i}E_sv_s = 0.$$

For $i \neq j$, we have

$$p_i(T)^{e_i}E_jv_j = p_j(T)^{e_j}E_jv_j = 0.$$

Since $p_i(x)$ and $p_j(x)$ are distinct primes, there exist polynomials $a(x)$ and $b(x)$ such that

$$1 = a(x)p_i(x)^{e_i} + b(x)p_j(x)^{e_j}.$$

Substituting T and applying both sides to E_jv, we have

$$E_jv_j = a(T)p_i(T)^{e_i}E_jv_j + b(T)p_j(T)^{e_j}E_jv_j = 0.$$

Therefore

$$v = E_1v_1 + \cdots + E_sv_s = E_iv_i \in E_iV,$$

and the theorem is proved.

As a first application of this theorem, we consider the question of when a basis for V can be chosen that consists of characteristic vectors of T.

(23.10) DEFINITION. A linear transformation $T \in L(V, V)$ is called *diagonable* if there exists a basis for V consisting of characteristic vectors of T. A matrix of T with respect to a basis of characteristic vectors is called a *diagonal matrix*; it has the form

$$\begin{pmatrix} \alpha_1 & & 0 \\ & \ddots & \\ 0 & & \alpha_n \end{pmatrix}, \qquad \alpha_i \in F,$$

with zeros except in the (i, i) positions, $1 \leq i \leq n$.

(23.11) THEOREM. *A linear transformation $T \in L(V, V)$ is diagonable if and only if the minimal polynomial of T has the form*

$$m(x) = (x - \xi_1) \cdots (x - \xi_s)$$

with distinct zeros $\xi_1, \ldots, \xi_s$ in F.

Proof. First suppose that T is diagonable and let $\{v_1, \ldots, v_n\}$ be a basis of T consisting of characteristic vectors belonging to characteristic roots $\xi_1, \ldots, \xi_n$ in F. Suppose the v_i are so numbered that $\xi_1, \ldots, \xi_s$ are distinct and every characteristic root ξ_j coincides with some ξ_i such that $1 \le i \le s$. Let

$$m(x) = (x - \xi_1) \cdots (x - \xi_s).$$

Since $T(v_i) = \xi_i v_i$, $1 \le i \le s$, we have

$$(T - \xi_i \cdot 1)v_i = 0$$

and hence

$$m(T)v_i = 0, \qquad 1 \le i \le n.$$

Therefore $m(T) = 0$ and by Lemma (22.3) the minimal polynomial is a factor of $m(x)$. But it is clear that if any prime factor of $m(x)$ is deleted we obtain a polynomial $m^*(x)$ such that $m^*(T) \neq 0$. For example, if $m^*(x) = (x - \xi_1) \cdots (x - \xi_{s-1})$, then

$$\begin{aligned} m^*(T)v_s &= (T - \xi_1) \cdots (T - \xi_{s-1})v_s \\ &= (\xi_s - \xi_1) \cdots (\xi_s - \xi_{s-1})v_s \neq 0. \end{aligned}$$

It follows that $m(x)$ is the minimal polynomial of T.

Now suppose that the minimal polynomial has the form

$$m(x) = (x - \xi_1) \cdots (x - \xi_s)$$

with distinct $\{\xi_i\}$ in F. By Theorem (23.9) we have

$$V = n(T - \xi_1 \cdot 1) \oplus \cdots \oplus n(T - \xi_s \cdot 1).$$

Let $\{v_{11}, \ldots, v_{1d_1}\}$ be a basis for $n(T - \xi_1 \cdot 1)$, $\{v_{21}, \ldots, v_{2d_2}\}$ a basis for $n(T - \xi_2 \cdot 1)$, etc. Then because V is the direct sum of the subspaces $n(T - \xi_i \cdot 1)$ it follows that

$$\{v_{11}, \ldots, v_{1d_1}, v_{21}, \ldots, v_{2d_2}, \ldots\}$$

is a basis for V. Finally, $w \in n(T - \xi_i \cdot 1)$ implies that $(T - \xi_i \cdot 1)w = 0$ or $T(w) = \xi_i \cdot w$, so that if $w \neq 0$ then w is a characteristic vector of T. Thus all the basis vectors v_{ij} are characteristic vectors of T, and the theorem is proved.

It is worthwhile to translate theorems on linear transformations into theorems on matrices. We have:

(23.12) COROLLARY. *A nececessary and sufficient condition for a matrix* $\mathbf{A} \in M_n(F)$ *to be similar to a diagonal matrix is that the minimal polynomial of* $\mathbf{A}$ *have the form*

$$m(x) = (x - \xi_1) \cdots (x - \xi_s)$$

with distinct $\xi_i \in F$.

We conclude this section with the following extension of Theorem (23.11), to sets of commuting diagonable linear transformations.

(23.13) THEOREM. *Let* $\{S_1, S_2, \ldots, S_k\}$ *be a set of diagonable linear transformations of* V, *such that* $S_iS_j = S_jS_i$ *for* $1 \le i, j \le k$. *Then there exists a basis of* V *such that the basis vectors are characteristic vectors simultaneously for the linear transformations* $S_1, \ldots, S_k$.

Proof. We use induction on dim V, the result being clear if dim $V = 1$. First suppose each S_i has only one characteristic root; then $S_i = \alpha_i 1$, $1 \le i \le k$, and any basis of V will consist of characteristic vectors for $\{S_1, \ldots, S_k\}$. We may now suppose that S_1, say, has more than one characteristic root, and that the theorem is true for sets of linear transformations acting on lower dimensional vector spaces and satisfying the hypothesis of the theorem. Let

$$m(x) = (x - \alpha_1) \cdots (x - \alpha_r), \qquad r > 1,$$

be the minimal polynomial of S_1. Then by Theorem (23.9),

$$V = V_1 \oplus \cdots \oplus V_r, \; V_i = n(S_1 - \alpha_i 1), \qquad 1 \le i \le r,$$

where each of the subspaces is different from zero and invariant relative to S_1. We shall prove that because the $\{S_i\}$ commute, each subspace V_i is invariant with respect to all the S_i, $1 \le i \le k$. By Theorem (23.9), we have $V_i = E_iV$, where $E_i = f_i(S_1)$ for some polynomial $f_i(x) \in F[x]$. Then $S_iS_j = S_jS_i$ implies $S_jE_i = E_iS_j$ and hence $S_jV_i = S_jE_iV = E_iS_jV \subset E_iV = V_i$, for $1 \le i \le r$, $1 \le j \le k$. Each subspace V_i has smaller dimension than V, because $r > 1$ and $V_i \ne 0$. Moreover, each S_j acts as a diagonable linear transformation on V_i, because the minimal polynomial of each S_j acting on V_i will divide the minimal polynomial of S_j on V, so that the conditions of Theorem (23.11) will be satisfied for the transformations $\{S_j\}$ acting on V_i, $1 \le i \le r$. By the induction hypothesis. each subspace $\{V_i\}$, $1 \le i \le r$, has a basis consisting of vectors which are characteristic vectors for all the $\{S_j\}$. These bases taken together form a basis of V with the required property, and the theorem is proved.

EXERCISES

1. Test the following matrices to determine whether or not they are similar to diagonal matrices in $M_2(R)$.

$$\begin{pmatrix} 2 & 1 \\ 0 & -1 \end{pmatrix}, \quad \begin{pmatrix} 1 & -2 \\ 1 & -1 \end{pmatrix}, \quad \begin{pmatrix} 3 & -2 \\ 2 & -1 \end{pmatrix}.$$

2. Show that the matrices

$$\begin{pmatrix} 1 & -2 \\ 1 & -1 \end{pmatrix}, \quad \begin{pmatrix} 0 & 1 \\ -1 & 0 \end{pmatrix}$$

are similar to diagonal matrices in $M_2(C)$, where C is the complex field, but not in $M_2(R)$. In each case find a basis of C_2 consisting of characteristic vectors for linear transformations defined by the matrices.

3. Show that the matrix

$$\begin{pmatrix} 0 & 0 & 1 \\ 1 & 0 & 0 \\ 0 & 1 & 0 \end{pmatrix}$$

is similar to a diagonal matrix in $M_3(C)$ but not in $M_3(R)$.

4. Show that the differentiation transformation $D: P_n \to P_n$ is not diagonable. (Recall that P_n has been used to denote the set of all polynomials $\alpha_0 + \alpha_1 x + \cdots + \alpha_n x^n$, with $\alpha_i \in R$.)

5. Let V_1 and V_2 be nonzero subspaces of a vector space V. Prove that $V = V_1 \oplus V_2$ if and only if $V = V_1 + V_2$ and $V_1 \cap V_2 = \{0\}$.

6. Let $T \in L(V, V)$ be a linear transformation such that $T^2 = 1$. Prove that $V = V_+ \oplus V_-$, where $V_+ = \{v \in V \mid T(v) = v\}$ and $V_- = \{v \in V \mid T(v) = -v\}$.

7. Show that the following converse of Lemma (23.6) holds. Let V be a direct sum, $V = V_1 \oplus \cdots \oplus V_s$, of nonzero subspaces V_i. Show that there exist linear transformations $E_1, \ldots, E_s$ such that $\sum E_i = 1$, $E_i E_j = E_j E_i = 0$, $i \neq j$, and $V_i = E_i V$, $1 \leq i \leq s$. (*Hint*: If $v = \sum v_i$, $v_i \in V_i$, set $E_i v = v_i$.)

8. Let $T \in L(V, V)$ have the minimal polynomial $m(x) \in F[x]$. Let $f(x)$ be an arbitrary polynomial in $F[x]$. Prove that

$$n[f(T)] = n[d(T)],$$

where $d(x)$ is the greatest common divisor of $f(x)$ and $m(x)$.

24. THE TRIANGULAR FORM THEOREM

There are two ways in which a linear transformation T can fail to be diagonable. One is that its minimal polynomial $m(x)$ cannot be factored into linear factors in $F[x]$ (for example, if $m(x) = x^2 + 1$ in $R[x]$

where R is the real field); the other is that $m(x) = (x - \xi_1)^{e_1} \cdots (x - \xi_s)^{e_s}$ with some $e_i > 1$. In the latter case it is desirable to have a theorem that applies to *all* linear transformations and that comes as close to the diagonal form theorem (23.11) as possible.

The main theorem is the following one.

(24.1) THEOREM (TRIANGULAR FORM THEOREM). *Let $T \in L(V, V)$, where V is a finite dimensional vector space over an arbitrary field F, and let the minimal polynomial of T be given by*

$$m(x) = (x - \alpha_1)^{e_1} \cdots (x - \alpha_s)^{e_s}$$

where the $\{e_i\}$ are positive integers and the $\{\alpha_i\}$ are distinct elements of F. Then there exists a basis of V such that the matrix $\mathbf{A}$ of T with respect to this basis has the form

$$\mathbf{A} = \begin{pmatrix} \mathbf{A}_1 & & & 0 \\ & \mathbf{A}_2 & & \\ & & \ddots & \\ 0 & & & \mathbf{A}_s \end{pmatrix}$$

where each $\mathbf{A}_i$ is a d_i-by-d_i block for some integer $d_i \geq e_i$, where $1 \leq i \leq s$, and each $\mathbf{A}_i$ can be expressed in the form

$$\mathbf{A}_i = \begin{pmatrix} \alpha_i & & * \\ & \ddots & \\ 0 & & \alpha_i \end{pmatrix}$$

where the matrix $\mathbf{A}_i$ has zeros below the diagonal and, possibly, nonzero entries $(*)$ above. All entries of $\mathbf{A}$ not contained in one of the blocks $\{\mathbf{A}_i\}$ are zero. To express it all in another way, given a square matrix $\mathbf{B}$ whose minimal polynomial is $m(x)$, there exists an invertible $\mathbf{S}$ such that $\mathbf{SBS}^{-1} = \mathbf{A}$ where $\mathbf{A}$ has the above-given form.

Remark. A very important observation is that the hypothesis of Theorem (24.1)—that the minimal polynomial of T can be factored into linear factors—is always satisfied if the field F is algebraically closed (see Section 21). The most common example of an algebraically closed field (and the only one we have discussed) is the field of complex numbers.

Proof. Let V_i be the null space of $(T - \alpha_i \cdot 1)^{e_i}$, for $1 \leq i \leq r$, and let d_i be the dimension of V_i. If we choose bases for the subspaces $V_1, V_2, \ldots, V_s$ separately, then, as we pointed out in the proof of Theorem (23.11), the totality of basis elements obtained form a basis of V because V is, by Theorem (23.9), the direct sum of the subspaces $\{V_i\}$. Then let us arrange a basis for V so that the first d_1 elements form a basis for V_1, the next d_2 elements form a basis for V_2, and so on. Since each subspace V_i is invariant relative to T, it is clear that the matrix of T relative to this basis has the form

$$\begin{pmatrix} \mathbf{A}_1 & & 0 \\ & \ddots & \\ 0 & & \mathbf{A}_s \end{pmatrix}.$$

It remains only to prove that the blocks $\mathbf{A}_i$ can be chosen to have the required form and that the inequalities $d_i \geq e_i$ hold, for $1 \leq i \leq s$.

Each space V_i is the null space of $(T - \alpha_i \cdot 1)^{e_i}$. In other words, if we let $N_i = T - \alpha_i \cdot 1$, then $N_i \in L(V_i, V_i)$ and we have $N_i^{e_i} = 0$.

Let us give a formal definition of this important concept.

DEFINITION. A linear transformation $T \in L(V, V)$ is called *nilpotent* if $T^k = 0$ for some positive integer k.

Returning to our situation, we have shown that T, viewed as a linear transformation on V_i, is the sum of a constant times the identity transformation, $\alpha_i \cdot 1$, and a nilpotent transformation N_i.

We now state a general result concerning nilpotent transformations, which will settle our problem.

(24.2) LEMMA. *Let N be a nilpotent transformation on a finite dimensional vector space W; then W has a basis $\{w_1, \ldots, w_t\}$ such that*

$$Nw_1 = 0, \qquad N(w_2) \in S(w_1), \ldots, N(w_i) \in S(w_1, \ldots, w_{i-1})$$

for $2 \leq i \leq t$.

We next present a proof of Lemma (24.2). Notice that the matrix of N with respect to the basis $\{w_1, \ldots, w_t\}$ has the form (illustrated for $t = 4$),

$$\begin{pmatrix} 0 & * & * & * \\ 0 & 0 & * & * \\ 0 & 0 & 0 & * \\ 0 & 0 & 0 & 0 \end{pmatrix}$$

so that the lemma is a special case of the triangular form theorem. The point is that this special case implies the whole theorem. We prove Lemma (24.2) by induction. First find $w_1 \neq 0$ such that $Nw_1 = 0$. Any vector will do for w_1 if $N = 0$ and, if $N^k \neq 0$ and $N^{k+1} = 0$, then let $w_1 = N^k(w) \neq 0$. Then $N(w_1) = N^{k+1}(w) = 0$. Suppose (as an induction hypothesis) that we have found linearly independent vectors $\{w_1, \ldots, w_i\}$ satisfying the conditions of the lemma and let $S = S(w_1, \ldots, w_i)$. If $S = W$, there is nothing more to prove. If $S \neq W$ and $N(W) \subset S$, then any vector not in S can be taken for w_{i+1}. Now suppose $N(W) \not\subset S$. Then there is an integer u such that $N^u(W) \not\subset S$, $N^{u+1}(W) \subset S$. Find $w_{i+1} \in N^u(W)$ such that $w_{i+1} \notin S$. Then $\{w_1, \ldots, w_{i+1}\}$ is a linearly independent set and $N(w_{i+1}) \in S$. This completes the proof of the lemma.

Let us apply Lemma (24.2) to the task of selecting an appropriate basis for the subspace V_i of V. Since $N_i = T - \alpha_1 \cdot 1$ is nilpotent on V_i, the lemma implies that V_i has a basis $\{v_{i,1}, \ldots, v_{i,d_i}\}$ such that

$$N_i(v_{i,1}) = 0, \quad N_i(v_{i,2}) \in S(v_{i,1}), \quad \ldots, \quad N_i(v_{i,k}) \subset S(v_{i,1}, \ldots, v_{i,k-1}) \quad \text{for } 2 \leq k \leq d_i.$$

Since $N_i = T - \alpha_i \cdot 1$, these equations yield the formulas

$$(T - \alpha_i \cdot 1)v_{i,1} = 0, \quad \ldots, \quad (T - \alpha_i \cdot 1)v_{i,k} \in S(v_{i,1}, \ldots, v_{i,k-1}), \quad \text{for } 2 \leq k \leq d_i.$$

These in turn yield

$$\begin{aligned} Tv_{i,1} &= \alpha_i v_{i,1} \\ Tv_{i,2} &= \alpha_{12} v_{i,1} + \alpha_i v_{i,2} \\ &\cdots\cdots\cdots\cdots\cdots\cdots \\ Tv_{i,k} &= \alpha_{1k} v_{i,1} + \cdots + \alpha_{k-1,k} v_{i,k-1} + \alpha_i v_{i,k} \end{aligned} \tag{24.3}$$

and we have shown that, relative to this basis, the matrix of T on the space V_i has the required form.

It remains to prove the inequalities $d_i \geq e_i$, $1 \leq i \leq r$. From Lemma (24.2) it follows that $N_i^{d_i} = 0$, where d_i is the dimension of V_i. Since $T = \alpha_i \cdot 1 + N_i$ on V_i, we have $(T - \alpha_i \cdot 1)^{d_i} = 0$ on V_i; and since V is the direct sum of the subspaces V_i, we have

$$(T - \alpha_1 \cdot 1)^{d_1} \cdots (T - \alpha_r \cdot 1)^{d_r} = 0. \tag{24.4}$$

Therefore, the minimal polynomial $m(x) = \Pi(x - \alpha_i)^{e_i}$ divides the polynomial $\Pi(x - \alpha_i)^{d_i}$, and from the theory of unique factorization in $F[x]$ we have $d_i \geq e_i$. This completes the proof of the theorem.

This result has many important corollaries. The first shows that the minimal polynomial has degree $\leq \dim V$, as we were led to expect in Examples A and B of Section 22.

(24.5) COROLLARY. *Let T be a linear transformation on a finite dimensional vector space V over an algebraically closed field F; then the minimal polynomial $m(x)$ of T has* degree $\leq \dim V$.

We recall that an element $\alpha \in F$ is called a characteristic root of T if there exists a nonzero vector $v \in V$ such that $Tv = \alpha v$ or, in other words, if $(T - \alpha \cdot 1)v = 0$. By Theorem (22.10), α is a characteristic root of T if and only if $D(\mathbf{A} - \alpha\mathbf{I}) = 0$, where $\mathbf{A}$ is the matrix of T with respect to some basis. This fact leads to the following definition.

(24.6) DEFINITION. Let $T \in L(V, V)$ and let $\mathbf{A}$ be a matrix of T with respect to some basis of V; then the matrix $x\mathbf{I} - \mathbf{A}$ has entries in the quotient field† (see Section 20) of the polynomial ring $F[x]$ and its determinant $h(x) = D(x\mathbf{I} - \mathbf{A})$ is called the *characteristic polynomial* of T.

We first prove several statements involving the definition.

Proposition. *The characteristic polynomial $h(x) \in F[x]$, and is independent of the choice of the matrix of T. The set of distinct zeros of $h(x)$ is identical with the set of distinct characteristic roots of T.*

Proof. Let $\mathbf{A}$ and $\mathbf{B}$ be matrices of T with respect to different bases. Then $\mathbf{B} = \mathbf{SAS}^{-1}$ for an invertible matrix $\mathbf{S}$ and we have

$$\begin{aligned} D(x\mathbf{I} - \mathbf{B}) &= D(x\mathbf{I} - \mathbf{SAS}^{-1}) \\ &= D[\mathbf{S}(x\mathbf{I} - \mathbf{A})\mathbf{S}^{-1}] \\ &= D(\mathbf{S})D(x\mathbf{I} - \mathbf{A})D(\mathbf{S})^{-1} \qquad [\text{by}(18.3)] \\ &= D(x\mathbf{I} - \mathbf{A}). \end{aligned}$$

The statement about the zeros of the characteristic polynomial is clear from the introductory remarks. The fact that $D(x\mathbf{I} - \mathbf{A}) \in F[x]$ follows from the formula for the complete expansion of $D(x\mathbf{I} - \mathbf{A})$ given in Section 17.

We now have the following basic corollary of Theorem (24.1).

(24.7) COROLLARY. *Let V be a vector space over an algebraically closed field F. Let $T \in L(V, V)$, let $m(x)$ be the minimal polynomial of T, and let $h(x)$ be the characteristic polynomial; then:*

1. $m(x) \mid h(x)$.
2. *Every zero of $h(x)$ is a zero of $m(x)$.*
3. $h(T) = 0$.

(The last statement is called the Cayley-Hamilton theorem.)

† The matrix $x\mathbf{I} - \mathbf{A}$ actually has entries in $F[x]$; the statement about the quotient field is needed because the determinant of a matrix has been defined only for matrices with entries in a field.

Proof. Let $\mathbf{A}$ be the matrix of T given in Theorem (24.1). Then

$$x\mathbf{I} - \mathbf{A} = \begin{matrix} \uparrow \\ d_1 \\ \downarrow \\ \uparrow \\ d_2 \\ \downarrow \end{matrix} \left(\begin{array}{c|c|c} \begin{matrix} x - \alpha_1 & & * \\ & \ddots & \\ 0 & & x - \alpha_1 \end{matrix} & & \\ \hline & \begin{matrix} x - \alpha_2 & & * \\ & \ddots & \\ 0 & & x - \alpha_2 \end{matrix} & \\ \hline & & \ddots \end{array} \right)$$

Therefore the characteristic polynomial is

$$h(x) = D(x\mathbf{I} - \mathbf{A}) = \Pi(x - \alpha_i)^{d_i}.$$

On the other hand, the minimal polynomial is given by

$$m(x) = \Pi(x - \alpha_i)^{e_i}.$$

By Theorem (24.1), $d_i \geq e_i$ for all i. All the statements of the corollary follow from these remarks.

We give now a restatement of the Cayley-Hamilton theorem for matrices.

(24.8) THEOREM (CAYLEY-HAMILTON). *Let F be a field contained in the field of complex numbers C, and let $\mathbf{A}$ be an n-by-n matrix with coefficients in F. Then $\mathbf{A}$ satisfies its characteristic equation*

$$D(x\mathbf{I} - \mathbf{A}) = 0.$$

In other words, letting $h(x) = D(x\mathbf{I} - \mathbf{A})$, we have $h(\mathbf{A}) = 0$.

Proof. Since $F \subset C$, $\mathbf{A}$ defines a linear transformation T on an n-dimensional vector space over C, such that $\mathbf{A}$ is the matrix of T with respect to some basis of the vector space. Since the field of complex numbers is algebraically closed, $h(T) = 0$ by part (3) of Corollary (24.7). Therefore $h(\mathbf{A}) = 0$, and the proof is completed.

The Cayley-Hamilton Theorem holds for matrices over an arbitrary field (see Exercise 8 in Section 25).

We conclude this section with two more applications of Theorem (24.1).

(24.9) THEOREM (JORDAN DECOMPOSITION). *Let $T \in L(V, V)$, where V is a finite dimensional vector space over an algebraically closed field F. There exist linear transformations D and N on V such that*

a. $T = D + N$.

b. *D is diagonable and N is nilpotent.*
c. *There exist polynomials $f(x)$ and $g(x) \in F[X]$ such that $D = f(T)$ and $N = g(T)$.*

The transformations D and N are uniquely determined in the sense that if D' and N' are diagonable and nilpotent transformations, respectively, such that $T = D' + N'$ and $D'N' = N'D'$, then $D' = D$ and $N' = N$.

Proof. Since F is algebraically closed, the minimal polynomial of T has the form

$$m(x) = (x - \alpha_1)^{e_1} \cdots (x - \alpha_s)^{e_s},$$

where $\alpha_1, \ldots, \alpha_s$ are the distinct characteristic roots. By Theorem (23.9), there exist polynomials $f_1(x), \ldots, f_s(x)$ such that if $E_i = f_i(T)$, $1 \leq i \leq s$, then

$$V = E_1V \oplus \cdots \oplus E_sV \qquad \text{and} \qquad E_iV = n((T - \alpha_i1)^{e_i}).$$

Let

$$D = \alpha_1E_1 + \cdots + \alpha_sE_s;$$

then $D = f(T)$ for some polynomial $f(x)$. Moreover D is a diagonable linear transformation. If $v_i \in V_i = E_iV$, $1 \leq i \leq s$, we have, setting $v_i = E_iv$,

$$\begin{aligned} Dv_i &= (\alpha_1E_1 + \cdots + \alpha_sE_s)E_iv \\ &= \alpha_iE_iE_iv = \alpha_iE_iv = \alpha_iv_i, \end{aligned}$$

using the properties of the transformations $\{E_i\}$ given in Theorem (23.9), and it follows that D is diagonable. Now let

$$N = T - D.$$

Then $N = g(T)$ for the polynomial $g(x) = x - f(x)$. We shall prove that N is nilpotent. It is sufficient to prove that $N^kv_i = 0$ for $v_i \in V_i$, $1 \leq i \leq s$, for some sufficiently large k. We have, setting $v_i = E_iv$, for $v \in V$,

$$Nv_i = (T - D)v_i = (T - \alpha_i)v_i$$

since $Dv_i = \alpha_iv_i$ by a previous computation. Then

$$N^{e_i}v_i = (T - \alpha_i)^{e_i}v_i = 0$$

since $v_i \in n((T - \alpha_i1)^{e_i})$, and hence N is nilpotent.

Now we come to the uniqueness part of the theorem. Suppose $T = D' + N'$, and D', N' satisfy the hypothesis of the last part of the theorem. Since $D'N' = N'D'$, we have $TD' = D'T$, and $TN' = N'T$. Therefore $D'D = DD'$ and $N'N = NN'$, since N and D are polynomials in T. From

$$T = D' + N' = D + N,$$

we obtain

$$D' - D = N - N'.$$

Since N and N' commute, we can apply the binomial theorem to show that

$$(N - N')^k = N^k - \binom{k}{1}N^{k-1}N' + \cdots (-1)^k(N')^k.$$

A typical term is $N^i(N')^j$, with $i + j = k$. If k is large enough, then it will follow that either $N^i = 0$ or $(N')^j = 0$, and hence $N - N'$ is nilpotent. On the other hand, by Theorem (23.13), there exists a basis of V consisting of characteristic vectors for D and D', and hence for $D' - D$. The matrix of $D' - D$ with respect to this basis will have the form

$$\begin{pmatrix} \beta_1 & & & 0 \\ & \beta_2 & & \\ & & \ddots & \\ 0 & & & \beta_n \end{pmatrix}.$$

It follows that since $D' - D = N - N'$ is nilpotent, some power of each β_i is zero. Therefore $\beta_1 = \cdots = \beta_n = 0$, and $D' - D = N - N' = 0$. This completes the proof of the theorem.

Of course, the hypothesis that F is algebraically closed isn't really used in the preceding theorem—all that is needed is that the characteristic roots of T all belong to F.

The Jordan decomposition of a linear transformation T is useful for computing the powers of T. The powers of T can be computed in terms of D and N using the binomial theorem, since the transformations D and N commute. Thus, if $T = D + N$, $T^2 = D^2 + ND + DN + N^2 = D^2 + 2ND + N^2$, and in general,

$$T^k = D^k + \binom{k}{1}D^{k-1}N + \cdots + \binom{k}{k-1}DN^{k-1} + N^k.$$

The powers of D are computed by taking the powers of the diagonal elements of a matrix corresponding to D, while all powers of N from some point on are zero. These remarks are applied in Section 34 to the problem of calculating the exponential of a matrix.

Finally we have the following result.

(24.10) THEOREM. *Let T be a linear transformation on an n-dimensional vector space V over an algebraically closed field F. Then the characteristic polynomial $h(x)$ of T can be factored in the form*

$$h(x) = (x - \alpha_1)^{d_1} \cdots (x - \alpha_s)^{d_s},$$

where $\{\alpha_1, \ldots, \alpha_s\}$ *are the distinct characteristic roots of T. Letting*

$$h(x) = x^n - \gamma_1 x^{n-1} + \cdots + (-1)^n \gamma_n, \qquad \gamma_i \in F,$$

be the expansion of $h(x)$ *in terms of the powers of x, we have*

$$\gamma_1 = d_1\alpha_1 + \cdots + d_s\alpha_s \qquad \text{and} \qquad \gamma_n = \alpha_1^{d_1} \cdots \alpha_s^{d_s}.$$

Moreover, if $\mathbf{B} = (\beta_{ij})$ *is the matrix of T with respect to an arbitrary basis of the vector space, then*

$$\gamma_1 = \sum_{i=1}^{n} \beta_{ii} \qquad \text{and} \qquad \gamma_n = D(\mathbf{B}).$$

Proof. The fact that $h(x)$ can be factored in the form

$$h(x) = (x - \alpha_1)^{d_1} \cdots (x - \alpha_s)^{d_s}$$

was proved in Corollary (24.7). The formulas

$$\gamma_1 = \sum_{i=1}^{s} d_i\alpha_i, \qquad \gamma_n = \prod_{i=1}^{s} \alpha_i^{d_i}$$

are obtained by expanding the right-hand side of the formula for $h(x)$ and comparing coefficients.

Now let $\mathbf{B} = (\beta_{ij})$ be the matrix of T with respect to an arbitrary basis. We have shown that the characteristic polynomial is independent of the choice of the basis. Therefore

$$h(x) = \begin{vmatrix} x - \beta_{11} & -\beta_{12} & \cdots & -\beta_{1n} \\ -\beta_{21} & x - \beta_{22} & \cdots & -\beta_{2n} \\ \cdots & \cdots & \cdots & \cdots \\ -\beta_{n1} & & \cdots & x - \beta_{nn} \end{vmatrix}.$$

The constant term γ_n of $h(x)$ is $h(0) = (-1)^n D(\mathbf{B})$. One way to obtain the fact that $\gamma_{n-1} = \beta_{11} + \cdots + \beta_{nn}$ is to use the complete expansion of the determinant [see (17.12) or (19.22)]. The terms in the complete expansion are products of elements from the first row, j_1 column, second row, j_2 column, etc., multiplied by $D(e_{j_1}, \ldots, e_{j_1})$. In order to have a nonzero term, all the columns must be different. The only nonzero term in the complete expansion of $h(x)$ which can contribute to the coefficient of x^{n-1} is

$$(x - \beta_{11})(x - \beta_{22}) \cdots (x - \beta_{nn}).$$

The sign associated with this term is $D(e_1, \ldots, e_n) = 1$. The coefficient of x^{n-1} in $(x - \beta_{11}) \cdots (x - \beta_{nn})$ is $-(\beta_{11} + \cdots + \beta_{nn})$. This completes the proof of the theorem.

The last part of the preceding theorem shows the importance of another numerical valued function of a matrix besides the determinant.

(24.11) DEFINITION. Let T be a linear transformation on a finite dimensional vector space V. The *trace* of T, $\mathrm{Tr}(T)$, is defined to be the sum of the diagonal elements $\sum \beta_{ii}$ of a matrix $\mathbf{B}$ of T with respect to some basis of the vector space. The trace $\mathrm{Tr}(T)$ is independent of the choice of the matrix $\mathbf{B}$. If $\mathbf{B} = (\beta_{ij})$ is an n-by-n matrix, then the trace of $\mathbf{B}$, $\mathrm{Tr}(\mathbf{B})$, is defined by $\mathrm{Tr}(\mathbf{B}) = \sum \beta_{ii}$.

The preceding theorem shows that $\mathrm{Tr}(T)$ is independent of the choice of the matrix $\mathbf{B}$, since $\mathrm{Tr}(T)$ is the coefficient of a power of x in the characteristic polynomial of T, which has already been shown to be independent of the matrix $\mathbf{B}$.

Example A. Let V be a three-dimensional vector space over the complex field C with basis $\{v_1, v_2, v_3\}$ and let $T \in L(V, V)$ be defined by the equations

$$\begin{aligned} Tv_1 &= -v_1 + 2v_3 \\ Tv_2 &= 3v_1 + 2v_2 + v_3 \\ Tv_3 &= -v_3. \end{aligned}$$

The matrix of T with respect to the basis $\{v_1, v_2, v_3\}$ is given by

$$\mathbf{A} = \begin{pmatrix} -1 & 3 & 0 \\ 0 & 2 & 0 \\ 2 & 1 & -1 \end{pmatrix}.$$

We shall show how to find a new basis for V with respect to which the matrix of T is in the form given in Theorem (24.1).

Step 1. Find the distinct characteristic roots of $\mathbf{A}$. This can be done either by finding the prime factors of the characteristic polynomial $h(x)$ or by finding the minimal polynomial $m(x)$ and determining its zeros since, by Corollary (24.7), $m(x)$ and $h(x)$ have the same set of distinct zeros.

The characteristic polynomial $h(x)$ is given by

$$h(x) = D(x\mathbf{I} - \mathbf{A}) = \begin{vmatrix} x+1 & -3 & 0 \\ 0 & x-2 & 0 \\ -2 & -1 & x+1 \end{vmatrix} = (x+1)^2(x-2).$$

At this point we know from Corollary (24.7) that the distinct characteristic roots of T are $\{-1, 2\}$ and that the minimal polynomial of T is either

$$(1+x)(2-x) \qquad \text{or} \qquad (1+x)^2(2-x).$$

Step 2. Find the null spaces of $T + 1$, $(T+1)^2$, $T - 2$. If V turns out to be the direct sum of the null spaces of $T + 1$ and $T - 2$, we will know that the minimal polynomial is $(x+1)(x-2)$ (why?) and if not then we will know that the minimal polynomial is $(x+1)^2(x-2)$ and will have to find the null space of $(T+1)^2$.

We have

$$\begin{aligned}(T+1)v_1 &= 2v_3\\ (T+1)v_2 &= 3v_1 + 3v_2 + v_3\\ (T+1)v_3 &= 0\end{aligned}$$

The rank of $T + 1$ can now be found by determining the maximal number of linearly independent vectors among $\langle 0, 0, 2\rangle, \langle 3, 3, 1\rangle, \langle 0, 0, 0\rangle$. In this case the number is obviously two and, by Theorem (13.9), the null space of $T + 1$ has dimension $3 - 2 = 1$.

Similarly, we have

$$\begin{aligned}(T-2)v_1 &= -3v_1 + 2v_3\\ (T-2)v_2 &= 3v_1 + v_3\\ (T-2)v_3 &= -3v_3\end{aligned}$$

and we find that rank $(T - 2) = 2$, so that the null space of $T - 2$ has dimension 1.

At this point we have shown that V is not the direct sum of $n(T + 1)$ and $n(T - 2)$. We may conclude that the minimal polynomial is

$$m(x) = (x + 1)^2(x - 2)$$

and that, by Theorem (23.11), it is impossible to find a basis of V with respect to which T has a diagonal matrix.

It remains to find the null spaces of $(T + 1)^2$ and $T - 2$. We have, from the computation of $T + 1$,

$$\begin{aligned}&(T+1)^2v_1 = (T+1)(2v_3) = 0,\\ &(T+1)^2v_2 = (T+1)(3v_1 + 3v_2 + v_3) = 9v_1 + 9v_2 + 9v_3,\\ &(T+1)^2v_3 = 0.\end{aligned}$$

Therefore $\{v_1, v_3\}$ is a basis for the null space of $(T + 1)^2$.

To find the null space of $T - 2$, we may suppose that $v = \xi_1 v_1 + \xi_2 v_2 + \xi_3 v_3 \in n(T - 2)$ and try to find ξ_1, ξ_2, ξ_3. We have

$$(T - 2)v = \xi_1(-3v_1 + 2v_3) + \xi_2(3v_1 + v_3) + \xi_3(-3v_3) = 0$$

and we have the following homogeneous system of equations to be solved for the ξ's:

$$\begin{aligned}-3\xi_1 + 3\xi_2 \qquad &= 0\\ 2\xi_1 + \xi_2 - 3\xi_3 &= 0.\end{aligned}$$

This system has the solution vector $\langle 1, 1, 1\rangle$. As a matter of fact, it is clear by inspection that $v_1 + v_2 + v_3$ is in $n(T - 2)$ and, since the dimension of $n(T - 2)$ is one, we know that $v_1 + v_2 + v_3$ generates $n(T - 2)$.

Step 3. Find the matrix of T with respect to the new basis. According to Theorem (24.1), we should find a basis $\{w_1, w_2\}$ for $n[(T + 1)^2]$ such that $(T + 1)w_1 = 0$, $(T + 1)w_2 \in S(w_1)$, and let w_3 be a basis of $n(T - 2)$.

We see that we should have

$$w_1 = v_3, \qquad w_2 = v_1, \qquad w_3 = v_1 + v_2 + v_3.$$

$$T(w_1) = -w_1, \qquad T(w_2) = -w_2 + 2w_1, \qquad T(w_3) = 2w_3.$$

The matrix relating these two bases is

$$\mathbf{S} = \begin{pmatrix} 0 & 1 & 1 \\ 0 & 0 & 1 \\ 1 & 0 & 1 \end{pmatrix}$$

and the matrix of T with respect to $\{w_1, w_2, w_3\}$ is, by the equations above,

$$\mathbf{B} = \begin{pmatrix} -1 & 2 & 0 \\ 0 & -1 & 0 \\ 0 & 0 & 2 \end{pmatrix}.$$

We should now recall that either $\mathbf{SB} = \mathbf{AS}$ or $\mathbf{BS} = \mathbf{SA}$ (to remember which of the two holds is much too hard!). Checking the multiplications we see that

$$\mathbf{SB} = \mathbf{AS}$$

or

$$\mathbf{B} = \mathbf{S}^{-1}\mathbf{AS}.$$

Example B. Find the Jordan decomposition of the *linear transformation* T given in Example C.

Following the proof of Theorem (24.9), we let $\mathbf{B}$ be the matrix of T with respect to $\{w_1, w_2, w_3\}$ given above. Then

$$B = \begin{pmatrix} -1 & 2 & 0 \\ 0 & -1 & 0 \\ 0 & 0 & 2 \end{pmatrix} = \begin{pmatrix} -1 & 0 & 0 \\ 0 & -1 & 0 \\ 0 & 0 & 2 \end{pmatrix} + \begin{pmatrix} 0 & 2 & 0 \\ 0 & 0 & 0 \\ 0 & 0 & 0 \end{pmatrix}.$$

Letting D and N be the linear transformations whose matrices with respect to the basis $\{w_1, w_2, w_3\}$ are

$$\begin{pmatrix} -1 & 0 & 0 \\ 0 & -1 & 0 \\ 0 & 0 & 2 \end{pmatrix} \quad \text{and} \quad \begin{pmatrix} 0 & 2 & 0 \\ 0 & 0 & 0 \\ 0 & 0 & 0 \end{pmatrix},$$

respectively, we have $T = D + N$, with D and N diagonable and nilpotent, respectively, and $DN = ND$. By the uniqueness part of Theorem (24.9), D and N give the Jordan decomposition of T.

We shall now give two examples which show how the trace function is useful in some other parts of mathematics.

Example C. *Incidence matrices and graphs.* A *graph* is a set of objects called *vertices* (or *nodes*), together with a set E of pairs of vertices $\{(v, v')\}$. The set E is assumed to satisfy the conditions

$$(v, v') \in E \text{ implies } v \neq v'.$$
$$(v, v') \in E \text{ implies } (v', v) \in E.$$

The pairs of points in E are called *edges* of the graph. A vertex v and an edge (v', v'') are said to be *incident* if $v = v'$ or $v' = v''$.

A graph can be represented by a diagram. For example, the figure below represents a graph with 5 vertices, and 5 edges {(1, 2), (2, 3), (2, 4), (3, 4), and (4, 5)}. (It is unnecessary to list the other pairs (2, 1), (3, 2), etc., because of our assumptions that $(2, 1) \in E$ if and only if $(1, 2) \in E$.)

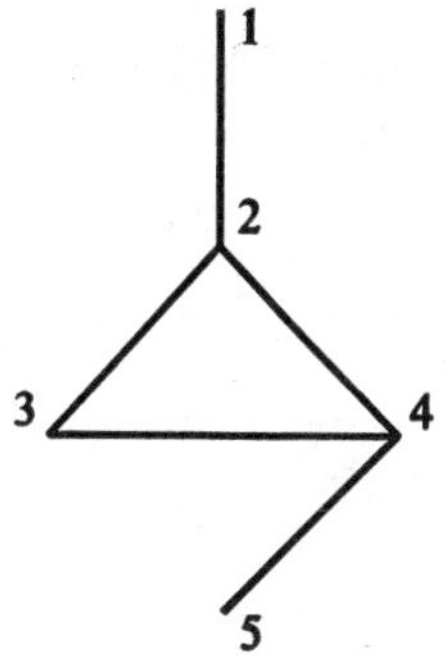

With each graph we can associate an *incidence matrix* $\mathbf{M} = (m_{ij})$ as follows. Assume the vertices are numbered $\{1, 2, \ldots, n\}$. Then $\mathbf{M}$ is an n-by-n matrix, whose entry in the (i, j) position is 1 if (i, j) is an edge of the graph and zero otherwise. The incidence matrix for the graph above is

$$\mathbf{M} = \begin{pmatrix} 0 & 1 & 0 & 0 & 0 \\ 1 & 0 & 1 & 1 & 0 \\ 0 & 1 & 0 & 1 & 0 \\ 0 & 1 & 1 & 0 & 1 \\ 0 & 0 & 0 & 1 & 0 \end{pmatrix}.$$

As an experiment, let's compute some powers of $\mathbf{M}$. We have

$$\mathbf{M}^2 = \begin{pmatrix} 1 & 0 & 1 & 1 & 0 \\ 0 & 3 & 1 & 1 & 1 \\ 1 & 1 & 2 & 1 & 1 \\ 1 & 1 & 1 & 3 & 0 \\ 0 & 1 & 1 & 0 & 1 \end{pmatrix};$$

$$\mathbf{M}^3 = \begin{pmatrix} 0 & 3 & 1 & 1 & 1 \\ 3 & 2 & 4 & 5 & 1 \\ 1 & 4 & 2 & 4 & 1 \\ 1 & 5 & 4 & 2 & 3 \\ 1 & 1 & 1 & 3 & 0 \end{pmatrix}.$$

Letting $\mathbf{M} = (m_{ij})$, we have

$$\mathrm{Tr}\,(\mathbf{M}^2) = \sum_{i=1}^{n} \sum_{j=1}^{n} m_{ij} m_{ji},$$

$$\mathrm{Tr}\,(\mathbf{M}^3) = \sum_{i=1}^{n} \sum_{j=1}^{n} \sum_{k=1}^{n} m_{ij} m_{jk} m_{ki}.$$

Using the definition of the incidence matrix, we know that

$$m_{ij} m_{ji} \neq 0$$

if and only if (i, j) is an edge of the graph, and in that case $m_{ij}m_{ji} = 1$. In case $m_{ij}m_{ji} = 1$, we shall also have $m_{ji}m_{ij} = 1$. Therefore,

$$\mathrm{Tr}\,(\mathbf{M}^2) = 2e,$$

where e is the number of edges in the graph. In our example, $\mathrm{Tr}\,(\mathbf{M}^2) =$ 10, and there are 5 edges.

Turning to $\mathbf{M}^3$, we see that $m_{ij}m_{jk}m_{ki} \neq 0$ only when there is a triangle in the graph, where (i, j, k) form a triangle if (ij), (jk) and (ki)

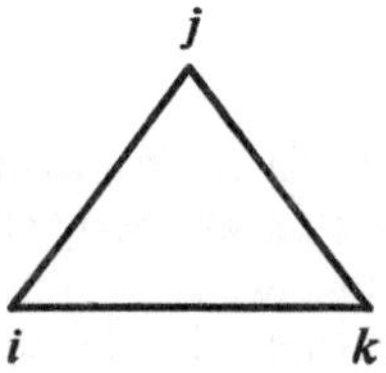

are all edges in the graph. It is easy to check that for each such triangle, the contribution to $\mathrm{Tr}\,\mathbf{M}^3 = 6$. Therefore,

$$\mathrm{Tr}\,(\mathbf{M}^3) = 6t,$$

where t is the number of triangles. In our example,

$$\mathrm{Tr}\,(\mathbf{M}^3) = 6,$$

and there is exactly one triangle.

Example D. Let V be a vector space with basis $\{v_1, \ldots, v_n\}$ over the field of real numbers. Let σ be a permutation of $\{1, 2, \ldots, n\}$, and let $\mathbf{A}_\sigma$ be the matrix of the linear transformation T_σ (defined in Section 19),

$$T_\sigma(v_i) = v_{\sigma(i)}.$$

This time, $\mathrm{Tr}\,(\mathbf{A}_\sigma)$ is equal to the number of basis vectors v_i such that $v_{\sigma(i)} = v_i$. In other words, $\mathrm{Tr}\,(\mathbf{A}_\sigma)$ counts the number of *fixed points* of the permutation σ, where a fixed point is defined to be an integer i, $1 \leq i \leq n$, such that $\sigma(i) = i$.

Examples C and D show the usefulness of the trace function for counting in situations where the data can be given in matrix form.

EXERCISES

1. Let T be a linear transformation on a vector space over the complex numbers such that

$$\begin{aligned} T(v_1) &= -v_1 - v_2 \\ T(v_2) &= v_1 - 3v_2 \end{aligned}$$

where $\{v_1, v_2\}$ is a basis for the vector space.

   a. What is the characteristic polynomial of T?
   b. What is the minimal polynomial of T?
   c. What are the characteristic roots of T?
   d. Does there exist a basis for the vector space consisting of characteristic vectors of T? Explain.
   e. Find a characteristic vector of T.
   f. Find a triangular matrix **B** and an invertible matrix **S** such that **SB** = **AS** where **A** is the matrix of T with respect to the basis $\{v_1, v_2\}$.
   g. Find the Jordan decomposition of T.

2. Answer the questions in Exercise 1 for the linear transformation $T \in L(C_2, C_2)$ defined by the equations

$$\begin{aligned} T(v_1) &= v_1 + iv_2 \\ T(v_2) &= -iv_1 + v_2, \end{aligned}$$

where $\{v_1, v_2\}$ is a basis for C_2.

3. Let V be a two-dimensional vector space over the real numbers R and let $T \in L(V, V)$ be defined by

$$\begin{aligned} T(v_1) &= -\alpha v_2 \\ T(v_2) &= \beta v_1 \end{aligned}$$

where α and β are positive real numbers. Does there exist a basis of V consisting of characteristic vectors of T? Explain.

Note: In Exercises 4 to 8, V denotes a finite dimensional vector space over the complex numbers C.

4. Let $T \in L(V, V)$ be a linear transformation whose characteristic roots are all equal to zero. Prove that T is nilpotent: $T^n = 0$ for some n.
5. Let $T \in L(V, V)$ be a linear transformation such that $T^2 = T$. Discuss whether or not there exists a basis of V consisting of characteristic vectors of T.
6. Answer the question of Exercise 5 for the case of a transformation T such that $T^r = 1$ for some positive integer r.
7. Let T be a linear transformation of rank 1, that is, $\dim T(V) = 1$. Then $T(V) = S(v_0)$ for some vector $v_0 \neq 0$. In particular, $T(v_0) = \lambda v_0$ for some $\lambda \in C$. Prove that

$$T^2 = \lambda T.$$

Does there exist a basis of V con sting of characteristic vectors of T? Explain.

8. Let $T = D + N$ be the Jordan decomposition of a linear transformation $T \in L(V, V)$. Prove that a linear transformation $X \in L(V, V)$ commutes with T, $XT = TX$, if and only if X commutes with D and N.

9. **a.** Show that if **A** and **B** are n-by-n matrices with coefficients in an arbitrary field F, then

$$\operatorname{Tr}(\mathbf{AB}) = \operatorname{Tr}(\mathbf{BA}).$$

b. Use (a) to show that if **A** and **B** are similar matrices, then $\operatorname{Tr}(\mathbf{A}) = \operatorname{Tr}(\mathbf{B})$.

c. Show that the matrices of trace zero form a subspace of $M_n(F)$ of dimension $n^2 - 1$. [*Hint:* The mapping $\operatorname{Tr}: M_n(F) \to F$ is a linear transformation.]

d. Prove that the subspace of $M_n(F)$ defined in (c) is generated by the matrices $\mathbf{AB} - \mathbf{BA}$, $\mathbf{A}, \mathbf{B} \in M_n(F)$.

10. A linear transformation $T \in L(V, V)$, where V is a finite dimensional vector space over an algebraically closed field F, is called *unipotent* if $T - 1$ is nilpotent.

a. Show that T is unipotent if and only if 1 is the only characteristic root of T.

b. Let $T \in L(V, V)$, and suppose that T is invertible. Prove that there exist linear transformations D and U such that $T = DU$, D is diagonable, U is unipotent, and D and U can both be expressed as polynomials in T.

c. Prove that D and U given in (*b*) are uniquely determined in the sense that if $T = D'U'$, with D' diagonable, U' unipotent, and $D'U' = U'D'$, then $D = D'$, and $U = U'$.

25. THE RATIONAL AND JORDAN CANONICAL FORMS

A basic question about matrices is the following one. Given two n-by-n matrices **A** and **B** with coefficients in F, how do we decide whether or not **A** is similar to **B**?

Example A. The following examples show some of the difficulties presented by this question.

(*i*) The matrices

$$\begin{pmatrix} 1 & 0 & 0 \\ 0 & 1 & 0 \\ 0 & 0 & 2 \end{pmatrix} \quad \text{and} \quad \begin{pmatrix} 1 & 0 & 0 \\ 0 & 2 & 0 \\ 0 & 0 & 2 \end{pmatrix}$$

have the same minimal polynomial $(x - 1)(x - 2)$ but are not similar.
(*ii*) The matrices

$$\begin{pmatrix} 1 & 0 \\ 0 & 1 \end{pmatrix} \quad \text{and} \quad \begin{pmatrix} 1 & 1 \\ 0 & 1 \end{pmatrix}$$

have the same characteristic polynomial and are not similar.

In this section we give a solution to the problem of deciding whether or not two matrices are similar. The main theorem, called the elementary divisor theorem, is rather difficult to prove, and we shall postpone the proof until Section 29. In this section we shall concentrate on proving some of the results leading up to the statement of the main theorem, and on the applications of the main theorem to particular cases.

Let V be a finite dimensional vector space over an arbitrary field F, and let T be a fixed linear transformation in $L(V, V)$. In order to exclude trivial cases, we shall assume that $V \neq 0$. It should also be emphasized that no assumption is made about F. Thus the discussion applies to vector spaces over the field of rational numbers Q or the real field R, and not only to an algebraically closed field, as has been the case earlier in this chapter.

The idea on which the whole discussion is based is the concept of a cyclic subspace of V relative to T.

(25.1) DEFINITION. A T-invariant subspace V_1 of V is called *cyclic* relative to T if $V_1 \neq 0$ and there exists a vector $v_1 \in V_1$ such that V_1 is generated by $\{v_1, Tv_1, T^2v_1, \ldots, T^kv_1\}$ for some k.

(25.2) LEMMA. *A subspace $V_1 \subset V$ is cyclic relative to T if for some vector $v_1 \in V_1$, $v_1 \neq 0$, and every vector in V_1 can be expressed in the form $f(T)v_1$, for some polynomial $f \in F[x]$.*

Proof. If V_1 is cyclic, then every vector $v \in V_1$ can be expressed as a linear combination

$$\begin{aligned} v &= \alpha_0 v_1 + \alpha_1(Tv_1) + \cdots + \alpha_k(T^kv_1) \\ &= f(T)v_1, \end{aligned}$$

where $f(x) = \alpha_0 + \alpha_1 x + \cdots + \alpha_k x^k$. Conversely, suppose $v_1 \in V_1$ is such that every $v \in V_1$ can be expressed in the form $f(T)v_1$ for some $f \in F[x]$. Then V_1 is generated by $\{v_1, Tv_1, T^2v_1, \ldots\}$ and by finite dimensionality there exists a k such that V_1 is generated by $\{v_1, Tv_1, \ldots, T^kv_1\}$ as required.

(25.3) DEFINITION. Let $v \in V$, $v \neq 0$, and let V_1 denote the subspace of V consisting of all vectors $\{f(T)v\}$, where f is an arbitrary polynomial in $F[x]$. Then V_1 is called the *cyclic subspace generated by v*, and will be denoted by $\langle v \rangle$.

We note that V_1 is a subspace of V. It is a cyclic subspace, by Lemma (25.2), and is also the uniquely determined smallest T-invariant subspace containing v (*why*?).

(25.4) Lemma. *Let $v \in V, v \neq 0$. There exists a nonzero polynomial in $F[x]$ with leading coefficient 1,†*

$$m_v(x) = x^d - \alpha_{d-1}x^{d-1} - \cdots - \alpha_0, \qquad \alpha_i \in F,$$

such that

$$m_v(T)v = 0$$

and, for every polynomial $f \in F[x]$,

$$f(T)v = 0 \qquad \textit{implies} \qquad m_v(x) \mid f(x).$$

Proof. By finite dimensionality, there exists an integer $d \geq 0$ such that $\{v, Tv, \ldots, T^{d-1}v\}$ are linearly independent (with the understanding that this set reduces to $\{v\}$ if $d = 0$), and such that $\{v, T(v), \ldots, T^d(v)\}$ are linearly dependent. Then there exist elements $\alpha_0, \ldots, \alpha_{d-1}$ in F such that

$$T^d v = \alpha_0 v + \alpha_1 Tv + \cdots + \alpha_{d-1}T^{d-1}v.$$

Letting

$$m_v(x) = x^d - \alpha_{d-1}x^{d-1} - \cdots - \alpha_0,$$

we have $m_v(T)v = 0$. Now suppose $f \in F[x]$ is such that $f(T)v = 0$. By the division algorithm,

$$f(x) = m_v(x)g(x) + r(x)$$

where either $r(x) = 0$ or $\deg r(x) < \deg m_v(x)$. Substitute T for x, and apply the formula to v to obtain

$$f(T)v = m_v(T)g(T)v + r(T)v.$$

Since $f(T)v = 0$ and $m_v(T)v = 0$, we have $r(T)v = 0$. If $r(x) \neq 0$, the equation $r(T)v = 0$ implies that the vectors $\{v, Tv, \ldots, T^{d-1}v\}$ are linearly dependent, contrary to assumption, and the lemma is proved.

(25.5) Definition. Let $v \in V, v \neq 0$. The unique polynomial $m_v(x)$ satisfying the conditions of Lemma (28.4) is called the *order* of v. It is uniquely determined as the polynomial $q(x)$ with leading coefficient 1, of least degree such that $q(T)v = 0$.

---

† The leading coefficient of a polynomial is the coefficient of the highest power of x appearing with a nonzero coefficient.

The fact that $m_v(x)$ is uniquely determined as the polynomial $q(x)$ of least degree such that $q(T)v = 0$ is shown as follows. If $q(T)v = 0$, then $m_v(x) \mid q(x)$. If $q(x)$ has the smallest degree among polynomials $f(x)$ such that $f(T)v = 0$, we must have $\deg m_v(x) = \deg q(x)$. Therefore $q(x) = \alpha m_v(x)$ for some $\alpha \in F$. Since both $q(x)$ and $m_v(x)$ have leading coefficient 1, we have $\alpha = 1$.

By now the reader should have an uneasy feeling that the order $m_v(x)$ is somehow related to the minimal polynomial of T, defined in Section 22. The following result settles this question.

(25.6) LEMMA. *Let $\langle v \rangle$ be a nonzero cyclic subspace of V, and let $T_{\langle v \rangle}$ be the restriction† of T to the subspace $\langle v \rangle$. Then the order $m_v(x)$ of v is equal to the minimal polynomial of $T_{\langle v \rangle}$ (up to a constant factor).*

Proof. A polynomial $f(x) \in F[x]$ has the property that $f(T)v = 0$ if and only if $f(T_{\langle v \rangle}) = 0$, because $\langle v \rangle$ is generated by $\langle v, Tv, T^2v, \dots \rangle$ and $f(T)T = Tf(T)$. Moreover, $f(T)v = f(T_{\langle v \rangle})v$. Letting $q(x)$ be the minimal polynomial of $T_{\langle v \rangle}$, we have $m_v(T) = 0$ on $\langle v \rangle$, and hence $q(x) \mid m_v(x)$. Conversely, $q(T)v = 0$, so that $m_v(x) \mid q(x)$, and hence $m_v(x)$ and $q(x)$ differ by a scalar factor. This completes the proof of the lemma.

Another worthwhile observation is the following one.

(25.7) LEMMA. *Let $m(x)$ be the minimal polynomial of T on V. For every nonzero vector $v \in V$, the order $m_v(x) \mid m(x)$.*

Proof. We have $m(T) = 0$ on V, and hence $m(T)v = 0$. The result now follows from Lemma (25.4).

The first main theorem of this Section can now be stated.

(25.8) THEOREM (ELEMENTARY DIVISOR THEOREM). *Let $T \in L(V, V)$, where V is a finite dimensional vector space over an arbitrary field, and $V \neq 0$. Then there exist nonzero vectors $\{v_1, v_2, \dots, v_r\}$ in V, whose orders are powers of prime polynomials in $F[x]$, $\{p_1(x)^{e_1}, \dots, p_r(x)^{e_r}\}$, and are such that V is the direct sum of the cyclic subspaces $\{\langle v_i \rangle, 1 \leq i \leq r\}$, that is,*

$$V = \langle v_1 \rangle \oplus \langle v_2 \rangle \oplus \cdots \oplus \langle v_r \rangle.$$

The number of summands r is uniquely determined, and the orders $\{p_i(x)^{e_i}\}$ of the vectors $\{v_i\}$ are uniquely determined up to a rearrangement.

† The restriction T_W of a linear transformation T to a T-invariant subspace W is the linear transformation $T_W \in L(W, W)$ defined by $T_W w = Tw$, $w \in W$.

The proof of the theorem will be given in Section 29. We shall spend the remainder of this section on what the theorem means and how it is used.

We remark first that the orders $\{p_1(x)^{e_1}, p_2(x)^{e_2}, \ldots\}$ may contain repetitions. The uniqueness part of the theorem, in more detail, states that if

$$V = \langle v_1' \rangle \oplus \cdots \oplus \langle v_s' \rangle,$$

for some other nonzero vectors $\{v_i'\}$ with prime power orders $\{q_1(x)^{f_1}, \ldots, q_s(x)^{f_s}\}$, then $r = s$, and the sets of prime powers,

$$\{p_1(x)^{e_1}, \ldots, p_r(x)^{e_r}\}, \qquad \{q_1(x)^{f_1}, \ldots, q_r(x)^{f_r}\}$$

are the same, up to a rearrangement.

(25.9) Definition. Let $T \in L(V, V)$, as in the theorem. The set of prime powers $\{p_1(x)^{e_1}, \ldots, p_r(x)^{e_r}\}$, with repetitions if necessary, is called the set of *elementary divisors* of T.

For example, to say that T has elementary divisors $\{x - 1, x + 1, x + 1, x^2 + 1\}$ over the field R means that $T \in L(V, V)$ for V a vector space over R, and that V is a direct sum of cyclic subspaces,

$$V = \langle v_1 \rangle \oplus \langle v_2 \rangle \oplus \langle v_3 \rangle \oplus \langle v_4 \rangle,$$

with the orders of $\{v_1, v_2, v_3, v_4\}$ equal to $\{x - 1, x - 1, x + 1, x^2 + 1\}$. As we shall see later, the elementary divisors describe the matrix of T with respect to a certain basis in a completely unambiguous way.

We now proceed to derive the form of the matrix of a linear transformation T with a given set of elementary divisors. The matrix obtained will be called the *rational canonical form* of T.

(25.10) Lemma. *Let $p(x) = x^d - \alpha_{d-1}x^{d-1} - \cdots - \alpha_0, \alpha_i \in F$, be a prime polynomial in $F[x]$, and suppose $\langle v \rangle$ is a cyclic subspace of V, and that $p(x)$ is the order of v. Then the matrix of $T_{\langle v \rangle}$ with respect to the basis $\{v, Tv, \ldots, T^{d-1}v\}$ of $\langle v \rangle$ is*

$$A_{p(x)} = \begin{pmatrix} 0 & & & \cdots & & \alpha_0 \\ 1 & 0 & & \cdots & & \alpha_1 \\ 0 & 1 & 0 & \cdots & & \\ \vdots & & & \ddots & & \vdots \\ & & & & 0 & \\ 0 & & \cdots & 0 & 1 & \alpha_{d-1} \end{pmatrix}.$$

Proof. By the proof of Lemma (25.4), a basis for $\langle v \rangle$ is given by $\{v, Tv, \ldots, T^{d-1}v\}$, and we have

$$T(T^{d-1}v) = \alpha_0 v + \alpha_1(Tv) + \cdots + \alpha_{d-1}(T^{d-1}v).$$

The form of the matrix is clear from these remarks.

(25.11) DEFINITION. The matrix $\mathbf{A}_{p(x)}$ is called the *companion matrix* of the polynomial $p(x)$.

Example B. What is the companion matrix of the polynomial $x^2 + 1$, over the rational field? We have to write $x^2 + 1$ in the form

$$x^2 + 1 = x^2 - 0x - (-1).$$

Then the companion matrix is

$$\mathbf{A}_{x^2+1} = \begin{pmatrix} 0 & -1 \\ 1 & 0 \end{pmatrix}.$$

(25.12) LEMMA. *Let* $p(x) = x^d - \alpha_{d-1}x^{d-1} - \cdots - \alpha_0$, $\alpha_i \in F$, *and let* $\langle v \rangle$ *be a cyclic subspace such that the order of* v *is* $p(x)^e$ *for some positive integer* e. *Then the matrix of* $T_{\langle v \rangle}$ *with respect to the basis*

$$\{p(T)^{e-1}v, Tp(T)^{e-1}v, \ldots, T^{d-1}p(T)^{e-1}v, p(T)^{e-2}v, Tp(T)^{e-2}v, \ldots, T^{d-1}p(T)^{e-2}v, \ldots, v, Tv, \ldots, T^{d-1}v\}$$

is

$$e \text{ blocks} \left\updownarrow \begin{pmatrix} \mathbf{A} & \mathbf{B} & 0 & \cdots & 0 \\ 0 & \mathbf{A} & \mathbf{B} & \ddots & \vdots \\ \vdots & & \ddots & & \mathbf{B} \\ 0 & & & \cdots & \mathbf{A} \end{pmatrix}\right.$$

where $\mathbf{A}$ *is the companion matrix* $\mathbf{A} = \mathbf{A}_{p(x)}$ *and* $\mathbf{B}$ *is the d-by-d matrix*

$$\mathbf{B} = \begin{pmatrix} 0 & \cdots & 0 & 1 \\ \vdots & & & 0 \\ \vdots & & & \vdots \\ 0 & \cdots & & 0 \end{pmatrix}.$$

Proof. The number of candidates given in the statement of the Lemma, for basis elements of $\langle v \rangle$, is $de = \deg p(x)^e$. Therefore the vectors will be a basis of $\langle v \rangle$ if we can show that they are linearly independent. A relation of linear dependence, however, would produce a nonzero polynomial $g(x)$ such that $g(T)v = 0$ and $\deg g(x) < \deg p(x)^e$, contrary to the assumption that the order of v is $p(x)^e$. We now compute the matrix of $T_{\langle v \rangle}$ with respect to the basis. We see that T maps each basis vector onto the next one, except for the basis vectors

$$T^{d-1}p(T)^{e-1}v, T^{d-1}p(T)^{e-2}v, \ldots, T^{d-1}v.$$

For these vectors we have

$$T(T^{d-1}p(T)^{e-1}) = \alpha_0 p(T)^{e-1}v + \cdots + \alpha_{d-1}T^{d-1}p(T)^{e-1}v,$$

since $p(T)p(T)^{e-1}v = 0$;

$$T(T^{d-1}p(T)^{e-2}v) = \alpha_0 p(T)^{e-2}v + \cdots + \alpha_{d-1}T^{d-1}p(T)^{e-2}v + p(T)^{e-1}v$$

since $p(T)p(T)^{e-2}v = p(T)^{e-1}v$, etc. The part of the matrix involving the columns labeled by

$$p(T)^i v, \ldots, T^{d-1}p(T)^i v, p(T)^{i-1}v, \ldots, T^{d-1}p(T)^{i-1}v$$

is given in detail below.

$$\begin{array}{l|c|c|c}
 & & p(T)^i v \cdots T^{d-1}p(T)^i v & p(T)^{i-1}v \cdots T^{d-1}p(T)^{i-1}v \\
 & & \vdots & \vdots \\
\hline
\begin{matrix} p(T)^i v \\ \vdots \\ T^{d-1}p(T)^i v \end{matrix} & \vdots & \begin{matrix} 0 & & \cdots & \alpha_0 \\ 1 & 0 & & \vdots \\ & 1 & 0 & \\ & & \ddots & \\ & & 1 & \alpha_{d-1} \end{matrix} & \begin{matrix} 0 & \cdots & 0 & 1 \\ & & & 0 \end{matrix} \\
\hline
\begin{matrix} p(T)^{i-1}v \\ \vdots \\ T^{d-1}p(T)^{i-1}v \end{matrix} & \vdots & \vdots & \begin{matrix} 0 & \cdots & & \alpha_0 \\ 1 & 0 & & \\ & 1 & \ddots & \\ & & 1 & \alpha_{d-1} \end{matrix}
\end{array}$$

This completes the proof of the lemma.

(25.13) Definition. The *de*-by-*de* matrix given in Lemma (25.12) is called the *companion matrix* of the polynomial $p(x)^e$.

We note that the polynomial $p(x)$ must be put in the form

$$x^d - \alpha_{d-1}x^{d-1} - \cdots - \alpha_0$$

in order for the formulas for the matrices to be valid.

Example C. What is the companion matrix of $(x^2 + 1)^2$, over the rational field? Putting $x^2 + 1$ in the required form, we have

$$x^2 + 1 = x^2 - 0x - (-1).$$

The companion matrix, by Lemma (25.12), is

$$\left(\begin{array}{cc|cc} 0 & -1 & 0 & 1 \\ 1 & 0 & 0 & 0 \\ \hline 0 & 0 & 0 & -1 \\ 0 & 0 & 1 & 0 \end{array}\right).$$

Assuming the truth of the elementary divisor theorem, we can now state the second main theorem of this section. We require two more definitions first.

(25.14) DEFINITION. Let $\mathbf{A}$ be an n-by-n matrix with coefficients in F. The *elementary divisors* of $\mathbf{A}$ are defined to be the elementary divisors of a linear transformation on an n-dimensional vector space whose matrix with respect to some basis is $\mathbf{A}$.

(25.15) DEFINITION. Let $T \in L(V, V)$, for an n-dimensional vector space over F, and suppose the elementary divisors of T are

$$\{p_1(x)^{e_1}, \ldots, p_r(x)^{e_r}\},$$

(with repetitions if necessary). Let $V = \langle v_1 \rangle \oplus \cdots \oplus \langle v_r \rangle$, where the order of v_i is $p_i(x)^{e_i}, 1 \le i \le r$. Choose a basis of V consisting of bases for the subspaces $\langle v_i \rangle$, $1 \le i \le r$, as in Lemmas (25.10) and (25.12). The *rational canonical form* of T is the matrix $\mathbf{C}$ of T with respect to this basis:

$$\mathbf{C} = \begin{pmatrix} \mathbf{C}_1 & 0 & \cdots & 0 \\ 0 & \mathbf{C}_2 & & \vdots \\ \vdots & & \ddots & 0 \\ 0 & & & \mathbf{C}_r \end{pmatrix},$$

where $\mathbf{C}_i$, $1 \le i \le r$, is the companion matrix of the elementary divisor $p_i(x)^{e_i}$. The *rational canonical form of an n-by-n matrix* $\mathbf{A}$ over F is defined to be the rational canonical form of a linear transformation T on an n-dimensional vector space over F, whose matrix with respect to some basis is $\mathbf{A}$.

Notice that except for the ordering of the basis vectors, every entry of $\mathbf{C}$ is determined by a knowledge of the $\{p_i(x)^{e_i}\}$.

(25.15) THEOREM. *Two n-by-n matrices* $\mathbf{A}$ *and* $\mathbf{B}$ *with coefficients in* F *are similar if they have the same rational canonical forms (up to a rearrangement of blocks on the diagonal).*

Proof. Let $\mathbf{C}$ be the common rational canonical form of $\mathbf{A}$ and $\mathbf{B}$. Since $\mathbf{A}$ and $\mathbf{C}$ are matrices of a linear transformation with respect to different bases, they are similar. Thus

$$\mathbf{A} = \mathbf{X}\mathbf{C}\mathbf{X}^{-1}$$

for some invertible n-by-n matrix $\mathbf{X}$. Similarly

$$\mathbf{B} = \mathbf{Y}\mathbf{C}\mathbf{Y}^{-1}$$

for some invertible n-by-n matrix $\mathbf{Y}$. Then

$$\mathbf{X}^{-1}\mathbf{A}\mathbf{X} = \mathbf{Y}^{-1}\mathbf{B}\mathbf{Y},$$

and

$$\mathbf{A} = \mathbf{X}\mathbf{Y}^{-1}\mathbf{B}(\mathbf{X}\mathbf{Y}^{-1})^{-1},$$

proving that $\mathbf{A}$ and $\mathbf{B}$ are similar.

(25.16) THEOREM. *The rational canonical form of a linear transformation $T \in L(V, V)$ is uniquely determined up to a rearrangement of the blocks on the diagonal.*

Proof. By the elementary divisor theorem, the elementary divisors of T are uniquely determined, up to a rearrangement. The proofs of Lemmas (25.10) and (25.12) show that the companion matrices of the elementary divisors are uniquely determined by the coefficients of the polynomials. This completes the proof. Probably the most useful form of the preceding theorems is the following one.

COROLLARY. *Two n-by-n matrices with entries in F are similar if and only if they have the same set of elementary divisors.*

The proof is immediate by Theorems (25.15) and (25.16).

(25.17) DEFINITION. The *Jordan normal form* (or *Jordan canonical form*) of a linear transformation (or a matrix) is defined to be the rational canonical form in case all the characteristic roots belong to the field F. In that case the elementary divisors all have the form $(x - \alpha_i)^e$, for $\alpha_i \in F$, and the companion matrix of $(x - \alpha_i)^e$ is the e-by-e matrix

$$\begin{pmatrix} \alpha_i & 1 & 0 & \cdots & 0 \\ 0 & \alpha_i & 1 & & \\ \alpha & & & & \\ \vdots & & & \ddots & 1 \\ 0 & & \cdots & & \alpha_i \end{pmatrix}.$$

The only statement in the definition to be checked is that the elementary divisors all have the form $(x - \alpha_i)^e$. By Lemma (25.7), the elementary divisors all divide the minimal polynomial, and from Section 24, the prime factors of the minimal polynomial have the form $x - \alpha_i$, where α_i is a characteristic root. The statement now follows from uniqueness of factorization of polynomials.

Example D. What is the rational canonical form of the matrix with rational coefficients with elementary divisors

$$(x - 1)^2, \qquad x^2 + 1?$$

The rational canonical form consists of two diagonal blocks, which are the companion matrices of $(x - 1)^2$ and $x^2 + 1$, respectively. The matrix is

$$\begin{pmatrix} 1 & 1 & 0 & 0 \\ 0 & 1 & 0 & 0 \\ 0 & 0 & 0 & -1 \\ 0 & 0 & 1 & 0 \end{pmatrix},$$

using Example C to find the companion matrix of $x^2 + 1$.

Example E. Find the rational or Jordan canonical form of the linear transformation T on a three-dimensional vector space over C given in Example A of Section 24, whose matrix with respect to a basis is

$$\mathbf{A} = \begin{pmatrix} -1 & 3 & 0 \\ 0 & 2 & 0 \\ 2 & 1 & -1 \end{pmatrix}.$$

Since C is algebraically closed, all the characteristic roots lie in C, and we shall be computing the Jordan normal form in this case. In the Example, it was shown that the minimal polynomial of $\mathbf{A}$ is $(x + 1)^2(x - 2)$. It was also shown that the underlying vector space is the direct sum of the null spaces of $(T + 1)^2$ and $T - 2$. It is left to the reader to show that these subspaces are both cyclic relative to T. It follows that the elementary divisors of T are

$$(x + 1)^2 \quad \text{and} \quad x - 2.$$

Therefore the Jordan canonical form is

$$\begin{pmatrix} -1 & 1 & 0 \\ 0 & -1 & 0 \\ 0 & 0 & 2 \end{pmatrix}.$$

EXERCISES

1. Find the rational canonical forms of linear transformations on a vector space over the field of rational numbers Q, whose elementary divisors are as follows:
   a. $(x - 1)^2, x^2 - x + 1$.
   b. $x + 1, (x^2 + x + 1)^2$.
   c. $x^2, (x + 1)^3$.

2. Find the rational canonical forms, over the field of rational numbers†, of the matrices

$$\begin{pmatrix} 2 & -1 \\ 1 & -1 \end{pmatrix}, \quad \begin{pmatrix} 0 & 0 & 1 \\ 1 & 0 & 0 \\ 0 & 1 & 0 \end{pmatrix}.$$

3. Find the rational canonical forms, over the field of real numbers, of the matrices given in Exercise 2.
4. Find the rational canonical forms, over the field of complex numbers, of the matrices given in Exercise 2.
5. Prove that V is cyclic relative to a linear transformation $T \in L(V, V)$ if and only if the minimal polynomial of T is equal to the characteristic polynomial.
6. Test to determine whether the following pairs of matrices are similar, over the complex field C.

   a. $\begin{pmatrix} 1 & 3 \\ 0 & 1 \end{pmatrix}, \quad \begin{pmatrix} 1 & 1 \\ 0 & 1 \end{pmatrix}.$

   b. $\begin{pmatrix} 0 & 1 \\ -1 & 0 \end{pmatrix}, \quad \begin{pmatrix} i & 0 \\ 0 & -i \end{pmatrix}.$

   c. $\begin{pmatrix} -1 & 0 & 0 & 0 \\ 0 & 1 & 0 & 0 \\ 0 & 0 & 1 & 0 \\ 0 & 0 & 0 & 2 \end{pmatrix}, \quad \begin{pmatrix} 1 & 0 & 0 & 0 \\ 0 & -1 & 0 & 0 \\ 0 & 0 & 2 & 0 \\ 0 & 0 & 0 & 1 \end{pmatrix}.$

7. Show that two n-by-n diagonal matrices with coefficients in F are similar if and only if the diagonal elements of one matrix are a rearrangement of the diagonal elements of the other.
8. Let F be an arbitrary field.

   a. Let $\mathbf{A}_{p(x)}$ be the companion matrix of a prime polynomial $p(x) \in F[x]$. Prove that

   $$D(x\mathbf{I} - \mathbf{A}_{p(x)}) = p(x).$$

   (*Hint:* Expand the determinant along the last column.)

   b. Let $\mathbf{A}$ be a matrix in block triangular form,

   $$\mathbf{A} = \begin{pmatrix} \mathbf{A}_1 & & & * \\ & \mathbf{A}_2 & & \\ & & \ddots & \\ 0 & & & \mathbf{A}_r \end{pmatrix},$$

   where the matrices $\mathbf{A}_1, \mathbf{A}_2, \ldots, \mathbf{A}_r$ are square matrices. Prove that

   $$D(x\mathbf{I} - \mathbf{A}) = D(x\mathbf{I} - \mathbf{A}_1) \cdots D(x\mathbf{I} - \mathbf{A}_r).$$

† The rational canonical form of a matrix $\mathbf{A}$ *over a field F* is the rational canonical form of a linear transformation corresponding to $\mathbf{A}$, on a vector space over F.

c. Let **A** be the companion matrix of a prime power $p(x)^e \in F[x]$. Prove that

$$D(x\mathbf{I} - \mathbf{A}) = p(x)^e.$$

d. Let **A** be a matrix whose elementary divisors are

$$\{p_1(x)^{e_1}, \ldots, p_r(x)^{e_r}\}.$$

Prove that

$$D(x\mathbf{I} - \mathbf{A}) = p_1(x)^{e_1} \cdots p_r(x)^{e_r}.$$

e. Prove the Cayley-Hamilton theorem for matrices with coefficients in F: If $h(x) = D(x\mathbf{I} - \mathbf{A})$, then $h(\mathbf{A}) = 0$.

Chapter 8

Dual Vector Spaces and Multilinear Algebra

This chapter begins with two sections on some important constructions on vector spaces, leading to quotient spaces and dual spaces. The section on dual spaces is based on the concept of a bilinear form defined on a pair of vector spaces. The next section contains the construction of the tensor product of two vector spaces and provides an introduction to the subject of what is called multilinear algebra. The last section contains an application of the theory of dual vector spaces to the proof of the elementary divisor theorem, which was stated in Section 25.

26. QUOTIENT SPACES AND DUAL VECTOR SPACES

It is useful to begin our discussion with some ideas from set theory. Let X be a set. A *relation* on X is an arbitrary set of ordered pairs $\mathcal{R} = \{(a, b)\}$, $a, b \in X$, where *ordered pair* means that we identify (a, b) with (a', b') only if $a = a'$ and $b = b'$. For example, the ordered pairs $(1, 2)$ and $(2, 1)$ are not equal. An example of a relation is the set of ordered pairs of integers $\mathcal{R} = \{(a, b)\}$, where $(a, b) \in \mathcal{R}$ if and only if $a < b$. It is often convenient to use the notation $a \mathcal{R} b$ to mean that the pair $(a, b) \in \mathcal{R}$.

(26.1) DEFINITION. A relation $\mathcal{R}$ on a set X is called an *equivalence relation* provided that

(*i*) $a\mathscr{R}a$ for all $a \in X$ (reflexive property).
(*ii*) $a\mathscr{R}b$ implies $b\mathscr{R}a$ (symmetric property).
(*iii*) $a\mathscr{R}b$ and $b\mathscr{R}c$ imply $a\mathscr{R}c$ (transitive property).

The relation of equality, $a = b$, is an example of an equivalence relation. A more interesting one is the following. Let X be a set, and let $f: X \to Y$ be a mapping of X into a set Y. Then define a relation $\mathscr{R}$ on X as follows:

$$a\mathscr{R}b \quad \text{if} \quad f(a) = f(b).$$

We leave it to the reader to check that $\mathscr{R}$ is an equivalence relation.

The main property of an equivalence relation is that it provides a partition of the set into nonoverlapping subsets, called *equivalence classes.* In our discussion of this fact, and others, it will be convenient to use the notation (introduced in Section 2),

$$\{a \in X \mid a \text{ has property } P\},$$

to denote the set of all elements in X which have some property P.

(26.2) THEOREM. *Let X be a set, and let $\mathscr{R}$ be an equivalence relation on X. For each $a \in X$, let $[a]$ be the subset of X defined by*

$$[a] = \{b \in X \mid b\mathscr{R}a\}.$$

Then the sets $\{[a], a \in X\}$ are called equivalence classes, and the set X is the union of all equivalence classes. In other words, every element of X belongs to at least one equivalence class. Moreover, if $[a]$ and $[a']$ are equivalence classes having at least one element in common, then they coincide, $[a] = [a']$.

Proof. The reflexive property of $\mathscr{R}$ states that $a\mathscr{R}a$ for all $a \in X$. Thus $a \in [a]$, and we have proved the first statement, that X is the union of the equivalence classes $[a]$. Now suppose $b \in [a] \cap [a']$. We have to prove that $[a] = [a']$. First let $c \in [a]$. Then $c\mathscr{R}a$. Since $b \in [a]$, we have $b\mathscr{R}a$, and hence $a\mathscr{R}b$ by the symmetric property of $\mathscr{R}$. Applying the transitivity, $c\mathscr{R}a$ and $a\mathscr{R}b$ imply that $c\mathscr{R}b$. Finally, $c\mathscr{R}b$ and $b\mathscr{R}a'$ imply $c\mathscr{R}a'$, and $c \in [a']$, again by transitivity and the fact that $[a] \subset [a']$. The same argument, with a and a' interchanged, shows that $[a'] \subset [a]$. Therefore $[a] = [a']$, and the theorem is proved.

For example, if $\mathscr{R}$ is the relation of equality, the equivalence class $[a]$ consists of the single element a. If $f: X \to Y$ is a function, and R the equivalence relation defined previously,

$$a\mathscr{R}b \quad \text{if} \quad f(a) = f(b),$$

then the equivalence class $[a]$ is the set of all $b \in X$ which are mapped by f onto the same element as a, that is, $[a] = \{b \in X \mid f(b) = f(a)\}$.

(26.3) DEFINITION. Let $\mathscr{R}$ be an equivalence relation on a set X. The set of equivalence classes $\{[a]\}$, $a \in X$, is called the *quotient set* of X by $\mathscr{R}$, and is denoted by $X/\mathscr{R}$.

We can now apply this construction to vector spaces.

(26.4) THEOREM. *Let V be a vector space over a field F, and let Y be a subspace of V. Then the relation $\mathscr{R}$ on the set V defined by*

$$v\mathscr{R}v' \quad \text{if} \quad v - v' \in Y$$

is an equivalence relation. The quotient set $V/\mathscr{R}$ consisting of all equivalence classes $\{[v] \mid v \in V\}$ is a vector space over F, with addition and scalar multiplication defined by

$$[v] + [v'] = [v + v'], \qquad v, v' \in V$$
$$\alpha[v] = [\alpha v], \qquad v \in V, \quad \alpha \in F.$$

If V is finite dimensional, then $V/\mathscr{R}$ is finite dimensional and we have

$$\dim V = \dim Y + \dim V/\mathscr{R}.$$

Proof. First we check that $\mathscr{R}$ is an equivalence relation. $v\mathscr{R}v$ for all v, since $v - v = 0 \in Y$ because Y is a subspace. If $v\mathscr{R}v'$, then $v - v' \in Y$. Since Y is a subspace, $-(v - v') = v' - v \in Y$, and $v'\mathscr{R}v$. Finally, if $v\mathscr{R}v'$ and $v'\mathscr{R}v''$, then $v - v' \in Y$ and $v' - v'' \in Y$. Then $v - v'' \in Y$, again using the fact that Y is a subspace.

The first thing to be checked in proving that $V/\mathscr{R}$ is a vector space is that the definitions of the operations make sense. This means that if $[v] = [v_1]$, and $[v'] = [v_1']$, then $[v + v'] = [v_1 + v_1]$. By assumption, we have $v - v_1 \in Y, v' - v_1' \in Y$, and hence $(v - v_1) + (v' - v_1') = v + v' - (v_1 + v_1') \in Y$, and $[v + v'] = [v_1 + v_1']$. Now let $[v] = [v_1]$, and let $\alpha \in F$. We have to check that $[\alpha v] = [av']$. Then $v - v' \in Y$, and since Y is a subspace $\alpha(v - v') = \alpha v - \alpha v' \in Y$. Therefore $[\alpha v] = [\alpha v']$. The verification of the vector space axioms is easily done using the definitions, and will be left to the reader.

Finally, suppose V is finite dimensional. Then the mapping

$$T : v \rightarrow [v]$$

is a linear transformation of V onto $V/\mathscr{R}$. By Theorem (13.9),

$$\dim T(V) + \dim n(T) = \dim V,$$

where $n(T)$ is the null space of T. We have $T(V) = V/\mathscr{R}$, and

$$n(T) = \{v \mid [v] = [0]\} = \{v \mid v - 0 \in Y\} = Y.$$

Substituting in the formula from (13.9), we have

$$\dim V/\mathscr{R} + \dim Y = \dim V,$$

and the theorem is proved.

(26.5) DEFINITION. Let V be a vector space over F, and let Y be a subspace of V. The vector space $V/\mathscr{R}$ defined in Theorem (25.4), is called the *quotient space* of V by Y and is denoted by V/Y.

Now let us see what this construction means in terms of linear transformations and matrices.

(26.6) DEFINITION. Let $T \in L(V, V)$, and let Y be a subspace of V which is invariant relative to T, $T(Y) \subset Y$. Let

$$T_Y : Y \to Y$$

be the mapping defined by

$$T_Y(y) = T(y), \quad y \in Y.$$

Let

$$T_{V/Y} : V/Y \to V/Y$$

be defined by

$$T_{V/Y}([v]) = [Tv].$$

Then $T_Y \in L(Y, Y)$ and $T_{V/Y} \in L(V/Y, V/Y)$. The transformation T_Y is called the *restriction* of I to Y, while $T_{V/Y}$ is called the transformation on V/Y *induced* by T.

There are a couple of points to be checked in the definition. It is clear that $T_Y \in L(Y, Y)$ since Y is invariant relative to T. We must check that the definition of $T_{V/Y}$ is legitimate, in other words that $[v] = [v']$ implies $[Tv] = [Tv']$. But $[v] = [v']$ implies $v - v' \in Y$, and $T(v - v') \in Y$. Then $Tv - Tv' \in Y$ and $[Tv] = [Tv']$. To show that $T_{V/Y} \in L(V/Y, V/Y)$, we have

$$\begin{aligned} T([v_1] + [v_2]) &= T([v_1 + v_2]) = [T(v_1 + v_2)] = [Tv_1 + Tv_2] \\ &= [Tv_1] + [Tv_2] = T[v_1] + T[v_2], \end{aligned}$$

and

$$\begin{aligned} T(\alpha[v]) &= T([\alpha v]) = [T(\alpha v)] = [\alpha(Tv)] = \alpha[Tv] \\ &= \alpha(T[v]). \end{aligned}$$

(26.7) THEOREM. *Let Y be a nonzero subspace of a finite dimensional vector space V which is invariant relative to a linear transformation*

$T \in L(V, V)$. Let $\{v_1, \ldots, v_n\}$ be a basis of V such that $\{v_1, \ldots, v_k\}$ is a basis for Y. If $Y \neq V$, then $k < n$, and $[v_{k+1}], \ldots, [v_n]$ is a basis of V/Y. Let $\mathbf{A} = (\alpha_{ij})$ be the matrix of T with respect to the basis $\{v_1, \ldots, v_n\}$. Then $\mathbf{A}$ has the form

$$\mathbf{A} = \begin{array}{c} \uparrow \\ k \\ \downarrow \\ \hline \uparrow \\ n-k \\ \downarrow \end{array} \overset{\leftarrow k \rightarrow \;|\; \leftarrow n-k \rightarrow}{\left(\begin{array}{c|c} \mathbf{A}_1 & \mathbf{A}_3 \\ \hline 0 & \mathbf{A}_2 \end{array}\right)}$$

where $\mathbf{A}_1$, $\mathbf{A}_2$ and $\mathbf{A}_3$ are k-by-k, $(n-k)$-by-$(n-k)$, and k-by-$(n-k)$ matrices, respectively. Then $\mathbf{A}_1$ is the matrix of T_Y with respect to the basis $\{v_1, \ldots, v_k\}$ and $\mathbf{A}_2$ is the matrix of $T_{V/Y}$ with respect to the basis $\{[v_{k+1}], \ldots, [v_n]\}$.

Proof. We first check that $\{[v_{k+1}], \ldots, [v_n]\}$ is a basis of V/Y. Let $[v] \in V/Y$; then

$$v = \alpha_1 v_1 + \cdots + \alpha_n v_n,$$

and
$$\begin{aligned} [v] &= \alpha_1 [v_1] + \cdots + \alpha_k [v_k] + \alpha_{k+1} [v_{k+1}] + \cdots + \alpha_n [v_n] \\ &= \alpha_{k+1} [v_{k+1}] + \cdots + \alpha_k [v_n] \end{aligned}$$

since $[v_1] = \cdots = [v_k] = 0$ in V/Y. Therefore V/Y is generated by $\{[v_{k+1}], \ldots, [v_n]\}$. Now suppose

$$\beta_{k+1} [v_{k+1}] + \cdots + \beta_n [v_n] = 0, \qquad \beta_i \in F.$$

Then

$$\beta_{k+1} v_{k+1} + \cdots + \beta_n v_n \in Y,$$

and hence

$$\beta_{k+1} v_{k+1} + \cdots + \beta_n v_n = \alpha_1 v_1 + \cdots + \alpha_k v_k,$$

for some $\alpha_i \in F$. Since $\{v_1, \ldots, v_n\}$ are linearly independent, we have $\beta_{n+1} = \cdots = \beta_n = 0$.

Now let us compute the matrix of T with respect to the basis $\{v_1, \ldots, v_n\}$. Since Y is invariant relative to T, $T(v_i) \in Y$ for $1 \leq i \leq k$. Therefore we have

$$\begin{aligned} T(v_1) &= \alpha_{11} v_1 + \alpha_{21} v_2 + \cdots + \alpha_{k1} v_k \\ &\cdots\cdots\cdots\cdots\cdots\cdots \\ T(v_k) &= \alpha_{1k} v_1 + \alpha_{2k} v_2 + \cdots + \alpha_{kk} v_k. \end{aligned}$$

The remaining equations are

$$T(v_{k+1}) = \alpha_{1,k+1}v_1 + \cdots + \alpha_{k,k+1}v_k + \alpha_{k+1,k+1}v_{k+1} + \cdots + \alpha_{n,k+1}v_n$$
$$\cdots\cdots\cdots\cdots\cdots\cdots\cdots\cdots\cdots\cdots$$
$$T(v_n) = \alpha_{1n}v_1 + \cdots + \alpha_{kn}v_k + \alpha_{k+1,n}v_{k+1} + \cdots + \alpha_{nn}v_n.$$

Then

$$T_{V/Y}([v_{k+1}]) = \alpha_{k+1,k+1}[v_{k+1}] + \cdots + \alpha_{n,k+1}[v_n]$$
$$\cdots\cdots\cdots\cdots\cdots\cdots\cdots\cdots$$
$$T_{V/Y}([v_n]) = \alpha_{k+1,n}[v_{k+1}] + \cdots + \alpha_{nn}[v_n].$$

Let

$$\mathbf{A}_1 = \begin{pmatrix} \alpha_{11} & \cdots & \alpha_{1k} \\ \cdots & \cdots & \cdots \\ \alpha_{k1} & \cdots & \alpha_{kk} \end{pmatrix}, \quad \mathbf{A}_2 = \begin{pmatrix} \alpha_{k+1,k+1} & \cdots & \alpha_{k+1,n} \\ \cdots & \cdots & \cdots \\ \alpha_{n,k+1} & \cdots & \alpha_{nn} \end{pmatrix},$$
$$\mathbf{A}_3 = \begin{pmatrix} \alpha_{1,k+1} & \cdots & \alpha_{1n} \\ \cdots & \cdots & \cdots \\ \alpha_{k,k+1} & \cdots & \alpha_{kn} \end{pmatrix}.$$

We have shown that the matrix $\mathbf{A}$ of T with respect to the basis $\{v_1, \ldots, v_n\}$ has the form stated in the theorem, and that $\mathbf{A}_1$ and $\mathbf{A}_2$ are the matrices of T_Y and $T_{V/Y}$, respectively. This completes the proof of the theorem.

Example A. Let $\mathbf{A}$ be the matrix

$$\mathbf{A} = \begin{pmatrix} 1 & 2 & -1 & 4 \\ -1 & 0 & 6 & 1 \\ 0 & 0 & 2 & 1 \\ 0 & 0 & -1 & 1 \end{pmatrix},$$

and let V be a vector space over R with basis $\{v_1, \ldots, v_4\}$. Let $T \in L(V, V)$ be the linear transformation whose matrix with respect to the basis is $\mathbf{A}$. By Theorem (26.7) we see that $Y = S(v_1, v_2)$ is invariant relative to T and that the matrices of T_Y and $T_{V/Y}$ with respect to the bases $\{v_1, v_2\}$ and $\{[v_3], [v_4]\}$ are

$$\begin{pmatrix} 1 & 2 \\ -1 & 0 \end{pmatrix} \quad \text{and} \quad \begin{pmatrix} 2 & 1 \\ -1 & 1 \end{pmatrix},$$

respectively.

We now turn to the important concept of dual vector space. The idea behind this construction is that for every vector space V and linear transformation $T \in L(V, V)$, there is a vector space V^*, called the *dual* of V, which is a kind of mirror to V, and a linear transformation T^* of V^* which mirrors the behavior of T.

(26.8) DEFINITION. Let V be a finite dimensional vector space over F. The *dual space* V^* of V is defined to be the vector space $L(V, F)$, where F is identified with the vector space of one-tuples over F. The elements of V^* are simply functions f from V into F such that $f(v_1 + v_2) = f(v_1) + f(v_2)$, $v_1, v_2 \in V$, and $f(\alpha v) = \alpha f(v)$, $\alpha \in F, v \in V$. Elements of V^* are called *linear functions* on V.

(26.9) LEMMA. *Let $\{v_1, \ldots, v_n\}$ be a basis for V over F. Then there exist linear functions $\{f_1, \ldots, f_n\}$ such that for each i,*

$$f_i(v_i) = 1, \qquad f_i(v_j) = 0, \qquad j \neq i.$$

The linear functions $\{f_1, \ldots, f_n\}$ form a basis for V^* over F, called the dual basis to $\{v_1, \ldots, v_n\}$.

Proof. First of all, the linear functions exist because of Theorem (13.1), which allows us to define linear transformations which map basis elements of a vector space onto arbitrary vectors in the image space. We next show that $\{f_1, \ldots, f_n\}$ are linearly independent. Suppose

$$\alpha_1 f_1 + \cdots + \alpha_n f_n = 0, \qquad \alpha_i \in F.$$

Then applying both sides to the vector v_1, and using the definition of the vector space operations in V^*, we have

$$\alpha_1 f_1(v_1) + \alpha_2 f_2(v_1) + \cdots + \alpha_n f_n(v_1) = 0.$$

Therefore $\alpha_1 = 0$ because $f_1(v_1) = 1$ and $f_2(v_1) = \cdots = f_n(v_1) = 0$. Similarly $\alpha_2 = \cdots = \alpha_n = 0$.

Finally we check that $\{f_1, \ldots, f_n\}$ form a set of generators for V^*. Let $f \in V^*$, and let $f(v_i) = \alpha_i$, $i = 1, 2, \ldots, n$. Then it is easily checked by applying both sides to the basis elements $\{v_1, \ldots, v_n\}$ in turn, that

$$f = \alpha_1 f_1 + \cdots + \alpha_n f_n.$$

This completes the proof of the lemma.

(26.10) THEOREM. *Let $T \in L(V, V)$. Define a mapping $T^*: V^* \to V^*$ by the rule*

$$(T^*f)(x) = f(Tx), \qquad x \in V,$$

for all $f \in V^*$. Then $T^* \in L(V^*, V^*)$, *and is called the* transpose *of the linear transformation T.*

Proof. There are several statements to be checked. First of all, T^*f is a linear function, because

$$\begin{aligned}(T^*f)(v_1 + v_2) &= f[T(v_1 + v_2)] = f[T(v_1)] + f[T(v_2)] \\ &= (T^*f)(v_1) + (T^*f)(v_2),\end{aligned}$$

for all $v_1, v_2 \in V$ and $f \in V^*$, and

$$(T^*f)(\alpha v) = f[T(\alpha v)] = f[\alpha(Tv)] = \alpha f(Tv)$$
$$= \alpha[T^*f(v)], \qquad v \in V, \alpha \in F.$$

We must then check that $T^* \in L(V^*, V^*)$. For $f_1, f_2 \in V^*$, we have

$$[T^*(f_1 + f_2)](v) = (f_1 + f_2)(Tv) = f_1(Tv) + f_2(Tv)$$
$$= (T^*f_1 + T^*f_2)(v).$$

Moreover,

$$[T^*(\alpha f)](v) = (\alpha f)(Tv) = f[\alpha(Tv)] = f[T(\alpha v)]$$
$$= (T^*f)(\alpha v) = [\alpha(T^*f)](v),$$

for all $v \in V$ and $\alpha \in F$. It is interesting to notice how the linearity of T and f, and the vector space operations in V^*, are used in these verifications.

(26.11) THEOREM. *Let $T \in L(V, V)$, let $\{v_1, \ldots, v_n\}$ be a basis of V, and $\{f_1, \ldots, f_n\}$ the dual basis of V^*, in the sense of Lemma (26.9). Let* **A** *be the matrix of T with respect to the basis $\{v_1, \ldots, v_n\}$. Then the matrix of T^* with respect to the basis $\{f_1, \ldots, f_n\}$ is the transpose matrix* ${}^t\mathbf{A}$. *(We recall that if α_{ij} is the element in the ith row, jth column of* **A**, *then the element in the ith row, jth column of* ${}^t\mathbf{A}$ *is α_{ji}.)*

Proof. We have

$$Tv_i = \sum_{j=1}^{n} \alpha_{ji} v_j.$$

Suppose

$$T^*f_i = \sum_{j=1}^{n} \beta_{ji} f_j,$$

where $\{f_1, \ldots, f_n\}$ is the dual basis of V^*, with respect to the basis $\{v_1, \ldots, v_n\}$ of V. We have to show that $\beta_{ji} = \alpha_{ij}$, for all i and j. Apply both sides of the second equation to an arbitrary basis vector v_k of V to obtain

$$T^*f_i(v_k) = \sum_{j=1}^{n} \beta_{ji} f_j(v_k).$$

The left side is

$$T^*f_i(v_k) = f_i(Tv_k) = f_i\left(\sum_{j=1}^{n} \alpha_{jk} v_j\right)$$
$$= \sum_{j=1}^{n} \alpha_{jk} f_i(v_j) = \alpha_{ik}.$$

The right-hand side is

$$\sum_{j=1}^{n} \beta_{ji} f_j(v_k) = \beta_{ki},$$

and we have shown that $\alpha_{ik} = \beta_{ki}$ for all i and k, completing the proof of the theorem.

EXERCISES

1. Show that the following concepts are equivalence relations:
   a. Equivalence of ordered pairs of points in the plane (see Examples A to C of Section 3), that is, $(A, B) \sim (C, D)$ if $B - A = D - C$
   b. Row equivalence of matrices (from Section 6)
   c. Similarity of matrices
2. Let Y be a subspace of a finite dimensional vector space V, and suppose that $V = Y \oplus Y'$ for some subspace Y'. Prove that the vector spaces Y' and V/Y are isomorphic [see Definition (11.14)]. Give an example to show that a given subspace Y of V may have more than one "complementary subspace" Y' such that $V = Y \oplus Y'$.
3. Let V be a finite dimensional vector space over F, let $T \in L(V, V)$, and let Y be a T-invariant subspace, with $Y \neq 0$, $Y \neq V$.
   a. Prove the minimal polynomials of the restriction T_Y and the induced mapping $T_{V/Y}$ divide the minimal polynomial of T.
   b. With reference to part (a), is the minimal polynomial of T equal to the product of the minimal polynomials of T_Y and $T_{V/Y}$?
4. Let V, Y, T be as in Exercise 3. Show that if both $T_Y = 0$ and $T_{V/Y} = 0$, then T is nilpotent, and in fact, $T^2 = 0$.
5. Let V, W be vector spaces over F, and let $T \in L(V, W)$. Let Y be a subspace of V. Show that

$$[v] \to Tv, \qquad [v] \in V/Y,$$

   defines a linear transformation of $V/Y \to W$ if and only if $Y \subset n(T)$.
6. Let V, W, T, Y be as in Exercise 5. Show that if $n(T) = Y$, then the induced linear transformation $T_{V/Y}$ is an isomorphism of V/Y onto the range $T(V)$ of T.
7. Let $T_1, T_2 \in L(V, V)$. Show that

$$(T_1T_2)^* = T_2^*T_1^*.$$

8. Prove that the minimal polynomial of $T \in L(V, V)$ is equal to the minimal polynomial of T^*.

27. BILINEAR FORMS AND DUALITY

We begin with a different and more general way of looking at the relationship between a vector space and its dual.

(27.1) Definition. A *bilinear form* on a pair of vector spaces V and V' over F is a function B which assigns to each ordered pair (v, v'), with

$v \in V$, $v' \in V'$, a uniquely determined element of F, $B(v, v')$, such that the following conditions are satisfied:

$$B(v_1 + v_2, v') = B(v_1, v') + B(v_2, v')$$
$$B(v, v'_1 + v'_2) = B(v, v'_1) + B(v, v'_2),$$
$$B(\alpha v, v') = B(v, \alpha v') = \alpha B(v, v'),$$

for $v, v_1, v_2 \in V$, $v', v'_1, v'_2 \in V'$, $\alpha \in F$.

Example A. Let V be a vector space over F, and V^* the dual vector space, $V^* = L(V, F)$. Let

$$B(v, f) = f(v), \qquad f \in V^*, \quad v \in V.$$

Then B is a bilinear form on the pair of vector spaces (V, V^*).

The proof is contained in the preceding section. The reader should notice that the vector space operations on V^* are defined in such a way as to force the mapping $(v, f) \to f(v)$ to be a bilinear form.

Example B. Let V and W be vector spaces over F with finite bases $\{v_1, \ldots, v_m\}$ and $\{w_1, \ldots, w_n\}$, respectively. Let $\mathbf{A} = (\alpha_{ij})$ be a fixed m-by-n matrix with coefficients in F. Let

$$v = \xi_1 v_1 + \cdots + \xi_m v_m \in V, \; w = \eta_1 w_1 + \cdots + \eta_n w_n \in W,$$

and define

$$B(v, w) = \sum_{i=1}^{m} \sum_{j=1}^{n} \alpha_{ij} \xi_i \eta_j .$$

Then $B(v, w)$ defines a bilinear form on (V, W). The verification of this fact is left to the exercises.

(27.2) DEFINITION. A bilinear form $B: (V, V') \to F$ is said to be *nondegenerate* provided that:

$$B(v, v') = 0 \text{ for all } v' \in V' \text{ implies } v = 0$$

and

$$B(v, v') = 0 \text{ for all } v \in V \text{ implies } v' = 0.$$

Since the usual arguments using the vector space axioms, applied to bilinear forms, show that

$$B(0, v') = B(v, 0) = 0$$

for all v, v' in V and V', respectively, the nondegeneracy condition asserts that the equation $B(v, v') = 0$ for all v', holds only in the unique case when it is forced to hold, that is, when $v = 0$.

(27.3) THEOREM. *Let (V, W) be a pair of finite dimensional vector spaces, and let $B(v, w)$ be a nondegenerate bilinear form on $(V, W) \to F$. For fixed $w \in W$, let $\varphi_w : V \to F$ be the mapping*

$$\varphi_w(v) = B(v, w), \qquad v \in V.$$

Then the mapping $\varphi_w \in V^*$, and the mapping

$$\Phi: W \to V^*$$

defined by

$$\Phi(w) = \varphi_w$$

is an isomorphism of W onto V^*. Similarly V is isomorphic to W^*.

Proof. The definition of a bilinear form implies that for each $w \in W$, φ_w belongs to the dual space V^*, and that the mapping

$$w \to \Phi(w) = \varphi_w$$

is a linear transformation of W into V^*. (The details of checking the assertions have already been presented several times before in other contexts, and will be omitted this time.) In order to show that Φ is an isomorphism, we have to show that Φ is one-to-one and onto. First suppose $\Phi(w) = \Phi(w')$; then

$$B(v, w) = B(v, w')$$

for all $v \in V$. Because of the bilinearity of B, it follows that

$$B(v, w - w') = 0$$

for all $v \in V$, and by the nondegeneracy of B, we have $w = w'$. Therefore Φ is one-to-one, and it follows that

$$\dim V^* \geq \dim W.$$

So far we have used only half of the definition of nondegeneracy. Using the other half, we obtain by the same argument a one-to-one linear transformation of V into W^*, and therefore

$$\dim W^* \geq \dim V.$$

Using the result of the preceding section that $\dim W = \dim W^*$, etc., we have

$$\dim V^* \geq \dim W = \dim W^* \geq \dim V.$$

It follows that

$$\dim V^* = \dim W,$$

and since $\Phi(W) \subset V^*$ and $\dim \Phi(W) = \dim V^*$, we conclude that Φ is onto. Therefore Φ is an isomorphism, and the theorem is proved.

(27.4) COROLLARY. *Let V, W, B be as in the theorem. Then* $\dim V = \dim W$.

This theorem makes the following definition rather natural.

(27.5) DEFINITION. A pair of finite dimensional vector spaces V and W are said to be *dual* with respect to a bilinear form $B: (V, W) \to F$ provided that B is nondegenerate.

In other words, V and W are dual with respect to B if, by the preceding theorem, each vector space is isomorphic to the dual space of the other, via the mappings defined by B.

(27.6) DEFINITION. Let V and V' be finite dimensional vector spaces which are dual with respect to a bilinear form B. Let $T \in L(V, V)$ and $T' \in L(V', V')$. Then T and T' are said to be *transposes* of each other provided that

$$B(v, T'(v')) = B(T(v), v').$$

for all $v \in V$, $v' \in V'$.

(27.7) THEOREM. *Let V and V' be finite dimensional vector spaces which are dual with respect to a bilinear form B. Let $T \in L(V, V)$; then there exists a uniquely determined linear transformation $T' \in L(V', V')$ such that T' and T are transposes of each other. Similarly each linear transformation $S \in L(V', V')$ has a unique transpose in $L(V, V)$.*

Proof. Because of the symmetry in the situation, it is sufficient to prove only the first statement. Let $T \in L(V, V)$; we have to show that for each $v' \in V'$ there exists a unique element $v_1' \in V'$ such that

$$B(v, v_1') = B(T(v), v'), \qquad v \in V, \tag{27.8}$$

so that we can define $T'(v')$ to be v_1'. Because T is a linear transformation, the mapping

$$v \to B(T(v), v')$$

is an element of the dual space of V. By Theorem (27.3) there exists a unique element $v_1' \in V$ such that (27.8) holds. Now we define $T': V' \to V'$ as $T'(v') = v_1'$. Then we have

$$B(v, T'(v')) = B(T(v), v') \tag{27.9}$$

for all $v \in V$, $v' \in V$. We have to prove that $T' \in L(V', V')$. The proof of this fact is a good illustration of how dual vector spaces are used. We have to show that for all $\alpha, \beta \in F$, $v_1', v_2' \in V'$,

$$T'(\alpha v_1' + \beta v_2') = \alpha T'(v_1') + \beta T'(v_2'). \tag{27.10}$$

The idea is to show that

$$B(v, T'(\alpha v_1' + \beta v_2')) = B(v, \alpha T'(v_1') + \beta T'(v_2'))$$

for all $v \in V$, and then use the nondegeneracy of the form to deduce (27.10). We have, from (27.9),

$$\begin{aligned} B(v, T'(\alpha v_1' + \beta v_2')) &= B(Tv, \alpha v_1' + \beta v_2') \\ &= \alpha B(Tv, v_1') + \beta B(Tv, v_2') \\ &= \alpha B(v, T'(v_1')) + \beta B(v, T'(v_2')) \\ &= B(v, \alpha T'(v_1') + \beta T'(v_2')), \end{aligned}$$

as required. We have now proved the existence of a transpose; it remains to prove uniqueness. But if there exist T' and T'' in $L(V', V')$ satisfying

$$B(Tv, v') = B(v, T'(v')) = B(v, T''(v'))$$

for all $v \in V$, $v' \in V'$, then $T' = T''$ by the nondegeneracy of B. This completes the proof of the theorem.

In a sense, we really haven't done anything that goes much beyond the concept of the dual of a vector space and the tranpose of a linear transformation. But the language of dual vector spaces is much simpler and more convenient to use in some applications, as we shall see later in the chapter.

(27.11) Definition. Let V and V' be dual vector spaces over F, with respect to a nondegenerate bilinear form B. Let V_1 and V_1' be subspaces of V and V', respectively. We define the *annihilator* $V_1^{\perp}$ of V_1 (with respect to B) to be the subspace of V' consisting of all v' such that $B(v_1, v') = 0$ for all $v_1 \in V_1$, or simply, $B(V_1, v') = 0$. Similarly, we define $(V_1')^{\perp} = \{v \in V \mid B(v, V_1') = 0\}$.

We note that $V_1^{\perp}$ and $(V_1')^{\perp}$ are really subspaces because of the fact that B is a bilinear form. The nondegeneracy of B is equivalent to the statements $V^{\perp} = 0$ and $(V')^{\perp} = 0$. We remark also that there should be no confusion from using the same symbol $^{\perp}$ for annihilators of subspaces of V or V', since the meaning will be clear from the situation.

The following result is the main theorem on annihilators. In the statement, reference is made to quotient spaces and induced linear transformations (introduced in Section 26).

(27.12) Theorem. *Let V and V' be finite dimensional vector spaces which are dual with respect to a nondegenerate bilinear form B, and let V_1 and V_1' be subspaces of V and V', respectively.*

a. $\dim V_1 + \dim V_1^{\perp} = \dim V$, $\dim (V_1')^{\perp} + \dim V_1' = \dim V'$.

b. $(V_1^{\perp})^{\perp} = V_1$, $((V_1')^{\perp})^{\perp} = V_1'$.

c. *The mapping* $V_1 \to V_1^{\perp}$ *is a one-to-one mapping of the set of subspaces of* V *onto the set of subspaces of* V', *such that* $V_1 \subset V_2$ *implies* $V_1^{\perp} \supset V_2^{\perp}$, *for all subspaces* V_1, V_2 *of* V.

d. *If* $V = V_1 \oplus V_2$, *then* $V' = V_1^{\perp} \oplus V_2^{\perp}$.

e. *The vector spaces* V_1 *and* $V'/V_1^{\perp}$ *are dual vector spaces with respect to the nondegenerate bilinear form* B_1 *defined by*

$$B_1(v_1, [v']) = B(v_1, v'), v_1 \in V_1, [v'] \in V'/V_1^{\perp}.$$

f. *Suppose* $T \in L(V, V)$ *and* $T' \in L(V', V')$ *are transposes of each other with respect to* B, *and suppose* V_1 *is a* T-*invariant subspace of* V. *Then* $V_1^{\perp}$ *is invariant relative to* T', *and the restriction* T_{V_1} *and the induced linear transformation* $(T')_{V'/V_1^{\perp}}$ *are transposes of each other with respect to the bilinear form* B_1 *defined in part* (**e**).

Proof. (a) Let $\{v_1, \ldots, v_k\}$ be a basis for V_1, and extend it to a basis $\{v_1, \ldots, v_n\}$ of V. Since V' is isomorphic to the dual space of V by (27.3), we can find a basis $\{v_1', \ldots, v_n'\}$ of V' corresponding to the dual basis of V^* relative to $\{v_1, \ldots, v_n\}$ (from Section 26). In other words, there is a basis $\{v_1', \ldots, v_n'\}$ of V' such that

$$B(v_i, v_j') = \begin{cases} 0, & i \neq j \\ 1, & i = j \end{cases}. \tag{27.13}$$

We assert that $\{v_{k+1}', \ldots, v_n'\}$ is a basis for $V_1^{\perp}$. Clearly these elements all belong to $V_1^{\perp}$ by (27.13). Now let $v' = \xi_1 v_1' + \cdots + \xi_n v_n' \in V_1^{\perp}$. Since $B(v_1, v') = \cdots = B(v_k, v') = 0$, we have $\xi_1 = \cdots = \xi_k = 0$ by (27.13), and the assertion is proved. We have shown that $\dim V_1^{\perp} = \dim V - \dim V_1$. The proof of the other half of (a) is exactly the same, and will be omitted.

(b) We have $V_1 \subset (V_1^{\perp})^{\perp}$, from the definition of annihilators. From part (a) we have

$$\begin{aligned} \dim (V_1^{\perp})^{\perp} &= \dim V' - \dim (V_1^{\perp}) \\ &= \dim V' - (\dim V - \dim V_1) \\ &= \dim V_1, \end{aligned}$$

since $\dim V' = \dim V$ by Corollary (27.4). Since the subspace $V_1 \subset (V_1^{\perp})^{\perp}$, they are equal. The second part of (b) is proved in exactly the same way.

(c) The definition of annihilator implies that $V_1 \subset V_2$ implies $V_1^{\perp} \supset V_2^{\perp}$. For the one-to-one property, suppose $V_1^{\perp} = V_2^{\perp}$. Applying (b), we have

$$V_1 = (V_1^{\perp})^{\perp} = ((V_2)^{\perp})^{\perp} = V_2.$$

The proof that every subspace of V^1 has the form $V_1^{\perp}$ for some subspace V_1 of V, is left to the reader. [*Hint*: Use the second statement in (b).]

(d) From (a) it follows that $\dim V_1^\perp + \dim V_2^\perp = \dim V'$. It is sufficient to prove that $V_1^\perp \cap V_2^\perp = 0$. Let $v' \in V_1^\perp \cap V_2^\perp$. Since $V = V_1 \oplus V_2$, we have $v' \in V^\perp = \{0\}$. by the nondegeneracy of B.

(e) We first have to show that B_1 is properly defined; in other words, if $v_1 = w_1$ in V_1, and $[v'] = [w']$ in $V'/V_1^\perp$, then we have to show that $B(v_1, v') = B(w_1, w')$. Since $[v'] = [w']$, we have $v' = w' + z'$, for $z' \in V_1^\perp$. Then

$$B(v_1, v') = B(w_1, w' + z') = B(w_1, w'),$$

since $w_1 \in V_1$ and $z' \in V_1^\perp$. The fact that B_1 is bilinear now follows at once from the bilinearity of B, and we shall leave the proof of this fact to the reader. Now we have to prove the nondegeneracy of B_1. Suppose, for $v_1 \in V_1$, that $B_1(v_1, V'/V_1^\perp) = 0$. Then $B(v_1, V') = 0$ by the definition of B_1, and $v_1 = 0$ by the nondegeneracy of B. Now suppose, for $v' \in V'$, that $B_1(V_1, [v']) = 0$. Then $B(V_1, v') = 0$ by the definition of V_1, and hence $v' \in V_1^\perp$. Therefore $[v'] = 0$ in $V'/V_1^\perp$, and part (e) is proved.

(f) We first have to show that $T'(V_1^\perp) \subset V_1^\perp$. Let $v' \in V_1^\perp$, and let $v_1 \in V_1$. Then

$$B(v_1, T'(v')) = B(T(v_1), v') = 0,$$

because $T(V_1) \subset V_1$, and we have $T'(V_1^\perp) \subset V_1^\perp$. It is now sufficient to prove that for $v_1 \in V_1, v' \in V'$,

$$B_1(T_{V_1}(v_1), [v']) = B_1(v_1, T'_{V'/V_1^\perp}([v'])).$$

From the definition of T_{V_1} and $T'_{V/V_1^\perp}$, this statement is equivalent to showing that

$$B_1(T(v_1), [v']) = B_1(v_1, [T'v']). \tag{27.14}$$

The definition of B_1 tells us that (27.14) is equivalent to

$$B(T(v_1), v') = B(v_1, T'v'),$$

which is exactly the condition that T and T' are transposes of each other. This completes the proof of the theorem.

EXERCISES

1. Check the statement made in Example B, that

$$B(v, w) = \sum \sum \alpha_{ij} \xi_i \eta_j$$

does define a bilinear form on the pair of vector spaces (V, W).

2. Let V and W be n-dimensional vector spaces over F, and let B be the bilinear form on (V, W) defined by an n-by-n matrix $\mathbf{A}$ as in Example B. Show that B is nondegenerate if and only if $\mathbf{A}$ has rank n.

3. A bilinear form $B: (V, V) \to F$ on a vector space V with itself, is called *symmetric* if $B(v, v') = B(v', v)$ for all v, v' in V, and *skew symmetric* if $B(v, v') = -B(v', v)$ for all v and v'. Prove that a bilinear form B on (V, V) is symmetric if and only if its matrix $\mathbf{A} = (\alpha_{ij})$ [where $\alpha_{ij} = B(v_i, v_j)$ for some fixed basis $\{v_i\}$ of V] satisfies ${}^t\mathbf{A} = \mathbf{A}$, and is skew symmetric if and only if ${}^t\mathbf{A} = -\mathbf{A}$.
4. Let V be a vector space over the field of real numbers, and let B be a nondegenerate skew symmetric bilinear form on (V, V). Prove that $\dim V$ is even: $\dim V = 2m$ for some positive integer m. [*Hint:* Let $\mathbf{A}$ be the matrix of B, as in Exercise 3. Then ${}^t\mathbf{A} = -\mathbf{A}$, and $D(\mathbf{A}) \neq 0$, by Exercise 2. What do these statements imply?]
5. Let B be a bilinear form on (V, V), where V is a vector space over the real numbers. Prove that B is skew symmetric if and only if $B(v, v) = 0$ for all $v \in V$.
6. Let V be a vector space, and V^* the dual vector space. Prove that for $v \in V$, the mapping $\lambda_v: f \to f(v), f \in V^*$, is an element of $(V^*)^*$. Prove that if V is finite dimensional, then $v \to \lambda_v$ is an isomorphism of V onto $(V^*)^*$.
7. Let V and V' be dual vector spaces with respect to a bilinear form B, and suppose T and T' are transposes of each other with respect to B. Show that T and T' have the same rank.
8. Let V, V', T, T' be as in Exercise 7. Show that α is a characteristic root of T if and only if α is a characteristic root of T'.

28. DIRECT SUMS AND TENSOR PRODUCTS

We are already familiar with direct sums of vector spaces from Chapter 7. We shall start out in this section with an approach to direct sums from a different point of view, to prepare the way for the more difficult concept of tensor product.

(28.1) DEFINITION. Let X and Y be sets. The *cartesian product*, $X \times Y$, of X and Y is the set of all *ordered pairs* (x, y), with $x \in X, y \in Y$. Two ordered pairs (x, y) and (x', y') are defined to be equal if and only if $x = x'$ and $y = y'$.

(28.2) DEFINITION. Let U and V be vector spaces over F. The *(external) direct sum* $U \dotplus V$ of U and V is the vector space whose underlying set is the cartesian product $U \times V$ of U and V, and with vector space operations defined by

$$(u, v) + (u_1, v_1) = (u + u_1, v + v_1),$$
$$\alpha(u, v) = (\alpha u, \alpha v),$$

for all vectors and scalars involved.

It is easily checked (and we omit the details) that $U \dotplus V$ is a vector space. We have taken pains to distinguish $U \dotplus V$ from the concept of an (internal) direct sum $U \oplus V$ defined in Chapter 7, although in a sense they are just different ways of looking at the same thing, as the following theorem will show.

(28.3) THEOREM. *Let U and V be finite dimensional vector spaces with bases $\{u_1, \ldots, u_k\}$, $\{v_1, \ldots, v_l\}$, respectively. Then:*

a. $\{(u_1, 0), \ldots, (u_k, 0), (0, v_1), \ldots, (0, v_l)\}$ *is a basis of* $U \dotplus V$.

b. $\dim (U \dotplus V) = \dim U + \dim V$.

c. *Letting U_1 and V_1 denote the sets $\{(u, 0) \mid u \in U\}$ and $\{(0, v) \mid v \in V\}$, respectively, U_1 and V_1 are subspaces of $U \dotplus V$, and $U \dotplus V$ is the (internal) direct sum, $U_1 \oplus V_1$.*

d. *Let $S \in L(U, U)$, $T \in L(V, V)$ and let $\mathbf{A}$ and $\mathbf{B}$ be the matrices of S and T with respect to the bases $\{u_1, \ldots, u_k\}$ and $\{v_1, \ldots, v_l\}$, respectively. Then $S \dotplus T: U \dotplus V \to U \dotplus V$, defined by*

$$(S \dotplus T)(u, v) = (S(u), T(v))$$

is a linear transformation of $U \dotplus V$, and its matrix with respect to the basis given in (a) is

$$\left(\begin{array}{c|c} \mathbf{A} & 0 \\ \hline 0 & \mathbf{B} \end{array}\right).$$

Proof. We shall give only a sketch of the proof, leaving some details to the reader.

(a) First, we show that the vectors are linearly independent. If

$$\alpha_1(u_1, 0) + \cdots + \alpha_k(u_k, 0) + \beta_1(0, v_1) + \cdots + \beta_l(0, v_l) = 0,$$

then

$$(\alpha_1 u_1 + \cdots + \alpha_k u_k, \beta_1 v_1 + \cdots + \beta_l v_l) = 0,$$

and

$$\alpha_1 u_1 + \cdots + \alpha_k u_k = 0, \qquad \beta_1 v_1 + \cdots + \beta_l v_l = 0.$$

Since $\{u_i\}$ and $\{v_i\}$ are linearly independent sets in U and V, we have $\alpha_1 = \cdots = \alpha_k = \beta_1 = \cdots = \beta_l = 0$. Now let $(u, v) \in U \dotplus V$. Then $u = \sum \alpha_i u_i$, $v = \sum \beta_j v_j$, and hence $(u, v) = \sum \alpha_i(u_i, 0) + \sum \beta_j(0, v_j)$, and part (a) is proved.

(b) Follows from part (a).

(c) The proof that U_1 and V_1 are subspaces is omitted. To show that $U \dotplus V$ is their direct sum, we have to check that every vector in $U \dotplus V$ is a sum of vectors in U_1 and V_1, and that $U_1 \cap V_1 = 0$. First, $(u, v) = (u, 0) + (0, v) \in U_1 + V_1$. Next, $(u, v) \in U_1 \cap V_1$ implies $u = 0$ and $v = 0$, and hence $(u, v) = 0$.

(d) The proof that $S + T$ is a linear transformation and has the matrix indicated is left as an exercise.

With the construction of the direct sum fresh in our minds, it is natural to ask for a vector space constructed from U and V whose dimension is the product (dim U)(dim V), when both of these are finite. At first, what seems to be required is simply to take the vector space having a basis consisting of all "products" $u_i \times v_j$, where u_i and v_j are bases of U and V, respectively. We would then have to impose restrictions on the behaviour of the basis elements, for example, how to interpret $u_i \times \alpha v_j$. Historically, this was the first approach to tensor products. Later various constructions were found which are independent of the choice of bases in U and V. For example, the tensor product $U \otimes V$ may be defined as $L(U^*, V)$, where U^* is the dual space of U, or as $B(U, V)^*$, the dual space of the vector space $B(U, V)$ of bilinear forms on U and V. It turns out that all these definitions are equivalent, and provide a vector space with the same basic properties. We shall take the approach of defining the tensor product in terms of these basic properties, and will give a construction of the tensor product starting with the vector space of functions on the cartesian product set $U \times V$. The other interpretations of the tensor product space, as well as the applications of the idea to linear transformations and matrices, then follow easily.

The proofs of the main theorems are fairly long, and it is probably a good idea for the reader to study the statements of Definitions (28.4), (28.9), Theorem (28.5), and Remark A following Definition (28.9) before tackling the proof of Theorem (28.5).

(28.4) DEFINITION. Let U, V, W be vector spaces over F. A *bilinear function* $\lambda: U \times V \to W$ is a mapping which assigns to each ordered pair $(u, v) \in U \times V$ a uniquely determined element $\lambda(u, v)$ in W, in such a way that the following conditions are satisfied:

$$\lambda(u_1 + u_2, v) = \lambda(u_1, v) + \lambda(u_2, v), \ \lambda(u, v_1 + v_2) = \lambda(u, v_1) + \lambda(u, v_2)$$
$$\lambda(\alpha u, v) = \lambda(u, \alpha v) = \alpha\lambda(u, v)$$

for all u's in U, v's in V, and α's in F.

Some examples of bilinear functions are already familiar. Bilinear forms are bilinear functions in which the third vector space W is the one-dimensional vector space F. Some other examples are the following ones.

Example A. Let V be a vector space over F, and let $\lambda: L(V, V) \times L(V, V) \to L(V, V)$ be the mapping $\lambda(T_1, T_2) = T_1T_2$, where T_1T_2 is the product of the linear transformations T_1 and T_2 in $L(V, V)$. The properties of multiplication of linear transformations show that λ is a bilinear function.

Example B. Let V be a vector space and V^* its dual. For $f \in V^*$ and $v \in V$, define a mapping $f \times v\colon V \to V$ by

$$(f \times v)(u) = f(u)v, \qquad u, v \in V, \quad f \in V^*.$$

Then it is left to the reader to check that (a) $f \times v \in L(V, V)$ for all f and v; and (b) the mapping $(f, v) \to f \times v$ is a bilinear mapping of $V^* \times V \to L(V, V)$.

We are going to show that there exists, for each pair of vector spaces U and V, a vector space $U \otimes V$, called their *tensor product*, with the property that for every bilinear map $\lambda\colon U \times V \to W$, there exists a *linear* transformation $L\colon U \otimes V \to W$ which in a certain sense is equivalent to the *bilinear* function λ. In order to make this idea more precise, it is convenient to introduce the notation $f \circ g$, for the *composite of two mappings of sets*, $g\colon X \to Y, f\colon Y \to Z$, given by

$$(f \circ g)(x) = f(g(x)).$$

This is the usual "function of a function" idea from calculus, and was already used in Chapter 3 to define the product of linear transformations.

(28.5) THEOREM. *Let U and V be vector spaces over a field F. There exists a vector space over F, called $U \otimes V$, and a fixed bilinear mapping $t\colon U \times V \to U \otimes V$, such that $U \otimes V$ is generated*† *by the elements $t(u, v)$, $u \in U$, $v \in V$. Moreover, for every bilinear mapping $\lambda\colon U \times V \to W$, there exists a linear transformation $L\colon U \otimes V \to W$, such that*

$$\lambda = L \circ t.$$

In other words, every bilinear mapping, $\lambda\colon U \times V \to W$, can be factored as a product of a fixed bilinear mapping $t\colon U \times V \to U \otimes V$ of $U \times V$ into a fixed vector space $U \otimes V$, followed by a linear transformation L: $U \otimes V \to W$.

Proof. Let $\mathscr{F}(U \times V)$ be the set of all functions $f\colon U \times V \to F$ such that $f(u, v) = 0$ except for a finite number of pairs $\{(u, v)\}$. Then $\mathscr{F}(U \times V)$ is a vector space over F, if we define the sum $f + f'$ and scalar product αf by the rules

$$(f + f')(u, v) = f(u, v) + f'(u, v),$$
$$(\alpha f)(u, v) = \alpha(f(u, v)).$$

It is clear that if both f and f' vanish except for a finite number of pairs (u, v), then the same is true for $f + f'$ and αf.

† A vector space is generated by a set of vectors $T = \{t\}$, where the set is possibly infinite, if every vector v is a linear combination $\alpha_1 t_1 + \cdots + \alpha_r t_r$, for some finite set $\{t_i\}$ in T, depending on v.

Let $u * v$ denote the function in $\mathscr{F}(U \times V)$ such that

$$(u * v)(u_1, v_1) = \begin{cases} 1 & \text{if } (u_1, v_1) = (u, v) \\ 0 & \text{if } (u_1, v_1) \neq (u, v). \end{cases}$$

We shall prove that an arbitrary function f in $\mathscr{F}(U \times V)$ is a linear combination of the functions $u * v$. Suppose $f(u, v) = 0$ except for $(u, v) \in \{(u_1, v_1), \ldots, (u_k, v_k)\}$; then

$$f = f(u_1, v_1)(u_1 * v_1) + \cdots + f(u_k, v_k)(u_k * v_k)$$

as we see by applying both sides to $(u_1, v_1), \ldots, (u_k, v_k)$ in turn, and using the definition of $u_i * v_i$. Moreover the coefficients in such a linear combination are uniquely determined, because if

$$\alpha_1(u_1 * v_1) + \cdots + \alpha_k(u_k * v_k) = \alpha_1'(u_1 * v_1) + \cdots + \alpha_k'(u_k * v_k);$$

then, applying both sides to (u_1, v_1), (u_2, v_2), etc., we see that $\alpha_1 = \alpha_1'$, $\alpha_2 = \alpha_2'$, etc.

Now we are ready to define $U \otimes V$. Let Y be the subspace of $\mathscr{F}(U \times V)$ generated by all functions

$$\begin{aligned} &(u_1 + u_2) * v - u_1 * v - u_2 * v, \\ &u * (v_1 + v_2) - u * v_1 - u * v_2, \\ &\alpha u * v - \alpha(u * v), \\ &u * \alpha v - \alpha(u * v), \end{aligned}$$

for $u, u_1, u_2 \in U$, $v, v_1, v_2 \in V$, and $\alpha \in F$. We now define $U \otimes V$ to be the quotient space $\mathscr{F}(U \times V)/Y$ (as defined in Section 26). Then $U \otimes V$ consists of all equivalence classes $[f]$, for $f \in \mathscr{F}(U \otimes V)$. Because each f is a linear combination of the functions $u * v$, it follows that every element of $U * V$ is a linear combination

$$\alpha_1 [u_1 * v_1] + \cdots + \alpha_k [u_k * v_k]$$

with $\alpha_i \in F$ and $u_i \in U$, $v_i \in V$.

We next define a bilinear function $t: U \times V \to U \otimes V$ by setting

$$t(u, v) = [u * v] \tag{28.6}$$

for all $u \in U$, $v \in V$. To show that t is bilinear, we start with the fact that since $U \otimes V = \mathscr{F}(U, V)/Y$, we have $[y] = 0$ in $U \otimes V$, for all $y \in Y$. Because of the definition of Y, we have

$$\begin{aligned} &[(u_1 + u_2) * v] - [u_1 * v] - [u_2 * v] = 0, \\ &[u * (v_1 + v_2)] - [u * v_1] - [u * v_2] = 0, \\ &[\alpha u * v] - [\alpha(u * v)] = [u * \alpha v] - [\alpha(u * v)] - 0, \end{aligned}$$

for all the vectors and scalars involved. The equations translate into the formulas

$$t(u_1 + u_2, v) = t(u_1, v) + t(u_2, v),$$
$$t(u, v_1 + v_2) = t(u, v_1) + t(u, v_2),$$
$$t(\alpha u, v) = \alpha t(u, v) = t(u, \alpha v),$$

which state precisely that t is a bilinear function.

Moreover, since every $f \in \mathscr{F}(U, V)$ is a linear combination of the functions $u * v$, it follows that $U \otimes V$ is generated by the elements $t(u, v)$, $u \in U, v \in V$.

Finally, we have to check that bilinear functions can be factored in the required way. Let $\lambda: U \times V \rightarrow W$ be an arbitrary bilinear function. Then we can define a linear transformation $\lambda_1: \mathscr{F}(U \times V) \rightarrow W$ by setting

(28.7) $$\lambda_1(\alpha_1(u_1 * v_1) + \cdots + \alpha_k(u_k * v_k)) = \alpha_1\lambda(u_1, v_1) + \cdots + \alpha_k\lambda(u_k, v_k),$$

for all $\alpha_i \in F$, and $u_i \in V$, $v_i \in V$. The definition of λ_1 is legitimate because we have shown that every element of $\mathscr{F}(U \times V)$ is a linear combination of the functions $u_i * v_i$, and that the coefficients in such a linear combination are uniquely determined. Checking that λ_1 is a linear transformation is by now a routine matter, and we shall omit the details.

The important point is to observe that because λ is bilinear, the subspace Y is contained in the nullspace of λ_1. For example, using the definition of λ_1 we have

$$\lambda_1((u_1 + u_2) * v - u_1 * v - u_2 * v) = \lambda(u_1 + u_2, v) - \lambda(u_1, v) - \lambda(u_2, v) = 0,$$

since λ is bilinear. Similarly the other generators of Y belong to the null space of λ_1.

Now we can define a linear transformation $L: U \otimes V \rightarrow W$ by setting

(28.8) $$L([f]) = \lambda_1(f), \quad f \in \mathscr{F}(U \times V).$$

The definition of L is justified because if $[f] = [f']$, then $f - f' \in Y$, and hence $\lambda_1(f - f') = \lambda_1(f) - \lambda_1(f') = 0$. The fact that L is linear is easily checked. (See also Exercise 5 of Section 26.)

We have to verify now that

$$L \circ t = \lambda.$$

For $u \in U$, $v \in V$, we have by (28.6), (28.7) and (28.8),

$$(L \circ t)(u, v) = L([u * v]) = \lambda_1(u * v) = \lambda(u, v),$$

and the theorem is proved.

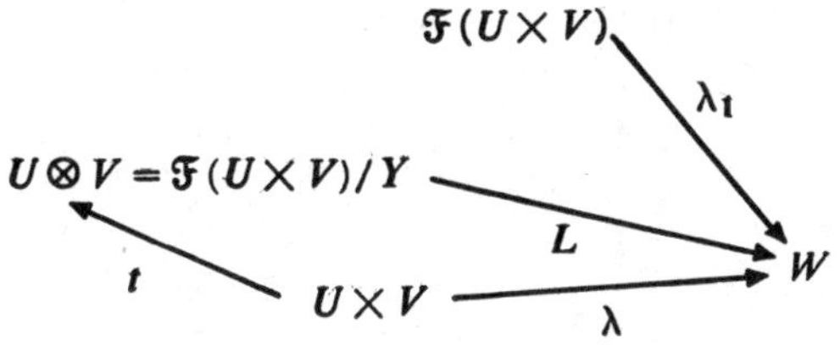

FIGURE 8.1

Figure 8.1 illustrates the various steps of the proof. This kind of a proof is sometimes called a proof by "general nonsense," which means that the proof is an application of the general ideas and constructions that we have made and is not easy to illustrate in terms of numerical examples. We shall now proceed to get a more concrete hold on $U \otimes V$.

(28.9) DEFINITION. A vector space $U \otimes V$ satisfying the conditions of the preceding theorem is called a *tensor product* of U and V. The bilinear map

$$t: U \times V \to U \otimes V$$

will be denoted by

$$t(u, v) = u \otimes v.$$

Remark. We remark that the product $u \otimes v$ has the following properties, which were all derived in the proof of the theorem.

(a) $u \otimes v$ is bilinear:

$$(\alpha u + \beta u') \otimes v = \alpha(u \otimes v) + \beta(u' \otimes v)$$
$$u \otimes (\alpha v + \beta v') = \alpha(u \otimes v) + \beta(u \otimes v')$$

for all vectors and scalars involved.

(b) The elements $u \otimes v$ generate $U \otimes V$: Every element of $U \otimes V$ can be expressed as a linear combination

$$\alpha_1(u_1 \otimes v_1) + \cdots + \alpha_k(u_k \otimes v_k), \qquad u_i \in U, \quad v_i \in V.$$

(c) For every bilinear function $\lambda: U \times V \to W$, there exists a linear transformation $L: U \otimes V \to W$ such that

$$L(u \otimes v) = \lambda(u, v),$$

for all $u \in U$, $v \in V$ (see Figure 8.2).

The next result gives the basic properties of $U \otimes V$ in case U and V are finite dimensional spaces.

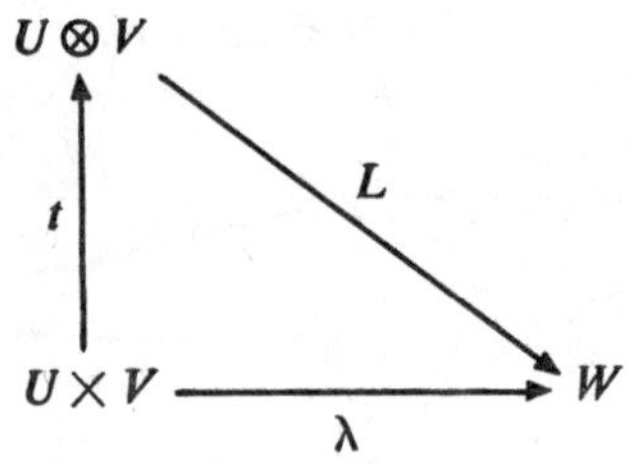

FIGURE 8.2

(28.10) THEOREM. *Let U and V be finite dimensional vector spaces over F with bases $\{u_1, \ldots, u_m\}$, $\{v_1, \ldots, v_n\}$, respectively.*

a. $\{u_i \otimes v_j\}$, *for* $1 \le i \le m$, $1 \le j \le n$, *is a basis of* $U \otimes V$.
b. $\dim (U \otimes V) = (\dim U)(\dim V)$.
c. *Let $S \in L(U, U)$, and let $T \in L(V, V)$, and let $\mathbf{A}$ and $\mathbf{B}$ be the matrices of S and T with respect to the bases $\{u_i\}$ and $\{v_j\}$, respectively. Then there exists a linear transformation*

$$S \otimes T: U \otimes V \to U \otimes V$$

such that for all $u \in U$, $v \in V$,

$$(S \otimes T)(u \otimes v) = S(u) \otimes T(v).$$

The matrix $\mathbf{C}$ of $S \otimes T$ with respect to the ordered basis

$$\{u_1 \otimes v_1, \ldots, u_1 \otimes v_n, \\ u_2 \otimes v_1, \ldots, u_2 \otimes v_n, \ldots, u_m \otimes v_1, \ldots, u_m \otimes v_n\}$$

is given as follows

$$\mathbf{C} = \begin{pmatrix} \alpha_{11}\mathbf{B} & \alpha_{12}\mathbf{B} & \alpha_{1m}\mathbf{B} \\ \cdots & \cdots & \cdots \\ \alpha_{m1}\mathbf{B} & \alpha_{m2}\mathbf{B} & \alpha_{mn}\mathbf{B} \end{pmatrix}.$$

Thus $\mathbf{C}$ is an mn-by-mn matrix, made up of m rows and columns of n-by-n blocks, with the block in the ith block row and jth block column being given by $\alpha_{ij}\mathbf{B}$. The matrix $\mathbf{C}$ will be denoted by

$$\mathbf{C} = \mathbf{A} \dot{\times} \mathbf{B},$$

and called the tensor product (*or* Kronecker product) *of* $\mathbf{A}$ *and* $\mathbf{B}$.

Proof. (*a*) Let $u = \sum \alpha_i u_i \in U$, $v = \sum \beta_j v_j \in V$. Then by the bilinearity of $u \otimes v$,

$$u \otimes v = (\sum \alpha_i u_i) \otimes (\sum \beta_j u_j) = \sum \alpha_i \beta_j (u_i \otimes v_j),$$

and we have proved that the vectors $\{u_i \otimes v_j\}$ generate $U \otimes V$. In order to prove that the vectors $\{u_i \otimes v_j\}$ are linearly independent, it will be sufficient to prove that $\dim U \otimes V \ge mn$ (why?). This will follow if we can find a linear transformation L of $U \otimes V$ *onto* some vector space W of dimension mn. This roundabout approach is necessary because all we really know about $U \otimes V$ is that we can construct linear transformations in $L(U \otimes V, W)$ under certain conditions. The construction of L uses the full force of the theory of dual vector spaces, from Section 27.

Let U^* be the dual space of U, and let $B(u, u^*)$ be the nondegenerate bilinear form on $U \times U^* \to F$, given by

$$B(u, u^*) = u^*(u), \qquad u \in U, \quad u^* \in U^*.$$

Let $u \in U$, $v \in V$, and define a linear transformation $u \times v \in L(U^*, V)$ by

$$(u \times v)(u^*) = B(u, u^*)v.$$

It is left to the reader to check that $u \times v$ is in fact a linear transformation and that, moreover, the mapping

$$\lambda: (u, v) \to u \times v$$

is a bilinear mapping of $U \times V \to L(U^*, V)$. By Theorem (28.5) there exists a linear transformation

$$L: U \otimes V \to L(U^*, V)$$

such that

$$L(u \otimes v) = u \times v.$$

It is now sufficient to prove that L maps $U \otimes V$ *onto* $L(U^*, V)$, because $\dim L(U^*, V) = mn$. Let $T \in L(U^*, V)$ and let $\{v_1, \ldots, v_n\}$ be a basis of V. Then for $u^* \in V^*$, we have

$$T(u^*) = \sum_{i=1}^{n} \xi_i v_i,$$

with uniquely determined coefficients $\xi_i \in F$. The coefficients ξ_i are functions of u^*,

$$T(u^*) = \sum_{i=1}^{n} \xi_i(u^*) v_i.$$

In fact, they are linear functions on U^* because, for example,

$$\begin{aligned} T(u_1^* + u_2^*) &= \sum_{i=1}^{n} \xi_i(u_1^* + u_2^*) v_i \\ &= T(u_1^*) + T(u_2^*) = \sum \xi_i(u_1^*) v_i + \sum \xi_i(u_2^*) v_i. \end{aligned}$$

Then

$$\sum \xi_i(u_1^* + u_2^*) v_i = \sum (\xi_i(u_1^*) + \xi_i(u_2^*)) v_i;$$

and comparing coefficients (since the $\{v_i\}$ are linearly independent), we have

$$\xi_i(u_1^* + u_2^*) = \xi_i(u_1^*) + \xi_i(u_2^*), \qquad 1 \le i \le n.$$

A similar argument shows that

$$\xi_i(\alpha u^*) = \alpha\xi_i(u^*), \qquad u^* \in U^*, \quad \alpha \in F.$$

We can now apply Theorem (27.3) to obtain elements $\{u_1, \ldots, u_m\}$ in U such that

$$B(u_i, u^*) = \xi_i(u^*), \qquad \text{for } 1 \leq i \leq n,$$

and hence

$$T(u^*) = \sum_{i=1}^{n} \xi_i(u^*)v_i = \sum_{i=1}^{n} B(u_i, u^*)v_i.$$

Therefore,

$$T = \sum_{i=1}^{n} u_i \times v_i.$$

Then

$$L(\sum u_i \otimes v_i) = \sum u_i \times v_i = T,$$

and we have proved that L is onto. Since $\dim L(U^*, V) = mn$, we have $\dim (U \otimes V) \geq mn$, and it follows that the vectors $\{u_i \otimes v_j\}$ are linearly independent. This completes the proof of part (a).

(*b*) Part (b) is an immediate consequence of part (a).

(*c*) We first use the fact that S and T are linear, and that $u \otimes v$ is a bilinear function, to show that the function

$$\mu: (u, v) \to S(u) \otimes T(v), \qquad u \in U, \quad v \in V,$$

is a bilinear mapping of $U \times V \to U \otimes V$. By Theorem (28.5), there exists a linear transformation in $L(U \otimes V, U \otimes V)$, which we shall denote by $S \otimes T$, such that

$$(S \otimes T)(u \otimes v) = S(u) \otimes T(v).$$

It remains to compute the matrix of $S \otimes T$ with respect to the given basis We have

$$S(u_i) = \sum_{k=1}^{m} \alpha_{ki}u_k, \qquad T(v_j) = \sum_{l=1}^{n} \beta_{lj}v_l,$$

where $\mathbf{A} = (\alpha_{ij})$, $\mathbf{B} = (\beta_{kl})$ are the matrices of S and T with respect to the given bases. Then

$$(S \otimes T)(u_i \otimes v_j) = \sum_{k=1}^{m} \sum_{l=1}^{n} \alpha_{ki}\beta_{lj}(u_k \otimes v_l).$$

To find the matrix of $S \otimes T$ with respect to the ordered basis of $U \otimes V$, $\{u_1 \otimes v_1, \ldots, u_1 \otimes v_n, u_2 \otimes v_1, \ldots, u_2 \otimes v_n, \text{etc.}\}$, we simply write down

	$u_i \otimes v_j$
$u_1 \otimes v_1$	$\alpha_{1i}\beta_{1j}$
$\vdots$	$\vdots$
$u_1 \otimes v_n$	$\alpha_{1i}\beta_{nj}$
$u_2 \otimes v_1$	$\alpha_{2i}\beta_{1j}$
$\vdots$	$\vdots$
$u_2 \otimes v_n$	$\alpha_{2i}\beta_{nj}$
$\vdots$	$\vdots$
$u_m \otimes v_1$	$\alpha_{mi}\beta_{1j}$
$\vdots$	$\vdots$
$u_m \otimes v_n$	$\alpha_{mi}\beta_{nj}$

FIGURE 8.3

the column of the matrix of $S \otimes T$ giving $(S \otimes T)(u_i \otimes v_j)$, arranged according to the basis we have chosen for $U \otimes V$. This is shown in Figure 8.3.

These remarks show that the matrix of $S \otimes T$ has the required form, and the proof of the theorem is completed.

In order to use tensor products, it is unnecessary to remember the details of the proofs of the last two theorems, which have undoubtedly been a strain to the reader's patience. The main facts are all contained in the statements of the theorems, and in Remark A, following the proof of Theorem (28.5).

We now state some easy but important facts about tensor products of linear transformations.

(28.11) THEOREM. *Let U and V be finite dimensional vector spaces over F.*

a. *Let $S_1, S_2 \in L(U, U)$, and let $T_1, T_2 \in L(V, V)$. Then*

$$(S_1 \otimes T_1)(S_2 \otimes T_2) = S_1S_2 \otimes T_1T_2.$$

b. *Let* $\mathbf{A}_1, \mathbf{A}_2$ *be m-by-m matrices, and let* $\mathbf{B}_1, \mathbf{B}_2$ *be n-by-n matrices. Then*

$$(\mathbf{A}_1 \dot{\times} \mathbf{B}_1)(\mathbf{A}_2 \dot{\times} \mathbf{B}_2) = \mathbf{A}_1\mathbf{A}_2 \dot{\times} \mathbf{B}_1\mathbf{B}_2 .$$

c. *Let* $S \in L(U, U), T \in L(V, V)$. *Then the trace of* $S \otimes T$ *is given by*

$$\operatorname{Tr}(S \otimes T) = (\operatorname{Tr} S)(\operatorname{Tr} T).$$

Proof. (a) Since the vectors $u \otimes v$ generate $U \otimes V$, it is sufficient to prove that the equation holds when both sides are applied to $u \otimes v$. Using the definitions of $S_1 \otimes T_1, S_2 \otimes T_2$ from (Theorem 28.10), we have

$$\begin{aligned}(S_1 \otimes T_1)(S_2 \otimes T_2)(u \otimes v) &= (S_1 \otimes T_1)(S_2(u) \otimes T_2(v)) \\ &= S_1S_2(u) \otimes T_1T_2(v)\end{aligned}$$

as required.

Part (b) follows from (a) since $\mathbf{A}_1 \dot{\times} \mathbf{B}_1$, $\mathbf{A}_2 \dot{\times} \mathbf{B}_2$ and $\mathbf{A}_1\mathbf{A}_2 \dot{\times} \mathbf{B}_1\mathbf{B}_2$ are the matrices of $S_1 \otimes T_1$, $S_2 \otimes T_2$ and $S_1S_2 \otimes T_1T_2$ with respect to a basis of $U \otimes V$. (The reader should note that we have at this point proved an interesting and new fact about matrix multiplication.)

In order to prove part (c), we use the facts that (i) the trace of linear transformation $S \otimes T$ is the trace of the matrix $\mathbf{A} \dot{\times} \mathbf{B}$ corresponding to the linear transformation with respect to a basis, and that (ii) the trace of a matrix is, from Section 24, the sum of the diagonal elements. From part (c) of Theorem (28.10), the diagonal blocks of $\mathbf{A} \dot{\times} \mathbf{B}$ are

$$\alpha_{11}\mathbf{B}, \ldots, \alpha_{mm}\mathbf{B}.$$

The diagonal elements of $\mathbf{A} \dot{\times} \mathbf{B}$ are therefore

$$\alpha_{11}\beta_{11}, \ldots, \alpha_{11}\beta_{nn}, \alpha_{22}\beta_{11}, \ldots, \alpha_{22}\beta_{nn}, \ldots, \alpha_{mm}\beta_{11}, \ldots, \alpha_{mm}\beta_{nn}.$$

Their sum is

$$(\alpha_{11} + \cdots + \alpha_{mm})(\beta_{11} + \cdots + \beta_{nn}) = (\operatorname{Tr} \mathbf{A})(\operatorname{Tr} \mathbf{B})$$

as we wished to show. This completes the proof of the theorem.

We conclude our introduction to tensor products with one important application. Other applications and examples are given in the exercises. The exposition for the rest of the section will be more condensed than usual, and several steps in the discussion will be left to the reader.

(28.12) THEOREM. (ASSOCIATIVITY OF THE TENSOR PRODUCT). *Let* V_1, V_2, *and* V_3 *be finite dimensional vector spaces over* F. *There exists an isomorphism of vector spaces between* $V_1 \otimes (V_2 \otimes V_3)$ *and* $(V_1 \otimes V_2) \otimes V_3$, *which carries* $v_1 \otimes (v_2 \otimes v_3) \to (v_1 \otimes v_2) \otimes v_3$, *for all* $v_1 \in V_1$, $v_2 \in V_2$, $v_3 \in V_3$.

Proof. Let $\{v_{1i}\}$ be a basis for V_1, $\{v_{2j}\}$ a basis for V_2, and $\{v_{3k}\}$ a basis for V_3. By Theorem (28.10), the vectors $\{v_{1i} \otimes (v_{2j} \otimes v_{3k})\}$ form a basis for $V_1 \otimes (V_2 \otimes V_3)$, while the vectors $\{(v_{1i} \otimes v_{2j}) \otimes v_{3k}\}$ form a basis for $(V_1 \otimes V_2) \otimes V_3$. There exists a linear transformation T which carries $v_{1i} \otimes (v_{2j} \otimes v_{3k})$ onto $(v_{1i} \otimes v_{2j}) \otimes v_{3k}$, for all i, j, k and T is an isomorphism of vector spaces. The fact that the isomorphism carries $v_1 \otimes (v_2 \otimes v_3)$ to $(v_1 \otimes v_2) \otimes v_3$ for all $v_1 \in V_1$, $v_2 \in V_2$ and $v_3 \in V_3$ is easily checked by expanding v_1, v_2, v_3 as linear combinations of the basis elements $\{v_{1i}\}$, $\{v_{2j}\}$, and $\{v_{3k}\}$, respectively.

We shall identify $V_1 \otimes (V_2 \otimes V_3)$ with $(V_1 \otimes V_2) \otimes V_3$, and call this vector space $V_1 \otimes V_2 \otimes V_3$ (without parentheses). We shall also write $v_1 \otimes v_2 \otimes v_3$ for the element in $V_1 \otimes V_2 \otimes V_3$ corresponding to $v_1 \otimes (v_2 \otimes v_3)$. Of course there is nothing special about a tensor product of three vector spaces; we can equally well form the tensor product vector space $V_1 \otimes V_2 \otimes \cdots \otimes V_m$ of m vector spaces $V_1, \ldots, V_m$. The elements of $V_1 \otimes V_2 \otimes \cdots \otimes V_m$ are sums

$$\sum v_{1i_1} \otimes v_{2i_2} \otimes \cdots \otimes v_{mi_m}$$

with $v_{1i_1} \in V_1$, $v_{2i_2} \in V_2, \ldots, v_{mi_m} \in V_m$.

(28.13) Definition. Let V be a finite dimensional vector space and k a positive integer. The *k-fold tensor product* $\bigotimes_k (V)$ (or the *vector space of k-fold tensors over* V) is the vector space $\underbrace{V \otimes \cdots \otimes V}_{k}$ (k factors). The elements of $\bigotimes_k (V)$ are called *k-fold tensors* and are sums of the form

$$t = \sum_{(i)} v_{i_1} \otimes v_{i_2} \otimes \cdots \otimes v_{i_k}, \qquad v_{i_j} \in V$$

where the sum is taken over some set of k-tuples $(i) = (i_1, \ldots, i_k)$, which are used to index the vectors from V which are involved in the tensor t.

We now show how to define symmetry operations on the tensor space $\bigotimes_k (V)$. Let σ be a permutation of the set $\{1, 2, \ldots, k\}$ (see Section 19). There exists a linear transformation $S_\sigma \in L(\bigotimes_k V, \bigotimes_k V)$ such that

$$S_\sigma(v_{i_1} \otimes \cdots \otimes v_{i_k}) = v_{i_{\sigma(1)}} \otimes \cdots \otimes v_{i_{\sigma(k)}},$$

for all $\{v_{i_1}, \ldots, v_{i_k}\} \in V$. Notice that S_σ does not change the *set* of vectors $\{v_{i_1}, \ldots, v_{i_k}\}$ in the tensor $v_{i_1} \otimes \cdots \otimes v_{i_k}$; it simply rearranges their order of occurrence. For example, in $\bigotimes_2 V$, let $\sigma = (12)$; then

$$S_\sigma(v \otimes w) = w \otimes v$$

for all vectors $v, w \in V$. In particular, $S_\sigma(v \otimes v) = v \otimes v$. The existence of S_σ is easily shown by defining S_σ on a basis of $\bigotimes_k V$.

(28.14) DEFINITION. A tensor $t \in \bigotimes_k V$ is called *symmetric* if $S_\sigma t = t$ for all permutations σ of $\{1, 2, \ldots, k\}$, and *skew symmetric* if $S_\sigma t = \epsilon(\sigma)t$ for all σ, where $\epsilon(\sigma)$ is the signature of σ [see Definition (19.16)].

Example A. In $\bigotimes_2 V$, the tensors of the form $\{v \otimes v \mid v \in V\}$ are symmetric, while those of the form $\{v \otimes w - w \otimes v \mid v, w \in V, v \neq w\}$ are skew symmetric.

Our goal is to study the space of skew symmetric tensors in $\bigotimes_k V$. We shall see that they have features which extend the most important properties of the determinant function.

(28.15) THEOREM.

a. *The set of skew symmetric tensors in $\bigotimes_k (V)$ form a subspace, which will be denoted by $\bigwedge_k (V)$ (read: the k-fold* wedge product *of V).*

b. *Let $v_1, \ldots, v_k$ be vectors in V. Then $v_1 \wedge \cdots \wedge v_k = \sum_\sigma \epsilon(\sigma)(v_{\sigma(1)} \otimes \cdots \otimes v_{\sigma(k)})$ is an element of $\bigwedge_k (V)$, for all $\{v_i\}$ in V.*

c. *The wedge product $v_1 \wedge \cdots \wedge v_k$ has the following properties:*

(*i*) *$v_1 \wedge \cdots \wedge v_k$ is linear, when viewed as a function of any one factor, for example,*

$$(\alpha v_1 + \beta v_1') \wedge v_2 \wedge \cdots \wedge v_k = \alpha(v_1 \wedge v_2 \wedge \cdots \wedge v_k) + \beta(v_1' \wedge v_2 \wedge \cdots \wedge v_k),$$

(*ii*) $\underbrace{v_1 \wedge \cdots \wedge \underset{i}{v_i} \wedge \cdots \wedge \underset{j}{v_j} \wedge \cdots \wedge v_k}_{} = -\underbrace{v_1 \wedge \cdots \wedge \underset{i}{v_j} \wedge \cdots \wedge \underset{j}{v_i} \wedge \cdots \wedge v_k}_{}$. *(Notice the similarity with the properties of the determinant function.)*

d. *$v_1 \wedge \cdots \wedge v_k \neq 0$ if and only if the vectors $v_1, \ldots, v_k$ are linearly independent.*

e. *Let $\{v_1, \ldots, v_n\}$ be a basis for V. Then the vectors*

$$\{v_{i_1} \wedge \cdots \wedge v_{i_k},\quad i_1 < i_2 < \cdots < i_n\}$$

form a basis for $\bigwedge_k V$.

Remark. Before giving the proof, we point out that (d) provides a far-reaching extension of the idea which led to the determinant function. The determinant provides a test for linear independence of a set of n vectors in an n-dimensional space. Part (d) provides a similar test which applies to k vectors in n-dimensional space, for $k = 1, 2, \ldots$.

Proof. (*a*) The fact that the skew symmetric tensors form a subspace of $\bigotimes_k V$ follows at once from the fact that the symmetry operators S_σ are linear transformations.

(*b*) Let τ be a permutation of $\{1, 2, \ldots, k\}$. Then

$$\begin{aligned} S_\tau(v_1 \wedge \cdots \wedge v_k) &= \sum_\sigma \epsilon(\sigma)(v_{\sigma\tau(1)} \otimes \cdots \otimes v_{\sigma\tau(k)}) \\ &= \epsilon(\tau) \sum_\sigma \epsilon(\sigma\tau)(v_{\sigma\tau(1)} \otimes \cdots \otimes v_{\sigma\tau(k)}) \\ &= \epsilon(\tau)(v_1 \wedge \cdots \wedge v_k) \end{aligned}$$

since $\epsilon(\sigma\tau) = \epsilon(\sigma)\epsilon(\tau)$ from Section 19, and since the mapping $\sigma \to \sigma\tau$ is simply a rearrangement of the permutations of $\{1, 2, \ldots, k\}$.

(*c*) Property (i) is clear, and (ii) follows by applying S_σ, with σ the transposition (ij), to $v_1 \wedge \cdots \wedge v_k$.

(*d*) If $v_1, \ldots, v_k$ are linearly dependent, then $v_1 \wedge \cdots \wedge v_k = 0$, by part (c) and exactly the reasoning given in Section 16 to prove the analogous result for determinants.

Now suppose $v_1, \ldots, v_k$ are linearly independent. Find vectors $v_{k+1}, \ldots, v_k$ such that $\{v_1, \ldots, v_n\}$ is a basis of V. Then the vectors appearing in the sum defining $v_1 \wedge \cdots \wedge v_k$ are distinct and form part of a basis of $\bigotimes_k V$. Hence $v_1 \wedge \cdots \wedge v_k \neq 0$.

(*e*) The vectors $\{v_{j_1} \otimes \cdots \otimes v_{j_k} \mid 1 \leq j_1, \ldots, j_k \leq n\}$ form a basis for $\bigotimes_k V$. Let

$$t = \sum \alpha_{j_1 \cdots j_k} (v_{j_1} \otimes \cdots \otimes v_{j_k}), \qquad \alpha_{j_1 \cdots j_k} \in F,$$

be an arbitrary skew symmetric tensor, and suppose for some set of indices $(j_1, \ldots, j_k)$, $\alpha_{j_1 \cdots j_k} \neq 0$. It follows from the definition of skew symmetric tensors, and by comparing coefficients, that

$$\alpha_{j_{\sigma(1)} \cdots j_{\sigma(k)}} = \epsilon(\sigma)\alpha_{j_1 \cdots j_k}$$

for all permutations σ of $\{1, 2, \ldots, k\}$. Therefore

$$t \pm \alpha_{j_1 \cdots j_k} (v_{j_1} \wedge \cdots \wedge v_{j_k})$$

a skew symmetric tensor with fewer nonzero summands than t. Continuing by induction, it follows that t is a linear combination of vectors of the form $\{v_{j_1} \wedge \cdots \wedge v_{j_k}\}$. Since $v_{j_1} \wedge \cdots \wedge v_{j_k}$ is $\pm\, v_{j_1'} \wedge \cdots \wedge v_{j_k'}$, with $j_1' < j_2' < \cdots < j_k'$, it follows that the vectors of the required form generate $\bigwedge_k (V)$. Finally, it is clear that the vectors $\{v_{j_1} \wedge \cdots \wedge v_{j_k} \mid j_1 < \cdots < j_k\}$ are linearly independent, because they involve different sets of basis vectors of $\bigotimes_k (V)$. This completes the proof of the theorem.

EXERCISES

In all the exercises, the vector spaces involved are assumed to be finite dimensional.

1. Compute $\mathbf{A}_1 \dot{\times} \mathbf{B}_1$, $\mathbf{A}_2 \dot{\times} \mathbf{B}_2$, and $(\mathbf{A}_1 \dot{\times} \mathbf{B}_1)(\mathbf{A}_2 \dot{\times} \mathbf{B}_2)$, where

$$\mathbf{A}_1 = \begin{pmatrix} 1 & 1 \\ 0 & 1 \end{pmatrix}, \quad \mathbf{B}_1 = \begin{pmatrix} 1 & 0 \\ 0 & 2 \end{pmatrix},$$
$$\mathbf{A}_2 = \begin{pmatrix} -1 & 0 \\ 0 & 1 \end{pmatrix}, \quad \mathbf{B}_2 = \begin{pmatrix} -1 & 0 \\ 1 & 0 \end{pmatrix}.$$

Check that the final result agrees with $\mathbf{A}_1\mathbf{A}_2 \dot{\times} \mathbf{B}_1\mathbf{B}_2$.

2. Let $\{u_1, \ldots, u_k\}$ be linearly independent vectors in U, and $\{v_1, \ldots, v_k\}$ arbitrary vectors in V. Show that $\sum u_i \otimes v_i = 0$ in $U \otimes V$ implies that $v_1 = \cdots = v_k = 0$.

3. Let $S \in L(U, U)$, $T \in L(V, V)$, and let α and β be characteristic roots of S and T, respectively, belonging to characteristic vectors $u \in U$ and $v \in V$. Prove that $u \otimes v \neq 0$ in $U \otimes V$ (using Exercise 2, for example) and that $u \otimes v$ is a characteristic vector of $S \otimes T$ belonging to the characteristic root $\alpha\beta$.

4. Suppose $\mathbf{A}$ and $\mathbf{B}$ are matrices in triangular form, with zeros above the diagonal. Show that $\mathbf{A} \dot{\times} \mathbf{B}$ has the same property, and hence that every characteristic root of $\mathbf{A} \dot{\times} \mathbf{B}$ can be expressed in the form $\alpha\beta$, where α is a characteristic root of $\mathbf{A}$ and β a characteristic root of $\mathbf{B}$.

5. Let $S \in L(U, U)$, $T \in L(V, V)$ and suppose the base field is algebraically closed. Prove that every characteristic root of $S \otimes T$ can be expressed in the form $\alpha\beta$, with α and β characteristic roots of S and T, respectively.

6. a. Prove that every linear transformation $T \in L(U, V)$ can be expressed in the form $\sum f_i \times v_i$, with $\{f_i\}$ in U^* and $\{v_i\}$ in V, where $f \times v$ is the linear transformation in $L(U, V)$ defined by $(f \times v)(u) = f(u)v$, $u \in U$, $f \in U^*$, $v \in V$. [*Hint:* Follow the proof of Theorem (28.10).]

   b. Let $X \in L(U, V)$ be expressed in the form $\sum f_i \times v_i$, $f_i \in U^*$, $v_i \in V$, according to part (*a*). Let $S \in L(U, U)$, $T \in L(V, V)$. Show that

$$(\textstyle\sum f_i \times v_i)S = \sum S^* f_i \times v_i,$$

   where S^* is the transpose of S. Also show that

$$T(\textstyle\sum f_i \times v_i) = \sum f_i \times Tv_i.$$

   c. Prove that $U^* \otimes V$ is isomorphic to $L(U, V)$, and that $\sum f_i \otimes v_i \to \sum f_i \times v_i$ gives the isomorphism. Letting S and T be as in part (b), $S^* \otimes T$ acts on $L(U, V)$ via the isomorphism of $L(U, V)$ with $U^* \otimes V$ in the following way:

$$(S^* \otimes T)(\textstyle\sum f_i \times v_i) = \sum S^* f_i \times Tv_i = T(\sum f_i \times v_i)S.$$

d. Prove that every $T \in L(U, V)$ can be expressed in the form $\sum f_i \times v_i$ with the $\{v_i\}$ linearly independent, and that in this case the rank of T is the number of nonzero summands in $\sum f_i \times v_i$.

e. Let $T \in L(U, V)$ be expressed in the form $\sum f_i \times v_i$. Prove that the trace of T is given by $\sum f_i(v_i)$.

7. Let $S \in L(U, U)$, $T \in L(V, V)$ and let the base field be algebraically closed. Suppose that S and T are invertible. Prove that if there exists a linear transformation $X \neq 0$ in $L(U, V)$ such that $XS = TX$, then S and T have a common characteristic root. [*Hint:* Using Exercise 6, put X in the form $X = \sum f_i \times v_i$, $f_i \in U^*$, $v_i \in V$. Then $XS = TX$ implies $T^{-1}XS = X$. Therefore, by part (c) of 5,

$$(S^* \otimes T^{-1})X = X,$$

and $S^* \otimes T^{-1}$ has a characteristic root equal to 1. Using Exercise 5, and properties of characteristic roots of S^* and T^{-1}, show that there exist characteristic roots α of S and β of T such that $\alpha\beta^{-1} = 1$.]

8. Let $V \otimes W$ and $V \otimes' W$ be two vector spaces, such that both satisfy the definition of the tensor product of the vector spaces V and W, relative to bilinear mappings $t: V \times W \to V \otimes W$ and $t': V \times W \to W \otimes' W$, respectively.

a. Prove that there exist linear transformations $S: V \otimes W \to V \otimes' W$ and $T: V \otimes' W \to V \otimes W$ such that $ST = 1$ and $TS = 1$.

b. Prove that both S and T are isomorphisms of vector spaces. (This problem shows that the tensor product vector space is uniquely determined by its defining properties.)

9. a. Let V and W be vector spaces over F, and let $B(V, W)$ be the set of all bilinear forms $f: V \times W \to F$. Show that $B(V, W)$ is a subspace of the vector space of functions $\mathscr{F}(V \times W)$.

b. Prove that the dual space $B(V, W)^*$ satisfies the definition of tensor product, with respect to the bilinear mapping

$$b: V \times W \to B(V, W)^*$$

defined by

$$b(v, w)(f) = f(v, w), \qquad f \in B(V, W), \quad v \in V, \quad w \in W.$$

10. Let $\bigwedge_k(V)$ be the k-fold wedge product of V, and let $\dim V = n$. Prove that $\dim \bigwedge_k(V) = \binom{n}{k}$ (binomial coefficient).

11. Let $T \in L(V, V)$. Extend Theorem (28.10) to prove that there exists a linear transformation $T^{(k)}$ of $\bigotimes_k(V)$ such that

$$T^{(k)}(v_1 \otimes \cdots \otimes v_k) = T(v_1) \otimes \cdots \otimes T(v_k) \qquad \text{for all } v_i \in V.$$

Prove that $\bigwedge_k(V)$ is an invariant subspace relative to $T^{(k)}$ for all $T \in L(V, V)$

12. Let $\dim V = n$, and let $\{v_1, \ldots, v_n\}$ be a basis for V. Prove that the tensor $v_1 \wedge \cdots \wedge v_n$ is a basis for $\bigwedge_n(V)$. Let $T \in L(V, V)$. Prove that

$$T^{(n)}(v_1 \wedge \cdots \wedge v_n) = D(T)(v_1 \wedge \cdots \wedge v_n),$$

for all $T \in L(V, V)$. (*Hint:* Use the uniqueness of the determinant function, proved in Section 17.)

29. A PROOF OF THE ELEMENTARY DIVISOR THEOREM

The theorem to be proved [Theorem (25.8)] states that if $T \in L(V, V)$, where V is a finite dimensional vector space, different from zero, over an arbitrary field F, then the following results hold.

(29.1) *There exist vectors $\{v_1, \ldots, v_r\}$, whose orders are prime powers $\{p_1(x)^{e_1}, \ldots, p_r(x)^{e_r}\}$ in $F[x]$, such that V is the direct sum of the cyclic subspaces relative to T generated by $\{v_1, \ldots, v_r\}$.*

(29.2) *Suppose*

$$V = \langle v_1 \rangle \oplus \cdots \oplus \langle v_r \rangle = \langle v_1' \rangle \oplus \cdots \oplus \langle v_s' \rangle,$$

where the vectors $\{v_i\}$ and $\{v_i'\}$ have prime power orders

$$\{p_1(x)^{e_1}, \ldots, p_r(x)^{e_r}\} \quad \text{and} \quad \{q_1(x)^{f_1}, \ldots, q_s(x)^{f_s}\},$$

respectively. Then $r = s$, and the polynomials

$$\{p_1(x)^{e_1}, \ldots, p_r(x)^{e_r}\} \quad \text{and} \quad \{q_1(x)^{f_1}, \ldots, q_s(x)^{f_s}\}$$

are the same, up to a rearrangement.

We shall first prove some preliminary results leading to the proof of (29.1). Let

$$m(x) = h_1(x)^{c_1} \cdots h_m(x)^{c_m}$$

be the minimal polynomial of T, factored in $F[x]$ as a product of powers of prime polynomials $\{h_1(x)^{c_1}, \ldots, h_m(x)^{c_m}\}$. By Theorem (23.9),

(29.3)
$$V = V_1 \oplus \cdots \oplus V_m,$$

where each subspace $V_i \neq 0$ and is the null space of $h_i(T)^{c_i}$, $1 \leq i \leq m$. Moreover, each V_i is invariant relative to T. Letting T_i denote the restriction of T to the subspace V_i, we have $h_i(T_i)^{c_i} = 0$ on V_i. Therefore the minimal polynomial of T_i is $h_i(x)^{b_i}$ for $b_i \leq c_i$. Suppose we were able to prove (29.1) for each of the linear transformations T_i, $1 \leq i \leq m$, so that

$$V_i = \langle v_{ij} \rangle \oplus \cdots \oplus \langle v_{ik} \rangle, \qquad 1 \leq i \leq m.$$

Then each cyclic T_i-space $\langle v_{ij}\rangle$ is also a cyclic subspace relative to T, whose generator v_{ij} has order a power of $h_i(x)$. Putting together these results for the spaces $\{V_1, \ldots, V_m\}$ in turn, and using (29.3), we obtain (29.1). We have shown that it is sufficient to prove (29.1) for a linear transformation $T \in L(V, V)$, whose minimal polynomial is a power of a prime.

(29.4) DEFINITION. A nonzero T-invariant subspace W of V is called *indecomposable* provided that W cannot be expressed as a direct sum of nonzero T-invariant subspaces.

(29.5) LEMMA. *Let V be a finite dimensional vector space different from zero, and let $T \in L(V, V)$. Then V is a direct sum of indecomposable, T-invariant subspaces.*

Proof. We use induction on $\dim(V)$. If V is already indecomposable, there is nothing to prove. If not, then $V = V_1 \oplus V_2$, where V_1 and V_2 are nonzero T-invariant subspaces. Letting T_1 and T_2 denote the restrictions of T to the subspaces V_1 and V_2, respectively, and the facts that $\dim(V_1) < \dim(V)$ and $\dim(V_2) < \dim(V)$, we can apply our induction hypothesis to conclude that

$$V_1 = V_{11} \oplus \cdots V_{1k}, \qquad V_2 = V_{21} \oplus \cdots \oplus V_{2k'},$$

where the subspaces V_{1i}, $1 \le i \le k$, are indecomposable and invariant relative to T_1, and the subspaces V_{2j}, $1 \le j \le k'$, are indecomposable and invariant relative to T_2. The subspaces $\{V_{ij}\}$ are all indecomposable and invariant relative to T, and V is their direct sum. This completes the proof of the lemma.

The proof of (29.1) has now been reduced, because of Lemma (29.5) and the preceding remarks, to giving a proof of the following result.

(29.6) THEOREM. *Let V be a finite dimensional vector space, different from zero, over an arbitrary field F, and let $T \in L(V, V)$ be a linear transformation whose minimal polynomial is a prime power $p(x)^a$, and such that V is indecomposable relative to T. Then V is cyclic relative to T.*

Proof. We first assert that there exists a vector $v_1 \in V$ whose order is equal to the minimal polynomial $p(x)^a$ of T. If this were not the case, then for every $v \neq 0$ in V, by Lemma (25.7), the order of v would be $p(x)^b$ for some $b < a$, and we would have $p(T)^{a-1}V = 0$, contrary to the definition of the minimal polynomial.

We shall prove Theorem (29.6) by showing that if the cyclic subspace $\langle v_1\rangle$ relative to T is not equal to V, then V is a direct sum,

$$V = \langle v_1\rangle \oplus W$$

for some nonzero T-invariant subspace W, contrary to the assumption that V is indecomposable relative to T. The proof is an application of the theory of dual vector spaces (Sections 26 and 27).

Let V^* be the dual space of V; then V and V^* are dual vector spaces with respect to the nondegenerate bilinear form $B(v, v^*)$, where (v, v^*) is the value,

$$B(v, v^*) = v^*(v),$$

of the linear function $v^* \in V^*$ at the vector $v \in V$. Let T^* be the transpose of T with respect to (v, v^*). Then for all polynomials $f(x) \in F[x]$, we have

$$B(f(T)v, v^*) = (v, f(T^*)v^*)$$

for all $v \in V$, $v^* \in V^*$. Therefore the minimal polynomial of T^* is also $p(x)^a$.

Now let $\langle v_1 \rangle^\perp$ be the annihilator of $\langle v_1 \rangle$ in V^*. From Section 27, $\langle v_1 \rangle^\perp$ is invariant relative to T^*, and the spaces $\langle v_1 \rangle$ and $V^*/\langle v_1 \rangle^\perp$ are dual with respect to the bilinear form

$$B_0(v, [v^*]) = B(v, v^*),$$

for $v \in \langle v_1 \rangle$ and $[v^*] \in V^*/\langle v_1 \rangle^\perp$. Moreover, the linear transformations $T_{\langle v_1 \rangle}$ and $T^*{}_{V^*/\langle v_1 \rangle^\perp}$ are transposes of each other with respect to the bilinear form B_0, by Theorem (27.12). The minimal polynomial of the restriction $T_{\langle v_1 \rangle}$ is equal to the order of v_1, by Lemma (25.6), and is therefore $p(x)^a$, by the choice of v_1. From our remark about the minimal polynomial of a transpose, the induced transformation $T_1^* = T_{V^*/\langle v_1 \rangle^\perp}$ also has the minimal polynomial $p(x)^a$. By the first part of the proof, there exists a vector $[v_1^*]$ in $V^*/\langle v_1 \rangle^\perp$ whose order is $p(x)^a$. Let W^* be the subspace generated by $\{v_1^*, \ldots, (T^*)^{d-1}v_1^*\}$ in V^*,

$$W^* = S(v_1^*, T^*v_1^*, \ldots, (T^*)^{d-1}v_1^*),$$

where $d = \deg p(x)^a$. Then the subspace W^* has the following properties.

(a) $\dim W^* = d$.
(b) $W^* \cap \langle v_1 \rangle^\perp = 0$.
(c) $T^*(W^*) \subset W^*$.

The statements (a) and (b) follow since

$$\alpha_1 v_1^* + \alpha_2(T^*v_1^*) + \cdots + \alpha_{d-1}(T^*)^{d-1}v_1^* \in \langle v_1 \rangle^\perp$$

implies

$$\alpha_1 [v_1^*] + \alpha_2 T_1^* [v_1^*] + \cdots + \alpha_{d-1}(T_1^*)^{d-1} [v_1^*] = 0$$

in $V^*/\langle v_1\rangle^\perp$, and hence $\alpha_1 = \alpha_2 = \cdots = \alpha_{d-1} = 0$ because $[v_1^*]$ has order $p(x)^a$. The statement (c) holds since T^* has the minimal polynomial $p(x)^a$ of degree d, and hence $T^*(T^*)^{d-1}v_1^*$ is a linear combination of $\{v_1^*, T^*v_1^*, \ldots, (T^*)^{d-1}v_1^*\}$.

Now, since $\dim \langle v_1\rangle^\perp = \dim V^* - d$, statements (a) to (c) imply that

$$V^* = W^* \oplus \langle v_1\rangle^\perp,$$

and that W^* is invariant relative to T^*. Then by Theorem (27.12),

$$V = (W^*)^\perp \oplus (\langle v_1\rangle^\perp)^\perp = (W^*)^\perp \oplus \langle v_1\rangle,$$

since $\langle v_1\rangle^{\perp\perp} = \langle v_1\rangle$. Moreover, $(W^*)^\perp$ is invariant relative to the transpose T of T^*, and we have proved that V is a direct sum of two nonzero T-invariant subspaces. This completes the proof of the first half of the elementary divisor theorem.

Now we are ready to prove the uniqueness part (29.2) of the elementary divisor theorem. Let

$$m(x) = h_1(x)^{c_1} \cdots h_t(x)^{c_t}$$

be the minimal polynomial of T, where the $\{h_i(x)\}$ are primes in $F[x]$, and we may assume that $m(x)$ and all the $h_i(x)$ have leading coefficients equal to one. The orders of the elements v_i and v_i' also have leading coefficients equal to one, and it follows that each $p_i(x)$ or $q_j(x)$ is equal to (not just a scalar multiple of) some prime $h_i(x)$. By Theorem (23.9),

$$V = n(h_1(T)^{c_1}) \oplus \cdots \oplus n(h_t(T)^{c_t}),$$

and each v_i or v_i' whose order is a power of $h_j(x)$ is contained in $n(h_j(T)^{c_j})$. It follows that for each j, $1 \le j \le t$,

$$n(h_j(T)^{c_j}) = \langle v_{j_1}\rangle \oplus \cdots \oplus \langle v_{j_{a_j}}\rangle = \langle v_{k_1}'\rangle \oplus \cdots \oplus \langle v_{k_{b_j}}'\rangle,$$

where $\{v_{j_1}, \ldots, v_{j_{a_j}}\}$ and $\{v_{k_1}', \cdots, v_{k_{b_j}}'\}$ are precisely the generators of the cyclic subspaces in (29.2) whose orders are powers of $h_j(x)$. Therefore it is sufficient to prove (29.2) for the case in which we have V expressed as a direct sum of cyclic spaces in two different ways,

$$V = \langle v_1\rangle \oplus \cdots \oplus \langle v_r\rangle = \langle v_1'\rangle \oplus \cdots \oplus \langle v_s'\rangle,$$

and the orders of the generators $\{v_i\}$ and $\{v_i'\}$ are all powers of single prime, which we shall call $p(x)$. We let the orders of the $\{v_i\}$ be $\{p(x)^{a_1}, \ldots, p(x)^{a_r}\}$, with $a_i > 0$, and the orders of the $\{v_i'\}$ will be denoted by $\{p(x)^{b_1}, \ldots, p(x)^{b_s}\}$ with all $b_j > 0$. To prove the uniqueness part of the theorem, it will be sufficient to prove that first, $r = s$, then that the number of a's and b's equal to one is the same, then that the number of a's and b's equal to two is the same, etc.

We begin by computing $n(p(T))$, using the two decompositions in turn. Let

$$v = f_1(T)v_1 + \cdots + f_r(T)v_r \in \langle v_1\rangle \oplus \cdots \oplus \langle v_r\rangle,$$

where the $f_i(x) \in F[x]$. Suppose $v \in n(p(T))$. Then

$$p(T)f_1(T)v_1 + \cdots + p(T)f_r(T)v_r = 0,$$

and because of the direct sum we have

$$p(T)f_1(T)v_1 = \cdots = p(T)f_r(T)v_r = 0.$$

Therefore

$$p(x)^{a_1} | p(x)f_1(x), \ldots, p(x)^{a_r} | p(x)f_r(x),$$

and we have, from uniqueness of factorization in $F[x]$,

$$p(x)^{a_1 - 1} | f_1(x), \ldots, p(x)^{a_r - 1} | f_r(x).$$

It follows that, putting $p(T)^{a_i - 1} = 1$ if $a_i = 1$, we have

$$n(p(T)) = \langle p(T)^{a_1 - 1}v_1\rangle \oplus \cdots \oplus \langle p(T)^{a_r - 1}v_r\rangle,$$

and the dimension of $n(p(T))$ is rd, where d is the degree of $p(x)$. A similar calculation using the subspaces $\langle v_i'\rangle$ shows that the dimension of $n(p(T))$ is sd, and we conclude that $r = s$.

Now let x_1 denote the number of a_i's equal to one, x_2 the number of a_i's equal to two, etc., Similarly, let y_1 denote the number of b_j's equal to one, y_2 the number equal to two, etc.

We can use the same argument used to compute $n(p(T))$ to find $n(p(T)^2)$ and obtain

$$\begin{aligned} n(p(T^2)) &= \sum_{a_i = 1} \langle v_i\rangle + \sum_{a_i \geq 2} \langle p_i(T)^{a_i - 2}v_i\rangle \qquad \text{(direct sum)} \\ &= \sum_{b_j = 1} \langle v_j'\rangle + \sum_{b_j \geq 2} \langle p_j(T)^{b_j - 2}v_j'\rangle \qquad \text{(direct sum)}. \end{aligned}$$

Computing the dimension of $n(p(T^2))$ using the facts that the dimension of $\langle v_i\rangle$ is d if $a_i = 1$ and the dimension of $\langle p_i(T)^{a_i - 2}v_i\rangle$ is $2d$, if $a_i \geq 2$, we have

$$x_1 d + (r - x_1)2d = y_1 d + (r - y_1)2d.$$

This equation implies that $x_1 - y_1 = 2(x_1 - y_1)$ and hence $x_1 - y_1 = 0$. Continuing in this way, we can show that $x_2 = y_2$, etc., and the theorem is proved.

EXERCISES

1. Prove that if **A** is an n-by-n matrix with coefficients in an arbitrary field F, then **A** is similar to ${}^t\mathbf{A}$. [*Hint:* Use the methods of this chapter to prove that if $T \in L(V, V)$, then the elementary divisors of T and T^* are the same.]
2. Let $T \in L(V, V)$. Prove that V is indecomposable relative to T if and only if V is cyclic and the minimal polynomial of T is a prime power, $p(x)^a$ for some prime $p(x) \in F[x]$. [See Theorem (29.6)].
3. Let $T \in L(V, V)$. V is called *irreducible* relative to T provided that the only invariant subspaces are V and $\{0\}$. Prove that a nonzero vector space V is irreducible relative to T if and only if V is cyclic and the minimal polynomial of T is a prime in $F[x]$.
4. Let $T \in L(V, V)$. V is called *completely reducible* relative to T provided that V is a direct sum, $V = \sum V_i$, of invariant subspaces V_i, such that each subspace V_i is irreducible relative to the restriction $T|_{V_i}$ of T to V_i. Assume that $V \neq \{0\}$. Prove that the following statements are equivalent.
   a. V is completely reducible relative to T.
   b. The minimal polynomial of T is a product of distinct monic prime factors in $F[x]$.
   c. The elementary divisors of T are all prime polynomials. [Note that this result is a generalization of Theorem (23.11).]

Chapter 9

Orthogonal and Unitary Transformations

In Chapters 7 and 8, general theorems were proved on the structure of a single linear transformation, including the Jordan decomposition, the elementary divisor theorem, and the rational and Jordan normal forms. For certain types of linear transformations on vector spaces over the fields of real or complex numbers, much more information about the nature of the characteristic roots, and sharper theorems on the structure of the transformations, can be proved. This chapter contains two sections on orthogonal transformations and quadratic forms on vector spaces over the field of real numbers. A final section is devoted to vector spaces over the complex numbers, and the theory of unitary, hermitian, and normal transformations. This chapter is independant of Chapter 8, and can be read immediately after Chapter 7.

30. THE STRUCTURE OF ORTHOGONAL TRANSFORMATIONS

In Section 14 we showed that every orthogonal transformation of the plane is either a rotation or a reflection. We wish to find in this section a geometrical description of orthogonal transformations on an n-dimensional real vector space V with an inner product (u, v). From the point of view of Chapter 7, given an orthogonal transformation T we look for an orthonormal basis of V with respect to which the matrix of T is as simple as possible.

The matrix

$$\mathbf{A} = \begin{pmatrix} \cos\frac{2\pi}{3} & -\sin\frac{2\pi}{3} \\ \sin\frac{2\pi}{3} & \cos\frac{2\pi}{3} \end{pmatrix} = \begin{pmatrix} -\frac{1}{2} & -\frac{\sqrt{3}}{2} \\ \frac{\sqrt{3}}{2} & -\frac{1}{2} \end{pmatrix}$$

is an orthogonal matrix such that $\mathbf{A}^3 = \mathbf{I}$. Its minimal polynomial is

$$x^2 + x + 1$$

which is a prime in the polynomial ring $R[x]$. Therefore, by Theorem (23.11), **A** cannot be diagonalized over the real field and cannot even be put in triangular form, since a triangular 2-by-2 orthogonal matrix can easily be shown to be diagonal. Thus the methods of Chapter 7 yield little new information, even in this simple case.

The difficulty is that the real field is not algebraically closed. It is here that, as Hermann Weyl remarked, Euclid enters the scene, brandishing his ruler and his compass. The ideas necessary to treat orthogonal transformations on a real vector space will throw additional light on the problems considered in Chapter 7, as well. The existence of an inner product allows us to get information about minimal polynomials, etc., that would have seemed almost impossible from the methods of Chapter 7 alone. We begin with some general definitions and theorems.

(30.1) DEFINITION. Let $T \in L(V, V)$ be a linear transformation on an arbitrary finite dimensional vector space V over an arbitrary field F. A nonzero invariant subspace $W \subset V$ (relative to T) is called *irreducible* if the only T-invariant subspaces contained in W are $\{0\}$ and W.

(30.2) THEOREM. **(a)** *If V is a vector space over an algebraically closed field F, then every irreducible invariant subspace W relative to $T \in L(V, V)$ has dimension* 1. **(b)** *If V is a vector space over the real field R and if W is an irreducible invariant subspace relative to $T \in L(V, V)$, then W has dimension* 1 *or* 2.

Proof. **(a)** Let W be an irreducible invariant subspace relative to T; then T defines a linear transformation T_W of W into itself, where

$$T_W(w) = T(w), \qquad w \in W.$$

From Section 24, W contains a characteristic vector w relative to T; then $S(w)$ is an invariant subspace contained in W, and hence $W = S(w)$ because W is irreducible. This proves part **(a)**

(b) Let W be an irreducible invariant subspace relative to T and let $m(x)$ be the minimal polynomial of T_W. From Theorem (23.9),

$m(x) = p(x)^e$ where $p(x)$ is a prime polynomial in $R[x]$; otherwise, W would be the direct sum of subspaces, contrary to the assumption that W is irreducible. If $m(x) = p(x)^e$ is the minimal polynomial of T_W, then $e = 1$; otherwise, the null space of $p(T)^{e-1}$ would be an invariant subspace different from $\{0\}$ and W. Thus we have, by Corollary (21.13), either

$$m(x) = x - \alpha \quad \text{or} \quad m(x) = x^2 + \alpha x + \beta, \quad \alpha^2 - 4\beta < 0.$$

Let w be a nonzero vector of W. Then, if $m(x) = x - \alpha$, w is a characteristic vector of T, and $W = S(w)$ as in part (**a**). If $m(x) = x^2 + \alpha x + \beta$ for $\alpha^2 - 4\beta < 0$, then $S[w, T(w)]$ is an invariant subspace, and hence $W = S[w, T(w)]$. Therefore W has dimension either 1 or 2, and the theorem is proved.

Now we apply this theorem to orthogonal transformations as follows.

(30.3) THEOREM. *Let T be an orthogonal transformation on a real vector space V with an inner product and let W be an irreducible invariant subspace relative to T; then either of the two following holds.*

a. $\dim W = 1$ *and, if $w \neq 0$ in W, then $T(w) = \pm w$.*
b. $\dim W = 2$ *and there is an orthonormal basis $\{w_1, w_2\}$ for W such that the matrix of T with respect to $\{w_1, w_2\}$ has the form*

$$\begin{pmatrix} \cos\theta & -\sin\theta \\ \sin\theta & \cos\theta \end{pmatrix}.$$

In other words, T_W is a rotation in the two-dimensional space W (see Section 14).

Proof. By Theorem (30.2), $\dim W$ is either 1 or 2. In case (1) of the theorem, $T(w) = \lambda w$ for some $\lambda \in R$, and

$$\|T(w)\| = \|w\|$$

implies $|\lambda| = 1$. Therefore, $\lambda = \pm 1$ and $T(w) = \pm w$.

If $\dim W = 2$, then the minimal polynomial of T has the form

$$x^2 + \alpha x + \beta, \quad \alpha^2 - 4\beta < 0.$$

Let $\{w_1, w_2\}$ be an orthonormal basis for W and let

$$T(w_1) = \lambda w_1 + \mu w_2.$$

Then $\lambda^2 + \mu^2 = 1$ and, since $(T(w_1), T(w_2)) = 0$, $T(w_2)$ is either $-\mu w_1 + \lambda w_2$ or $\mu w_1 - \lambda w_2$. In the first case, the matrix of T is

$$\begin{pmatrix} \lambda & -\mu \\ \mu & \lambda \end{pmatrix}$$

and we can find θ such that $\cos\theta = \lambda$, $\sin\theta = \mu$, since $\lambda^2 + \mu^2 = 1$. In the latter case, the matrix is

$$\begin{pmatrix} \lambda & \mu \\ \mu & -\lambda \end{pmatrix}$$

which satisfies the equation $x^2 - 1 = 0$, and since $x^2 - 1 = (x + 1)(x - 1)$ is not a prime this case cannot occur. This completes the proof of the theorem.

The last theorem becomes of especial interest when combined with the following result.

(30.4) THEOREM. *Let T be an orthogonal transformation on a real vector space V with an inner product; then V is a direct sum of irreducible invariant subspaces $\{W_1, \ldots, W_s\}$, for $s \geq 1$, such that vectors belonging to distinct subspaces W_i and W_j are orthogonal.*

Proof. We prove the theorem by induction on dim V, the result being obvious if dim $V = 1$. Assume the theorem for subspaces of dimenson $<$ dim V and let T be an orthogonal transformation on V. Let W_1 be a nonzero invariant subspace of least dimension; then W_1 is an irreducible invariant subspace. Let $W_1^\perp$ be the subspace consisting of all vectors orthogonal to the vectors in W_1. We prove that

$$W = W_1 \oplus W_1^\perp.$$

Clearly, $W_1 \cap W_1^\perp = \{0\}$, since $(w, w) \neq 0$ if $w \neq 0$. Now let $w \in W$ and let $\{w_1, \ldots, w_s\}$ be an orthonormal basis for W_1 where $s = 1$ or 2. Then

$$w = \sum_{i=1}^{s} (w, w_i)w_i + \left(w - \sum_{i=1}^{s} (w, w_i)w_i\right)$$

and since $\sum_1^s (w, w_i)w_i \in W_1$ and $w - \sum_1^s (w, w_i)w_i \in W_1^\perp$, we have $W = W_1 + W_1^\perp$. This fact together with the result that $W_1 \cap W_1^\perp = \{0\}$ implies that $W = W_1 \oplus W_1^\perp$.

Now we prove the key result that $W_1^\perp$ is also an invariant subspace. Let $w' \in W_1^\perp$. Since T is orthogonal, $T(W_1) = W_1$ and $(W_1, w') = (T(W_1), T(w')) = (W_1, T(w')) = 0$. Therefore $T(w')$ is also orthogonal to all the vectors in W_1.

Since $T(W_1^\perp) \subset W_1^\perp$, T is an orthogonal transformation of $W_1^\perp$ and, by the induction hypothesis, $W_1^\perp$ is a direct sum of pairwise orthogonal irreducible invariant subspaces. Since $V = W_1 \oplus W_1^\perp$, the same is true for V, and the theorem is proved.

Combining Theorems (30.3) and (30.4), we obtain the following determination of the form of the matrix of an orthogonal transformation.

(30.5) THEOREM. *Let T be an orthogonal transformation of a real vector space V with an inner product; then there exists an orthonormal basis for V such that the matrix of T with respect to this basis has the form*

$$\begin{pmatrix} 1 & & & & & & & & & & 0 \\ & \ddots & & & & & & & & & \\ & & 1 & & & & & & & & \\ & & & -1 & & & & & & & \\ & & & & \ddots & & & & & & \\ & & & & & -1 & & & & & \\ & & & & & & \cos\theta_1 & -\sin\theta_1 & & & \\ & & & & & & \sin\theta_1 & \cos\theta_1 & & & \\ & & & & & & & & \cos\theta_2 & -\sin\theta_2 & \\ & & & & & & & & \sin\theta_2 & \cos\theta_2 & \\ 0 & & & & & & & & & & \ddots \end{pmatrix}$$

with zeros except in the 1-*by*-1 *or* 2-*by*-2 *blocks along the diagonal.*

Proof. The proof is immediate by Theorems (30.3) and (30.4) since we can choose orthonormal bases for the individual subspaces W_i in Theorem (30.4) which, when taken together, will form an orthonormal basis of V.

EXERCISES

1. Prove that, if T is an orthogonal transformation on R_2 such that $D(T) = -1$, there exists an orthonormal basis for R_2 such that the matrix of T with respect to this basis is

$$\begin{pmatrix} -1 & 0 \\ 0 & 1 \end{pmatrix}.$$

2. An orthogonal transformation T of R_3 is called a *rotation* if $D(T) = 1$. Prove that if T is a rotation in R_3 there exists an orthonormal basis of R_3 such that the matrix of T with respect to this basis is

$$\begin{pmatrix} 1 & 0 & 0 \\ 0 & \cos\theta & -\sin\theta \\ 0 & \sin\theta & \cos\theta \end{pmatrix}$$

for some real number θ.

3. Prove that an orthogonal transformation T in R_m has 1 as a characteristic root if $D(T) = 1$ and m is odd. What can you say if m is even?

31. THE PRINCIPAL-AXIS THEOREM

In analytic geometry the following problem is considered. Let

$$f(x_1, x_2) = ax_1^2 + bx_1x_2 + cx_2^2 + dx_1 + ex_2 + f = 0$$

be an equation of the second degree in x_1 and x_2. The problem is to find a new coordinate system

$$\begin{aligned} X_1 &= (\cos\theta)x_1 + (\sin\theta)x_2 + c_1 \\ X_2 &= (-\sin\theta)x_1 + (\cos\theta)x_2 + c_2 \end{aligned}$$

obtained by a rotation and a translation from the original one, such that in the new coordinate system the equation becomes either of the following.

(31.1) $$f(X_1, X_2) = AX_1^2 + BX_2^2 + C = 0.$$

(31.2) $$f(X_1, X_2) = AX_2^2 - DX_1 = 0.$$

The graph of $f(X_1, X_2) = 0$ can then be classified as a circle, ellipse, hyperbola, parabola, etc. It is clear that, if we can first find a rotation of axes

$$\begin{aligned} X_1 &= (\cos\theta)x_1 + (\sin\theta)x_2 \\ X_2 &= (-\sin\theta)x_1 + (\cos\theta)x_2 \end{aligned}$$

such that the second-degree terms $ax_1^2 + bx_1x_2 + cx_2^2$ become

(31.3) $$AX_1^2 + BX_2^2,$$

then the new equation has the form

$$f(X_1, X_2) = AX_1^2 + BX_2^2 + CX_1 + DX_2 + D$$

and can be put in the form (31.1) or (31.2) by a translation of axes: $X_1' = X_1 + c_1$, $X_2' = X_2 + c_2$.

The problem we shall consider in this section is a generalization of the problem of finding a rotation of axes that will put the second-degree terms of $f(x_1, x_2)$ in the form (31.3).

(31.4) Definition. Let V be an n-dimensional vector space over the real numbers R. A *quadratic form* on V is a function Q which assigns to each vector $a \in V$ a real number $Q(a)$ such that the following conditions are satisfied.

(*i*) $Q(\alpha a) = \alpha^2 Q(a)$, $\alpha \in R$, $a \in V$.

(*ii*) If we define $B(a, b) = \frac{1}{2}[\,Q(a + b) - Q(a) - Q(b)\,]$, then B is a *bilinear form* on V, that is,

$$\begin{aligned} B(\alpha a_1 + \beta a_2, b) &= \alpha B(a_1, b) + \beta B(a_2, b), \\ B(a, \alpha b_1 + \beta b_2) &= \alpha B(a, b_1) + \beta B(a, b_2), \end{aligned}$$

for all $a, b \in V$ and $\alpha, \beta \in R$.

(31.5) THEOREM. *Let Q be a quadratic form on V and let $\{e_1, \ldots, e_n\}$ be a basis of V over R. Define an n-by-n matrix $\mathbf{S} = (\sigma_{ij})$ by setting*

$$\sigma_{ij} = B(e_i, e_j), \qquad 1 \le i,\ j \le n,$$

where B is the bilinear form defined in part (ii) of Definition (31.4). Then ${}^t\mathbf{S} = \mathbf{S}$, and for all $a = \sum \alpha_i e_i \in V$ we have

$$(31.6) \qquad Q(a) = B(a, a) = \sum_{i,j=1}^{n} \alpha_i\alpha_j\sigma_{ij} = \sum_{i=1}^{n} \alpha_i^2\sigma_{ii} + 2\sum_{i<j} \alpha_i\alpha_j\sigma_{ij}.$$

Conversely, if $\mathbf{S} = (\sigma_{ij})$ is an arbitrary n-by-n real matrix such that ${}^t\mathbf{S} = \mathbf{S}$, then (31.6) defines a quadratic form Q such that if we set $B(a, b) = \frac{1}{2}[\,Q(a + b) - Q(a) - Q(b)\,]$ then $B(e_i, e_j) = \sigma_{ij}$.

Remark. Assuming the truth of the theorem, the function

$$f(x_1, x_2) = ax_1^2 + bx_1x_2 + cx_2^2$$

defines a quadratic form on R_2 such that if $\{x_1, x_2\}$ are the coordinates of a vector x with respect to the basis $\{e_1, e_2\}$ then the matrix of f defined in the theorem is given by

$$\begin{pmatrix} a & \frac{1}{2}b \\ \frac{1}{2}b & c \end{pmatrix}.$$

Proof of Theorem (31.5) The first part of the theorem is immediate from Definition (31.4). For the converse, let $\mathbf{S} = {}^t\mathbf{S}$ be given and define a function

$$B(a, b) = \sum_{i,j=1}^{n} \alpha_i\beta_j\sigma_{ij}$$

for $a = \sum \alpha_i e_i$, $b = \sum \beta_i e_i$. Then B is a bilinear form on V such that $B(e_i, e_j) = \sigma_{ij}$ and $B(a, b) = B(b, a)$, for $a, b \in V$. It is also clear that if we define

$$Q(a) = B(a, a),$$

then Q is a quadratic form such that $B(a, b) = \frac{1}{2}[\,Q(a + b) - Q(a) - Q(b)\,]$, since $B(a, b) = B(b, a)$. This completes the proof.

(31.7) DEFINITION. A real n-by-n matrix $\mathbf{S}$ is called *symmetric* if ${}^t\mathbf{S} = \mathbf{S}$. The *matrix* $\mathbf{S} = (\sigma_{ij})$ of a *quadratic form* Q with respect to the basis $\{e_1, \ldots, e_n\}$ is defined by

$$\sigma_{ij} = B(e_i, e_j)$$

where

$$B(a, b) = \tfrac{1}{2}[\,Q(a + b) - Q(a) - Q(b)\,]$$

is the bilinear form associated with Q.

(31.8) THEOREM. *Let Q be a quadratic form on V whose matrix with respect to the basis $\{e_1, \ldots, e_n\}$ is $\mathbf{S} = (\sigma_{ij})$. Let $\{f_1, \ldots, f_n\}$ be another basis of V such that*

$$f_i = \sum_{j=1}^{n} \gamma_{ji} e_j, \qquad 1 \leq i \leq n.$$

Then the matrix of Q with respect to the basis $\{f_1, \cdots, f_n\}$ is given by

$$\mathbf{S}' = {}^t\mathbf{C}\mathbf{S}\mathbf{C}$$

where $\mathbf{C} = (\gamma_{ij})$.

Proof. Let $\mathbf{S}' = (\sigma'_{ij})$. Then

$$\begin{aligned} \sigma'_{ij} = B(f_i, f_j) &= B\left(\sum_{k=1}^{n} \gamma_{ki} e_k, \sum_{l=1}^{n} \gamma_{lj} e_l\right) \\ &= \sum_{k=1}^{n} \sum_{l=1}^{n} \gamma_{ki} \gamma_{lj} B(e_k, e_l) \\ &= \sum_{k=1}^{n} \sum_{l=1}^{n} \gamma_{ki} \sigma_{kl} \gamma_{lj}, \end{aligned}$$

and the theorem is proved.

Now we can state the main theorem of this section.

(31.9) PRINCIPAL-AXIS THEOREM.† *Let V be a vector space over R with an inner product (a, b) and let $\{e_1, \ldots, e_n\}$ be an orthonormal basis of V. Let Q be a quadratic form on V whose matrix with respect to $\{e_1, \ldots, e_n\}$ is $\mathbf{S} = (\sigma_{ij})$. Then there exists an orthonormal basis $\{f_1, \ldots, f_n\}$ such that the matrix of Q with respect to $f_1, \ldots, f_n$ is*

$$\mathbf{S}' = \begin{pmatrix} \sigma_1 & & & 0 \\ & \sigma_2 & & \\ & & \ddots & \\ 0 & & & \sigma_n \end{pmatrix}$$

where the σ_i are the characteristic roots of $\mathbf{S}$. If

$$f_i = \sum_{j=1}^{n} \gamma_{ji} e_j, \qquad 1 \leq i \leq n,$$

then $\mathbf{C} = (\gamma_{ij})$ is an orthogonal matrix. If $a \in V$ is expressed in terms of the new basis $\{f_1, \ldots, f_n\}$ by $a = \sum_{i=1}^{n} \alpha_i f_i$, then $Q(a) = \sum_{i=1}^{n} \alpha_i^2 \sigma_i$. The vectors $f_1, \ldots, f_n$ are called principal axes of Q.

† For a different proof and an application of this theorem to mechanics, see Synge and Griffith, p. 318 (listed in the Bibliography).

(**31.10**) DEFINITION. A linear transformation T on a real vector space with an inner product (a, b) is called a *symmetric transformation* if

$$(Ta, b) = (a, Tb), \qquad a, b \in V.$$

(**31.11**) *Let $\{e_1, \ldots, e_n\}$ be an orthonormal basis of V. A linear transformation $T \in L(V, V)$, whose matrix with respect to $\{e_1, \ldots, e_n\}$ is $\mathbf{S} = (\sigma_{ij})$, is a symmetric transformation if and only if $\mathbf{S}$ is a symmetric matrix.*

The proof is similar to the proof of part (4) of Theorem (15.11) and will be omitted.

(**31.12**) THEOREM. *Let T be a symmetric transformation on a real vector space V; then there exists an orthonormal basis of V consisting of characteristic vectors of V.*

We shall first prove that Theorem (31.12) implies Theorem (31.9), and then we shall prove Theorem (31.12).

Proof that (31.12) implies (31.9). In the notation of Theorem (31.9) let Q be a quadratic form on V with the matrix $\mathbf{S}$ with respect to the orthonormal basis $\{e_1, \ldots, e_n\}$. By Theorems (31.8) and (31.5) and the results of Section 15 it is sufficient to construct an *orthogonal* matrix $\mathbf{C}$ such that

$$ {}^t\mathbf{C}\mathbf{S}\mathbf{C} = \begin{pmatrix} \sigma_1 & & 0 \\ & \ddots & \\ 0 & & \sigma_n \end{pmatrix}$$

where $\sigma_1, \ldots, \sigma_n$ are the characteristic roots of S. For an *orthogonal* matrix $\mathbf{C}$, ${}^t\mathbf{C} = \mathbf{C}^{-1}$ and therefore it is sufficient to find an orthogonal matrix $\mathbf{C}$ such that

$$\mathbf{C}^{-1}\mathbf{S}\mathbf{C} = \begin{pmatrix} \sigma_1 & & 0 \\ & \ddots & \\ 0 & & \sigma_n \end{pmatrix}.$$

This is a problem on the linear transformation T whose matrix with respect to $\{e_1, \ldots, e_n\}$ is $\mathbf{S}$. By (31.11), T is a symmetric transformation and from the results of Section 15 the task of finding $\mathbf{C}$ is exactly the problem stated in Theorem (31.12).

Proof of Theorem (31.12) We first prove that V is a direct sum of pairwise orthogonal irreducible invariant subspaces relative to T. The argument is the same as the proof of Theorem (30.4); we have only to check that, if W is invariant relative to T, then $W^\perp$ is invariant relative to T. Let $w' \in W^\perp$ and $w \in W$; then

$$(w, Tw') = (Tw, w') = 0$$

since $Tw \in W$ and $w' \in W^{\perp}$. We may now conclude that V is the direct sum of pairwise orthogonal irreducible invariant subspaces $\{W_1, \ldots, W_s\}$.

It is now sufficient to prove that for all i, where $1 \le i \le s$, $\dim W_i = 1$. Since $\dim W_i = 1$ or 2, it is sufficient to prove that a symmetric transformation T on a two-dimensional real vector space W always has a characteristic vector. Let $\{w_1, w_2\}$ be an orthonormal basis for W; then, relative to $\{w_1, w_2\}$, T has a symmetric matrix

$$\begin{pmatrix} \lambda & \xi \\ \xi & \mu \end{pmatrix}, \qquad \lambda, \mu, \xi \in R.$$

This matrix satisfies the equation

$$x^2 - (\lambda + \mu)x + (\lambda\mu - \xi^2) = 0.$$

Letting $A = -(\lambda + \mu)$ and $B = \lambda\mu - \xi^2$, we have

$$\begin{aligned} A^2 - 4B &= (\lambda + \mu)^2 - 4(\lambda\mu - \xi^2) \\ &= \lambda^2 - 2\lambda\mu + \mu^2 + 4\xi^2 = (\lambda - \mu)^2 + 4\xi^2 \ge 0 \end{aligned}$$

for all real numbers λ, ξ, μ. Therefore the polynomial $x^2 + Ax + B$ factors into linear factors in $R[x]$, and it follows from Section 24 that W contains a characteristic vector. This completes the proof of the theorem.

EXERCISES

1. Consider the symmetric matrix

$$\mathbf{X} = \begin{pmatrix} 0 & 0 & 1 \\ 0 & 1 & 0 \\ 1 & 0 & 0 \end{pmatrix}.$$

   a. Find the characteristic roots of $\mathbf{X}$.

   b. Define a symmetric transformation T in R_3 whose matrix with respect to the orthonormal basis of unit vectors $\{e_1, e_2, e_3\}$ is $\mathbf{X}$. Find by the methods of Chapter 7 a basis of R_3 consisting of characteristic vectors of T [we know that can be done, by Theorem (31.12)]. Modify this basis to obtain an orthonormal basis $\{f_1, f_2, f_3\}$ of R_3 consisting of characteristic vectors of T. Let

$$f_i = \sum_{j=1}^{3} \mu_{ji} e_j$$

   and

$$Tf_i = \alpha_i f_i, \qquad \alpha_i \in R, \quad i = 1, 2, 3.$$

   Then $\mathbf{M} = (\mu_{ij})$ is an orthogonal matrix, such that $\mathbf{M}^{-1}\mathbf{X}\mathbf{M}$ is diagonal. Since $\mathbf{M}$ is orthogonal, $\mathbf{M}^{-1} = {}^t\mathbf{M}$, so this computational procedure also can be applied to find the principal axes of a vector space with respect to a quadratic form.

2. Find orthonormal bases of R_2 which exhibit the principal axes of the quadratic forms:
   a. $8x_1^2 + 8x_1x_2 + 2x_2^2$,
   b. $x_1^2 + 6x_1x_2 + x_2^2$.
3. Find an orthonormal basis of R_4 which exhibits the principal axes of the quadratic form $x_1x_2 + x_3x_4$.
4. A quadratic form $Q(x)$ is called *positive definite* if $Q(x) > 0$ for every vector $x \neq 0$. Prove that Q is positive definite if and only if all the characteristic roots of the matrix **S** of Q are positive. Define a *negative definite* quadratic form, and state and prove a similar result.
5. Let Q be a quadratic form on a vector space V over the field of real numbers, and let B be the bilinear form associated with Q. Prove that there exist subspaces V_+, V_-, and V_0 of V such that $B(V_+, V_-) = B(V_+, V_0) = B(V_-, V_0) = 0$, $V = V_+ \oplus V_- \oplus V_0$, and the quadratic forms Q_+, Q_-, and Q_0 defined on V_+, V_-, V_0, respectively, by the rule $Q_+(x) = Q(x)$, $x \in V_+$, etc., are positive definite, negative definite, and identically zero, respectively. (See Exercise 4 for definitions of positive definite and negative definite quadratic forms.)
6. Let Q be a quadratic form on V, and let $V = V_+ \oplus V_- \oplus V_0$ be a decomposition of V as in Exercise 5. Prove that V_0 is the set of all vectors $v \in V$ such that $B(v, V) = 0$, where B is the bilinear form associated with Q.
7. Let Q be a quadratic form on V, and let $V = V_+ \oplus V_- \oplus V_0$ and $V = V'_+ \oplus V'_- \oplus V'_0$ be two decompositions of V satisfying the conditions of Exercise 5. Prove that $\dim V_+ = \dim V'_+$, $\dim V_- = \dim V'_-$, and that $V_0 = V'_0$. [*Hint:* $V_0 = V'_0$ by Exercise 6. Define a linear transformation $T\colon V_+ \to V'_+$ as follows. Let $v \in V_+$, and let $v = v'_+ + v'_- + v'_0$ be the decomposition of v according to the second decomposition of the vector space. Set $T(v) = v'_+$. Let v belong to the null space of T. Show that $Q(v) \geq 0$ and $Q(v) \leq 0$, so that $v = 0$. Conclude that $\dim V_+ \leq \dim V'_+$, etc.]
8. Let Q be a quadratic form on V, and let $V = V_+ \oplus V_- \oplus V_0$ be a decomposition satisfying the conditions of Exercise 5. Prove that $\dim V_+$ is equal to the number of characteristic roots of the matrix of Q which are positive, $\dim V_-$ the number of negative ones, and $\dim V_0$ the number of characteristic roots which are equal to zero.
9. †Let Q be a quadratic form on V, viewed as a function $Q(x_1, \ldots, x_n)$ of n variables $x_1, \ldots, x_n$. The set of points $x = \langle x_1, \ldots, x_n \rangle$ such that $Q(x_1, \ldots, x_n) = 1$ defines a surface in R_n. Suppose that Q is positive definite, and that the characteristic roots of the matrix of Q are all different. Prove that if $\{v_1, \ldots, v_n\}$ are a set of principal axes of Q, then each v_i is normal to the surface $(Qx)=1$ at $(Q(v_i))^{-1/2}v_i$. (See Figure 9.1).

---

† Exercises 9 and 10 require some familiarity with the calculus of functions of several variables.

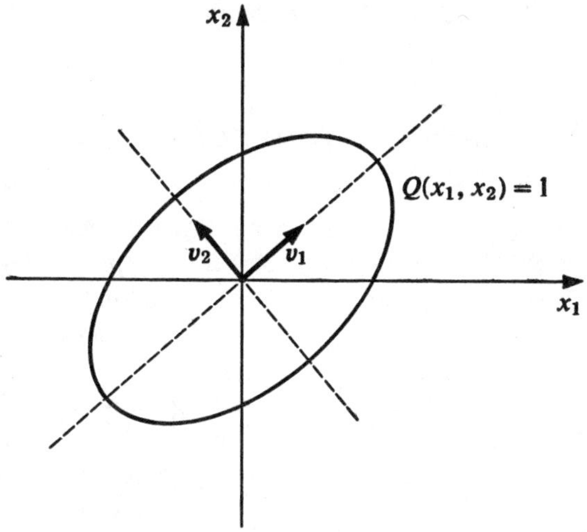

FIGURE 9.1

[*Hint:* First consider the case where $Q(x_1, \ldots, x_n) = \sum \alpha_i x_i^2$. A vector $v = \langle \xi_1, \ldots, \xi_n \rangle$ is normal to the surface at x if and only if it is a scalar multiple of $\langle \partial Q/\partial x_1, \ldots, \partial Q/\partial x_n \rangle$ evaluated at x. Use this information to verify the assertion in the given case, and settle the general case by making a change of variables by an orthogonal transformation.]

10. Let Q be viewed as a function of n variables, as in the preceding exercise. Show that if $Q(x_1, \ldots, x_n)$ is a positive definite quadratic form with matrix $\mathbf{S}$, then

$$I = \int_0^\infty \cdots \int_0^\infty e^{-Q(x_1, \ldots, x_n)}\, dx_1\, dx_2 \cdots dx_n = 2^{-n} \sqrt{\frac{\pi^n}{\det \mathbf{S}}}.$$

[*Hint:* First show that if $\lambda > 0$,

$$\int_0^\infty e^{-\lambda x^2} = \left(\frac{\pi}{4\lambda}\right)^{1/2}.$$

(See R. C. Buck, *Advanced Calculus*, 2nd ed., McGraw-Hill, 1965, Chap. 3.) Then use the principal-axis theorem to find an orthogonal matrix $\mathbf{C}$ such that the transformation

$$x_i = \sum_{j=1}^{n} \gamma_{ij} y_j, \qquad \mathbf{C} = (\gamma_{ij}),$$

carries $Q(x_1, \ldots, x_n)$ into $\lambda_1 y_1^2 + \cdots + \lambda_n y_n^2$, where the λ_i are the (positive) characteristic roots of $\mathbf{S}$. Then use the formula for changing the variable in a multiple integral (see R. C. Buck, loc. cit.) to show that

$$I = \int_0^\infty \cdots \int_0^\infty e^{-(\lambda_1 y_1^2 + \cdots + \lambda_n y_n^2)} |J|\, dy_1 \cdots dy_n,$$

where J is the jacobian of the transformation, and is ± 1. The integral can now be done by the one-variable case.]

32. UNITARY TRANSFORMATIONS AND THE SPECTRAL THEOREM

In this section $V = \{u, v, \ldots\}$ denotes a finite dimensional vector space over the complex field $C = \{a, b, \ldots\}$. $R = \{\alpha, \beta, \ldots\}$ denotes the field of real numbers, and each $a \in C$ can be expressed in the form $a = \alpha + \beta i$, $\alpha, \beta \in R$, and $i^2 = -1$. For $a = \alpha + \beta i \in C$, we let $\bar{a} = \alpha - \beta i$ denote the complex conjugate of a.

From Section 21, we recall that if a complex number a is viewed as a vector in R_2, then its length is given by $|a| = \sqrt{a\bar{a}}$. We use this idea to introduce the concept of length of vectors in a complex vector space.

First let $V = C_n$, the vector space of n-tuples over C.

(32.1) THEOREM. *Let $u = \langle a_1, \ldots, a_n\rangle$, $v = \langle b_1, \ldots, b_n\rangle$ belong to the vector space $V = C_n$, and define*

$$(u, v) = \sum_{i=1}^{n} a_i\overline{b_i}.$$

Then the scalar product (u, v) has the following properties.

a. $(u_1 + u_2, v) = (u_1, v) + (u_2, v)$,
$(u, v_1 + v_2) = (u, v_1) + (u, v_2)$,
$(au, v) = a(u, v)$, $\quad (u, av) = \bar{a}(u, v)$, $\quad (u, v) = \overline{(v, u)}$,
for all vectors u, v, etc., in V and $a \in C$.

b. *For all $u \in V$, (u, u) is real and nonnegative. Moreover, $(u, u) = 0$ if and only if $u = 0$. We define the* length $\|u\|$ *of $u \in V$ by*

$$\|u\| = \sqrt{(u, u)}.$$

Proof. There is no difficulty with the behavior of (u, v) as far as addition is concerned. We have

$$(au, v) = \sum (aa_i)\overline{b_i} = a(\sum a_i\overline{b_i}) = a(u, v)$$

and

$$(u, av) = \sum a_i(\overline{ab_i}) = \sum a_i\overline{a}\overline{b_i} = \bar{a}\sum a_i\overline{b_i} = \bar{a}(u,v),$$

using the fact from Section 21 that $\overline{ab} = \bar{a}\bar{b}$. Finally, $(v, u) = \sum b_i\overline{a_i} = \overline{(u, v)}$, since $\bar{\bar{a}} = a$ for all $a \in C$.

For part (ii), we have

$$(u, u) = \sum a_i\overline{a_i},$$

and from Section 21, $a_i\overline{a_i}$ is real and nonnegative, so that the same is true for (u, u). If $(u, u) = 0$, then because all $a_i\bar{a}_i \geq 0$, we have $a_i\bar{a}_i = 0$, and $a_i = 0$ for all i. Therefore $(u, u) = 0$ if and only if $u = 0$, and the theorem is proved.

The reader should note that because of the property $(u, av) = \bar{a}(u, v)$, the function (u, v) is not a bilinear form on the vector space V.

(32.2) DEFINITION. Let V be an arbitrary vector space over C. A mapping (u, v) which assigns to each pair of vectors $\{u, v\}$ a complex number (u, v) is called a *hermitian scalar product* (after the French mathematician C. Hermite) if (u, v) has the properties (a) and (b) of Theorem (32.1). A set of vectors $\{u_1, \ldots, u_s\}$ in V is called an *orthonormal set* [with respect to (u, v)] if $\|u_i\| = 1$ for all i, and $(u_i, u_j) = 0$ if $i \neq j$. Two vectors u and v are called *orthogonal* if $(u, v) = 0$.

(32.3) THEOREM. *Let (u, v) be a hermitian scalar product on V. Every subspace $W \neq 0$ of V has an orthonormal basis. If $\{u_1, \ldots, u_s\}$ is an orthonormal basis for W, and $v = \sum a_iu_i$ and $w = \sum b_iu_i$ are arbitrary vectors in W, then*

$$(v, w) = \sum a_i\bar{b}_i.$$

Proof. We imitate the Gram-Schmidt process given in Chapter 4 for real vector spaces with an inner product. If dim $W = 1$, and $W = S(w)$, $w \neq 0$, then $w_1 = \|w\|^{-1}w$ has length 1 and is an orthonormal basis of W. Now let $W = S(w_1, \ldots, w_s)$, and assume, as an induction hypothesis, that $S(w_1, \ldots, w_{s-1})$ has an orthonormal basis $\{u_1, \ldots, u_{s-1}\}$. Let $w_s \notin S(w_1, \ldots, w_{s-1})$, and put

$$w = w_s - \sum_{i=1}^{s-1} (w_s, u_i)u_i.$$

Then $w \neq 0$, and for $1 \leq j \leq s - 1$,

$$\begin{aligned}(w, u_j) &= (w_s, u_j) - \sum_{i=1}^{s-1} ((w_s, u_i)u_i, u_j)\\ &= (w_s, u_j) - (w_s, u_j) = 0.\end{aligned}$$

Setting $u_s = \|w\|^{-1}w$, we have $\|u_s\| = 1$ and $(u_s, u_j) = 0$ for $1 \leq j \leq s - 1$. Since $(u, v) = \overline{(v, u)}$, we have also $(u_j, u_s) = 0$ for $1 \leq j \leq s - 1$, and the first part of the theorem is proved.

Now let $v = \sum a_iv_i$, and $w = \sum b_iu_i$ be vectors in W, expressed in terms of the orthonormal basis $\{u_1, \ldots, u_s\}$. Then

$$\begin{aligned}(v, w) &= (\sum a_iu_i, \sum b_ju_j) = \sum a_i\bar{b}_j(u_i, u_j)\\ &= \sum a_i\bar{b}_i\end{aligned}$$

as required. This completes the proof of the theorem.

(32.4) DEFINITION. Let (u, v) be a hermitian scalar product on V. Let W be a subspace of V, and let $W^\perp = \{v \in V \mid (v, w) = 0 \text{ for all } w \in W\}$. Then $W^\perp$ is a subspace of V called the *orthogonal complement* of W.

Note that since $(v, w) = 0$ implies $(w, v) = 0$, the definition could have been worded just as well

$$W^\perp = \{v \in V \mid (w, v) = 0 \quad \text{for all } w \in W\}.$$

The properties of (u, v) imply that $W^\perp$ is a subspace. Since $W^\perp$ is defined in terms of a particular scalar product on V, there is no danger of confusion with the annihilators $V_1^\perp$ defined in Chapter 8, with reference to a bilinear form on a pair of vector spaces.

(32.5) THEOREM. *The mapping $W \to W^\perp$ of the set of subspaces of a vector space V with a hermitian scalar product has the following properties:*

a. $W_1 \subset W_2$ *implies* $W_1^\perp \supset W_2^\perp$.
b. $W^{\perp\perp} = W$.
c. $V = W \oplus W^\perp$, *for all subspaces W of V.*

Proof. Statement (a) is clear from the definition of $W^\perp$. Statement (c) follows by the same argument used in the proof of Theorem (30.4) to show the corresponding result for real vector spaces, and we shall not repeat the details. From part (b), we have

$$\dim W^\perp = \dim V - \dim W,$$

and hence

$$\dim W^{\perp\perp} = \dim W.$$

Since $W \subset W^{\perp\perp}$ from the definition of $W^\perp$, we conclude that $W = W^{\perp\perp}$, and part (b) is proved. This completes the proof of the theorem.

(32.6) DEFINITION. Let V be a vector space with a hermitian scalar product (u, v). A linear transformation $U \in L(V, V)$ is called a *unitary transformation* on V if $\|U(v)\| = \|v\|$ for all $v \in V$.

The unitary transformations on complex vector spaces with hermitian scalar products are the counterparts of the orthogonal transformations on real inner product spaces; they are the distance-preserving transformations.

(32.7) THEOREM. *Let V be a complex vector space with a hermitian scalar product (u, v). A linear transformation $U \in L(V, V)$ is unitary if and only if any one of the following equivalent conditions is satisfied.*

a. *$(U(u), U(v)) = (u, v)$ for all $u, v \in V$.*
b. *U carries orthonormal bases of V onto orthonormal bases, that is, whenever $\{v_1, \ldots, v_n\}$ is an orthonormal basis, so is $\{U(v_1), \ldots, U(v_n)\}$.*
c. *Letting $\mathbf{A} = (a_{ij})$ be the matrix of U with respect to an orthonormal basis $\{v_1, \ldots, v_n\}$, $Uv_i = \sum a_{ji}v_j$, the matrix $\mathbf{A}$ has the property that*

$$\mathbf{A} \cdot {}^t\overline{\mathbf{A}} = \mathbf{I},$$

where ${}^t\overline{\mathbf{A}}$ is the matrix whose (i, j) entry is $\overline{a_{ji}}$. A matrix $\mathbf{A}$ satisfying the condition (c) is called a unitary matrix.

Proof. Once again the proof is similar to the corresponding result for orthogonal transformations. First suppose U is unitary. Applying the condition $\|Uv\| = \|v\|$ to $v + w$ and $v + iw$ yields

$$(Uv, Uw) + (Uw, Uv) = (v, w) + (w, v)$$

and

$$(iU(w), U(v)) + (Uv, iU(w)) = (iw, v) + (v, iw).$$

Using the fact that $\bar{i} = -i$, the second equation yields

$$i(U(w), U(v)) - i(U(v), U(w)) = i(w, v) - i(v, w).$$

Upon canceling i, we can add the equations to obtain

$$(U(v), U(w)) = (v, w),$$

which is (a).

Now assume (a). If $\{v_1, \ldots, v_n\}$ is an orthonormal basis of V, then (a) implies that $\{U(v_1), \ldots, U(v_n)\}$ is an orthonormal set, and hence an orthonormal basis since vectors in an orthonormal set are linearly independent. Therefore (a) implies (b).

Next assume (b), and let $\{v_1, \ldots, v_n\}$ be an orthonormal basis. Letting $\mathbf{A} = (a_{ij})$ be the matrix of U with respect to the basis, we have

$$\begin{aligned} (U(v_i), U(v_j)) &= \left(\sum_{k=1}^{n} a_{ki}v_k, \sum_{l=1}^{n} a_{lj}v_l\right) \\ &= \sum_{k=1}^{n}\sum_{l=1}^{n} a_{ki}\overline{a_{lj}}(v_k, v_l) \\ &= \sum_{k=1}^{n} a_{ki}\overline{a_{kj}} = \begin{cases} 1 & \text{if } i = j \\ 0 & \text{if } i \neq j. \end{cases} \end{aligned}$$

These equations imply that the matrix of $\mathbf{A}$ satisfies (c).

Finally the equations for $(U(v_i), U(v_j))$ in terms of the coefficients of $\mathbf{A}$ show that if $\mathbf{A}$ satisfies the condition $\mathbf{A}{}^t\overline{\mathbf{A}} = \mathbf{I}$, then $\{U(v_1), \ldots, U(v_n)\}$ is an orthonormal basis, and hence that U is unitary. This completes the proof of the theorem.

(32.8) THEOREM. *Let $U \in L(V, V)$ be a unitary transformation. Then the characteristic roots of U all have absolute value equal to 1. Moreover, there exists an orthonormal basis of V consisting of characteristic vectors of U.*

Proof. First let v be a characteristic vector belonging to a characteristic root a of U. Then $Uv = av$ implies

$$(Uv, Uv) = (v, v) = a\bar{a}(v, v)$$

and hence $a\bar{a} = 1$, proving that $|a| = 1$. We prove the second part of the theorem by induction on dim V, the result being clear if dim $V = 1$. Suppose dim $V > 1$. Since C is an algebraically closed field, there exists a characteristic vector v_1 belonging to some characteristic root of U, and we may assume that $\|v_1\| = 1$. Let $W = S(v_1)$; then by Theorem (32.5), $V = W \oplus W^{\perp}$, and dim $W^{\perp} =$ dim $V - 1$. If we can show that $W^{\perp}$ is invariant relative to U, then the restriction of U to $W^{\perp}$ will be a unitary transformation of $W^{\perp}$, and the theorem will follow from the induction hypothesis. Let $(v_1, w) = 0$; we have to show that $(v_1, U(w)) = 0$. Since $Uv_1 = av_1$ for some $a \neq 0$, we have

$$0 = (v_1, w) = (U(v_1), U(w)) = (av_1, U(w)) = a(v_1, U(w))$$

and $(v_1, U(w)) = 0$ as required.

(32.9) COROLLARY. *Let $\mathbf{A}$ be a unitary n-by-n matrix, that is, $\mathbf{A}^t\bar{\mathbf{A}} = \mathbf{I}$. Then there exists a unitary matrix $\mathbf{B}$ such that $\mathbf{BAB}^{-1}$ is a diagonal matrix whose diagonal entries are all complex numbers of absolute value* 1.

Proof. Let V be an n-dimensional space with a hermitian scalar product (for example C_n) and let $\{v_1, \ldots, v_n\}$ be an orthonormal basis of V. Let U be the linear transformation whose matrix with respect to the basis $\{v_1, \ldots, v_n\}$ is $\mathbf{A}$. By Theorem (32.7), U is a unitary transformation. Apply Theorem (32.8) to obtain an orthonormal basis $\{w_1, \ldots, w_n\}$ such that the corresponding matrix is diagonal, with diagonal entries of absolute value equal to one. Then the diagonal matrix is equal to $\mathbf{BAB}^{-1}$, where $\mathbf{B}$ is the matrix carrying one orthonormal basis into the other. By Theorem (32.7) again, $\mathbf{B}$ is a unitary matrix, and the corollary is proved.

We now introduce the analog of the transpose of a linear transformation, adapted to vector spaces with hermitian scalar products.

(32.10) DEFINITION. Let V be a vector space with a hermitian scalar product (u, v). Let $T \in L(V, V)$. A linear transformation $T' \in L(V, V)$ is called an *adjoint* of T if

$$(Tu, v) = (u, T'v)$$

for all $u, v \in V$.

The next theorem shows that the adjoint always exists, and is uniquely determined.

(32.11) THEOREM. *Let $T \in L(V, V)$. Then T has a uniquely determined adjoint T'. If the matrix of T with respect to an orthonormal basis is* **A**, *then the matrix of T' is* ${}^t\bar{\mathbf{A}}$. *We have*

$$(aT)' = \bar{a}T', \qquad (T_1 + T_2)' = T_1' + T_2',$$
$$(T_1T_2)' = T_2'T_1', \qquad \text{and} \qquad T'' = T$$

for all linear transformations T, T_1, T_2 and $a \in C$.

Proof. The proof of the existence of T' can be shown by the methods of Section 27. Instead, we shall give a more down-to-earth approach using matrices. Let $\mathbf{A} = (a_{ij})$ be the matrix of T with respect to an orthonormal basis $\{v_1, \ldots, v_n\}$, and let T' be the linear transformation whose matrix with respect to this basis is ${}^t\bar{\mathbf{A}}$. Then

$$Tv_i = \sum_k a_{ki}v_k, \qquad T'v_j = \sum_k \bar{a}_{jk}v_k,$$

and

$$(Tv_i, v_j) = \left(\sum_k a_{ki}v_k, v_j\right) = \sum_k a_{ki}(v_k, v_j) = a_{ji}$$

while

$$(v_i, T'v_j) = (v_i, \sum \bar{a}_{jk}v_k) = \sum_k a_{jk}(v_i, v_k) = a_{ji}.$$

Therefore T' is an adjoint of T, and has the matrix ${}^t\bar{\mathbf{A}}$ with respect to the basis. The uniqueness of T' follows from the fact that if T'' is another adjoint, then

$$(v, (T' - T'')w) = 0$$

for all v and w, and hence $T' = T''$. The proofs of the formulas for $(aT)'$, $(T_1 + T_2)'$, $(T_1T_2)'$ and T'' follow directly from the uniqueness of the adjoint and are left to the reader as exercises. This completes the proof.

(32.12) DEFINITION. A linear transformation $T \in L(V, V)$ is *self-adjoint*, or *hermitian*, if $T = T'$; *normal* if $TT' = T'T$.

Example A. Let **A** be a real symmetric matrix, and let T be a linear transformation whose matrix with respect to an orthonormal basis is **A**. Then T is a self-adjoint transformation.

Self-adjoint transformations are always normal. Examples of normal transformations which are not in general self-adjoint are the unitary transformations.

The main theorem of this section is the result that normal transformations are always diagonable, and that characteristic vectors of a normal transformation belonging to distinct characteristic roots are orthogonal. We shall first prove some preliminary results. The main result is stated in a particularly useful and elegant form, called the *spectral theorem*, which exhibits the connection between the *spectrum* of T, that is defined to be the set of characteristic roots of T, and the action of T, on the vector space V.

(32.13) LEMMA. *Let T be a normal transformation on V. There exist common characteristic vectors for T and T'. For such a vector v, $Tv = av$ and $T'v = \bar{a}v$.*

Proof. There exists a characteristic root a of T because C is algebraically closed. Let V_1 be the set of all vectors $v \in V$ such that $Tv = av$. If $v \in V_1$, then the fact that $TT' = T'T$ implies that $TT'v = T'Tv = aT'v$, and V_1 is invariant relative to T'. Now find in V_1 a characteristic vector for T'; this vector will have the required property.

Now let $Tv = av$, $T'v = bv$. Then

$$a(v, v) = (Tv, v) = (v, T'v) = (v, bv) = \bar{b}(v, v),$$

and we have $a = \bar{b}$ since $(v, v) \neq 0$. This completes the proof.

(32.14) LEMMA. *Let T be normal, and let v and v' be characteristic vectors for T and T' simultaneously such that v and v' belong to distinct characteristic roots for T. Then $(v, v') = 0$.*

Proof. Let $Tv = av$, $Tv' = bv'$, $a \neq b$. Since v and v' satisfy the hypothesis of Lemma (32.13), we have $T'v = \bar{a}v$, $T'v' = \bar{b}v'$. Then

$$\begin{aligned} a(v, v') &= (Tv, v') = (v, T'v') = (v, \bar{b}v') \\ &= b(v, v'), \end{aligned}$$

and $(v, v') = 0$ since $a \neq b$.

(32.15) LEMMA. *Let $\{E_1, \ldots, E_s\}$ be a set of linear transformations of V such that $1 = \sum E_i$, and $E_iE_j = 0$ if $i \neq j$. Then the $\{E_i\}$ are self-adjoint, for $1 \leq i \leq s$, if and only if the subspaces $\{E_iV\}$ are mutually orthogonal.*

Proof. First suppose the $\{E_i\}$ are self-adjoint. Then letting $v = E_i(w)$, $v' = E_j(w')$ for $i \neq j$, we have

$$(v, v') = (E_i(w), E_j(w')) = (w, E_iE_j(w')) = 0$$

since $E_i' = E_i$, and $E_iE_j = 0$.

Conversely, suppose $(E_iV, E_jV) = 0$ if $i \neq j$. Then $(V, E_i'E_jV) = 0$ and hence $E_i'E_j = 0$ if $i \neq j$. Using Theorem (32.11), we have also

$$1 = E_1' + \cdots + E_s', \qquad E_i'E_j' = 0, \quad i \neq j.$$

Then

$$E_i = 1E_i = (E_1' + \cdots + E_s')E_i = E_i'E_i$$

and

$$E_i' = E_i' \cdot 1 = E_i'(E_1 + \cdots + E_s) = E_i'E_i.$$

Therefore $E_i = E_i'$, and the lemma is proved.

Now we are ready to state the main theorem on normal transformations.

(32.16) THEOREM (SPECTRAL THEOREM FOR NORMAL TRANSFORMATIONS). *Let T be a normal transformation on a vector space V over C with a hermitian scalar product, and let $\{\alpha_1, \ldots, \alpha_s\}$ be the distinct characteristic roots of T. Then there exist polynomials $\{f_1(x), \ldots, f_s(x)\}$ in $C[x]$ such that the transformations $\{E_i = f_i(T)\}$, $1 \leq i \leq s$, are self-adjoint and satisfy the conditions $1 = E_1 + \cdots + E_s$, and $E_iE_j = 0$ if $i \neq j$. Moreover T can be expressed in the form*

$$T = \alpha_1 E_1 + \cdots + \alpha_s E_s.$$

Such a decomposition of T is called a spectral decomposition of T.

Proof. We first prove, by induction on dim V, that T is diagonable. The result is clear if dim $V = 1$, and we may assume dim $V > 1$. By Lemma (32.13), there exists a common characteristic vector w for T and T'. Let W be the subspace $S(w)$ generated by w. By Theorem (32.5), $V = W \oplus W^{\perp}$. We prove that $W^{\perp}$ is invariant relative to T and T'. Let $v \in W^{\perp}$. Then $(w, Tv) = (w, T'v) = 0$ since w is a characteristic vector for both T and T', and hence $Tv \in W^{\perp}$ and $T'v \in W^{\perp}$. The restriction of T to $W^{\perp}$ is easily shown to be normal (the details are left as an exercise). By induction, it follows that T is diagonable.

We can now apply Theorem (23.9) to find polynomials $\{f_1(x), \ldots, f_s(x)\}$ in $C[x]$ such that the transformations $E_i = f_i(T)$, $1 \leq i \leq s$, satisfy the conditions

$$1 = \sum E_i, \; E_iE_j = 0 \qquad \text{if } i \neq j$$

and

$$T = \alpha_1 E_1 + \cdots + \alpha_s E_s.$$

Moreover V is the direct sum, $V = E_1V \oplus \cdots \oplus E_sV$, and $E_iV = \{v \in V \mid Tv = \alpha_i v\}$, for $1 \leq i \leq s$. It remains to prove that the E_i are self-adjoint, and for this it is sufficient, by Lemma (32.15), to show that the subspaces E_iV are mutually orthogonal. Since T and T' commute, each subspace E_iV is invariant relative to both T and T'. Therefore the restriction of T' to E_iV is normal and is diagonable on E_iV. Moreover all the characteristic roots of T' on E_iV are equal to $\bar{a}_i$ by Lemma (32.13). Therefore every nonzero vector in E_iV is a characteristic vector for both T and T'. We can now apply Lemma (32.14) to conclude that $(E_iV, E_jV) = 0$ if $i \neq j$, and the theorem is proved.

(32.17) COROLLARY. *Let T be a normal transformation on V. There exists an orthonormal basis of V consisting of characteristic vectors of T.*

This result has essentially been proved along with Theorem (32.16), and we leave the proof of the corollary as an exercise.

We conclude this section with one of the striking applications of the Spectral Theorem. We recall that each complex number $z = \alpha + \beta i \neq 0$ can be expressed in polar form, $z = r(\cos\theta + i\sin\theta)$, where r is a positive real number, and $\cos\theta + i\sin\theta$ is a complex number of absolute value one. Our application is a far-reaching generalization of this fact about complex numbers, to transformations on a vector space with a hermitian scalar product. The analog of a complex number of absolute value one is certainly a unitary transformation. We now define the analog of a positive real number.

(32.18) DEFINITION. A linear transformation on V is called *positive* if T is self-adjoint and (Tv, v) is real and positive for all vectors $v \neq 0$.

(32.19) LEMMA. *The characteristic roots of a self-adjoint transformation T are all real. T is positive if and only if they are all positive.*

Proof. Let T be self-adjoint, and let a be a characteristic root of T belonging to a characteristic vector v. Then

$$(Tv, v) = (v, Tv) = \overline{(Tv, v)}$$

so that (Tv, v) is real. Moreover, $Tv = av$ implies that $(Tv, v) = a(v, v)$, and

$$a = \frac{(Tv, v)}{(v, v)}$$

is real. If T is positive, the formula shows that a is positive. Conversely, suppose all characteristic roots of the self-adjoint transformation T are positive. By the Spectral Theorem (32.16),

$$T = \sum a_i E_i,$$

where the a_i are real and positive. Let $v \in V$ be expressed in the form

$$v = \sum E_i v_i,$$

for some $v_i \in V$, and suppose $v \neq 0$. Then

$$(Tv, v) = \sum a_i(E_iv_i, E_jv_j) > 0$$

since $v \neq 0$, and $(E_iv_i, E_jv_j) = 0$ if $i \neq j$.

Now we are ready to prove our application of the Spectral Theorem. Notice that in the proof, the Spectral Theorem is used to construct a square root of a certain transformation, using the fact that if $T = \sum \alpha_i E_i$ is the decomposition of T according to Theorem (32.16), then $T^2 = \sum \alpha_i^2 E_i$, $T^3 = \sum \alpha_i^3 E_i$, etc., because of the properties of the E_i's.

(32.20) THEOREM (POLAR DECOMPOSITION). *Let T be an invertible linear transformation on a vector space V over C with a hermitian scalar product. Then T can be expressed in the form $T = US$, where U is unitary and S is positive.*

Proof. Since T is invertible, (Tv, Tv) is real and positive for all $v \neq 0$. Then $(T'Tv, v) = (Tv, Tv)$ is also real and positive for $v \neq 0$, and since $(T'T)' = T'T'' = T'T$ by Theorem (32.11), $T'T$ is a positive transformation. Then by Lemma (32.19), the spectral decomposition of $T'T$ will have the form

$$T'T = \sum a_i E_i,$$

with real and positive characteristic roots $\{a_i\}$. Put

$$S = \sum \sqrt{a_i} E_i;$$

then since $E_i^2 = E_i$ and $E_iE_j = 0$ if $i \neq j$, we have $S^2 = T'T$. Moreover S is clearly self-adjoint and positive, since the characteristic roots are $\{\sqrt{a_i}\}$ and the $\{E_i\}$ are self-adjoint. Since all the characteristic roots of S are $\neq 0$, S is invertible. Put $U = TS^{-1}$. Then $T = US$, where S is positive. Moreover,

$$U'U = (S^{-1})'T'TS^{-1} = S^{-1}S^2S^{-1} = 1,$$

since S^{-1} is also self-adjoint, and $S^2 = T'T$. Therefore U is unitary (see Exercise 5 below), and the theorem is proved.

EXERCISES

Throughout the exercises, V denotes a finite dimensional vector space over C with a hermitian scalar product (u, v).

1. Prove that the unitary transformations on V form a group with respect to the operation of multiplication.

2. Show that a unitary matrix in triangular form (with zeros below the diagonal) must be in diagonal form.

3. Show that

$$\mathbf{A} = \begin{pmatrix} 0 & i \\ -i & 0 \end{pmatrix}$$

is a unitary matrix. Find a unitary matrix **B** such that $\mathbf{BAB}^{-1}$ is a diagonal matrix.

4. Let V^* be the dual space of V. Prove that if $f \in V^*$, then there exists a unique vector $w \in V$ such that $f(v) = (v, w)$ for all $v \in V$.

5. Show that $U \in L(V, V)$ is unitary if and only if $UU' = 1$, where U' is the adjoint of U.

6. Show that if $T \in L(V, V)$ is normal, then W is a T-invariant subspace of V if and only if $W^{\perp}$ is a T'-invariant subspace.

7. Prove that if T is a normal transformation with spectral decomposition $T = \sum \alpha_i E_i$ according to Theorem (32.16), then $T' = \sum \bar{\alpha}_i E_i$ is a spectral decomposition of T'.

8. Show that if U is a unitary transformation, then U is normal.

9. Show that a normal transformation T with spectral decomposition $T = \sum \alpha_i E_i$ is unitary if and only if $|\alpha_i| = 1$ for all characteristic roots α_i of T.

10. Prove that if $T = \sum \alpha_i E_i$ is the spectral decomposition of a normal transformation T, then for every polynomial $f(x) \in C[x]$, $f(T)$ is a normal transformation with the spectral decomposition $f(T) = \sum f(\alpha_i) E_i$.

11. Let T be a normal transformation with spectral decomposition $T = \sum \alpha_i E_i$. Show that a linear transformation $X \in L(V, V)$ commutes with T if and only if X commutes with all the idempotents E_i.

12. Let $(u, v)_1$ be a second hermitian scalar product on V. Prove that there exists a positive transformation T with respect to the given scalar product (u, v) such that

$$(u, v)_1 = (Tu, v)$$

for all $u, v \in V$.

Chapter 10

Some Applications of Linear Algebra

We shall give in this chapter three applications of linear algebra, each to a different part of mathematics. The first is to the geometrical problem of classifying finite symmetry groups in three dimensions, completing to some extent the work begun in Section 14 on symmetry groups in the plane. The second shows how the language of vectors and matrices is used in analysis to translate a system of linear first-order differential equations with constant coefficients, to a single vector differential equation, and then to solve the equation in an elegant way using the exponential function of a matrix. Finally, we give an application to a problem in classical algebra, on sums of squares.

33. FINITE SYMMETRY GROUPS IN THREE DIMENSIONS

We begin with some general remarks about orthogonal transformations on an n-dimensional real vector space V with an inner product (u, v). If $S \subset V$, we denote by $S^{\perp}$ the set of all vectors $v \in V$ such that $(s, v) = 0$ for all $s \in S$. Then $S^{\perp}$ is always a subspace and, as we have shown in Section 30, if S is a subspace then

$$V = S \oplus S^{\perp}.$$

In particular, if x is a nonzero vector, then $(x)^{\perp}$ is an $(n-1)$-dimensional subspace such that

$$V = S(x) \oplus (x)^{\perp}.$$

In the language of Section 10, $(x)^\perp$ is a hyperplane passing through the origin. Perhaps the simplest kind of orthogonal transformation on V is a transformation T such that, for some nonzero vector x,

$$Tx = -x$$
$$Tu = u, \qquad u \in (x)^\perp.$$

If H denotes the hyperplane $(x)^\perp$, then T is called a *reflection with respect to H*.† Geometrically, T sends each point of V onto its mirror image with respect to H, leaving the elements of H fixed. The matrix of a reflection T, with respect to a basis containing a basis for the hyperplane left fixed by T and a vector orthogonal to the hyperplane, is given by

$$\begin{pmatrix} 1 & & & & 0 \\ & 1 & & & \\ & & \ddots & & \\ & & & 1 & \\ 0 & & & & -1 \end{pmatrix}$$

so that we have

$$T^2 = 1, \qquad D(T) = -1$$

for all reflections T. Our first important result is the following theorem, due to E. Cartan and J. Dieudonné, which asserts that every orthogonal transformation is a product of reflections.

(33.1) THEOREM. *Every orthogonal transformation on an n-dimensional vector space V with an inner product is a product of at most n reflections.*

Proof.‡ We use induction on n, the result being clear if $n = 1$, since the only orthogonal transformations on a one-dimensional space are ± 1. Suppose $\dim V > 1$ and that the theorem is true for orthogonal transformations on an $(n - 1)$-dimensional space. Let T be a given orthogonal transformation on V and let $x \neq 0$ be a vector in V.

Case 1. Suppose $Tx = x$; then if $u \in H = (x)^\perp$, we have

$$(Tu, x) = (Tu, Tx) = (u, x) = 0,$$

so that $Tu \in H$, and T defines an orthogonal transformation on the $(n - 1)$-dimensional space H. By the induction hypothesis there exist reflections $T_1, \ldots, T_s$, for $s \leq n - 1$ of H, such that

(33.2) $$Tu = T_1 \cdots T_s u, \qquad u \in H.$$

† The question of the uniqueness of a reflection with respect to a given hyperplane is settled in the exercises.

‡ A proof of a more general form of the theorem is given by Artin (listed in the Bibliography).

Extend each T_i to a linear transformation T_i' on V by defining $T_i'x = x$ and $T_i'u = T_iu$, for $u \in H$, $i = 1, \ldots, s$. We show now that each T_i' is a reflection on V. There exists an $(n - 2)$-dimensional subspace $H_i \subset H = (x)^{\perp}$ whose elements are left fixed by T_i; then T_i' leaves fixed the vectors in the hyperplane $H_i' = H_i + S(x)$ of V. Moreover, if x_i generates $H_i^{\perp}$ in H, then $x_i \in (H_i')^{\perp}$ and $T_i'x_i = T_ix_i = -x_i$. These remarks show that T_i' is a reflection with respect to H_i'. Finally, from the definition of the transformations T_i' and (33.2) it follows that

$$T = T_1' \cdots T_s'.$$

Thus we have shown that an orthogonal transformation which leaves a vector fixed is a product of at most $n - 1$ reflections.

Case 2. Suppose $Tx \neq x$; then $u = Tx - x \neq 0$. Let $H = (u)^{\perp}$ and let U be a reflection with respect to H. We have

$$(Tx + x, Tx - x) = (Tx, Tx) + (x, Tx) - (Tx, x) - (x, x) = 0$$

because T is orthogonal and the form is symmetric. Therefore, $Tx + x \in H = (u)^{\perp}$ where $u = Tx - x$, and we have

$$U(Tx + x) = Tx + x.$$

Since U is a reflection with respect to $H = (Tx - x)^{\perp}$, we have

$$U(Tx - x) = -(Tx - x).$$

Adding these equations, we obtain

$$2UT(x) = 2x$$

and hence

$$UT(x) = x.$$

By Case 1 there exist $s \leq n - 1$ reflections $T_1, \ldots, T_s$ such that

$$UT = T_1 \cdots T_s.$$

Since $U^2 = 1$ we have

$$T = U(UT) = UT_1 \cdots T_s$$

which is a product of at most n reflections. This completes the proof.

As a corollary, we obtain some geometrical insight into the following result, which was proved in another way in the exercises of Section 30.

(33.3) COROLLARY. *Let T be an orthogonal transformation of a three-dimensional real vector space such that $D(T) = +1$. Then there exists a nonzero vector $v \in V$ such that $Tv = v$.*

Proof. We may assume $T \neq 1$. By Theorem (33.1), T is a product of one, two, or three reflections of determinant -1. It follows that T is a product of exactly two reflections, $T = T_1T_2$, where T_i is a reflection with respect to a two-dimensional space H_i for $i = 1, 2$. By Theorem (7.5),

$$\dim (H_1 + H_2) + \dim (H_1 \cap H_2) = \dim H_1 + \dim H_2 = 4$$

and since $\dim (H_1 + H_2) \leq 3$ we have $\dim (H_1 \cap H_2) \geq 1$. Let x be a nonzero vector in $H_1 \cap H_2$; then $Tx = x$, and the corollary is proved.

We define a *rotation* in a three-dimensional space as an orthogonal transformation of determinant $+1$. Then the corollary asserts that every rotation in three dimensions is a rotation about an axis, the axis being the line determined by the vector left fixed. This fact is of fundamental importance in mechanics and was proved with the use of an interesting geometrical argument by Euler.† We shall see that it is also the key to the classification of finite symmetry groups in R_3. Our presentation of this material is based on the introductory discussion in Section 14 and the proof of the main theorem is taken from Weyl's book (see the Bibliography). Apart from its geometrical interest, Weyl's argument gives a penetrating introduction to the theory of finite groups.

Let us make the problem precise. By a *finite symmetry group* in three dimensions we mean a finite group of orthogonal transformations in R_3. For simplicity we shall determine the *finite groups of rotations* in R_3 and indicate in the exercises the connection with the general problem. Let us begin with a list of some examples of finite rotation groups in R_3. By the *order* of a finite group we mean the number of elements in the group. The cyclic group $\mathscr{C}_n$ of order n and the dihedral group $\mathscr{D}_n$ of order $2n$, discussed in Section 14, are the first examples.

For our purposes in this section it is better to think of $\mathscr{D}_n$ as the symmetry group of a regular n-sided polygon. Although the transformation S in $\mathscr{D}_n$ which flips the polygon over along a line of symmetry is a reflection when viewed as a transformation in the plane, S is a rotation when viewed as a transformation in R_3.

The next examples are the symmetry groups of the regular polyhedra in R_3. There are exactly five of these, which are listed in the following table along with the number of vertices V, edges E, and faces F.

	F	E	V
tetrahedron	4	6	4
cube	6	12	8
octahedron	8	12	6
dodecahedron	12	30	20
icosahedron	20	30	12

† See Synge and Griffith, pp. 279–280 (listed in the Bibliography).

(For a derivation of this list, based on Euler's formula for polyhedra "without holes," $F - E + V = 2$, see the book by Courant and Robbins and also those by Weyl and by Coxeter, listed in the Bibliography.) The faces are equilateral triangles in the case of the tetrahedron, squares in the case of the cube, equilateral traingles in the case of the octahedron, pentagons in the dodecahedron, and equilateral triangles in the case of the icosahedron; see Figure 10.1.

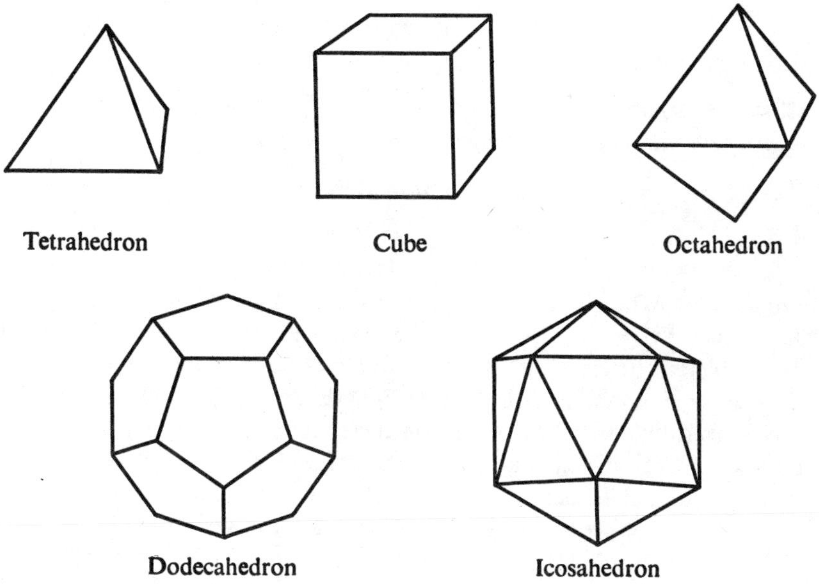

FIGURE 10.1

Apparently, we should obtain five different groups of rotations from these figures, the group in each case being the set of all rotations that carry the figure onto itself. On closer inspection we see that this is not the case. For example, the cube and the octahedron are dual in the sense that, if we take either figure and connect the centers of the faces with line segments, these line segments are the edges of the other; see Figure 10.2. Thus, any rotation carrying the cube onto itself will also be a symmetry of the octahedron, and conversely. Similarly, the dodecahedron and the icosahedron are dual figures. So, from the regular polyhedra we obtain only three additional groups: the group $\mathscr{T}$ of rotations of the tetrahedron the group $\mathscr{O}$ of rotations of the octahedron, and the group $\mathscr{I}$ of rotations of the icosahedron.

Our main theorem can now be stated.

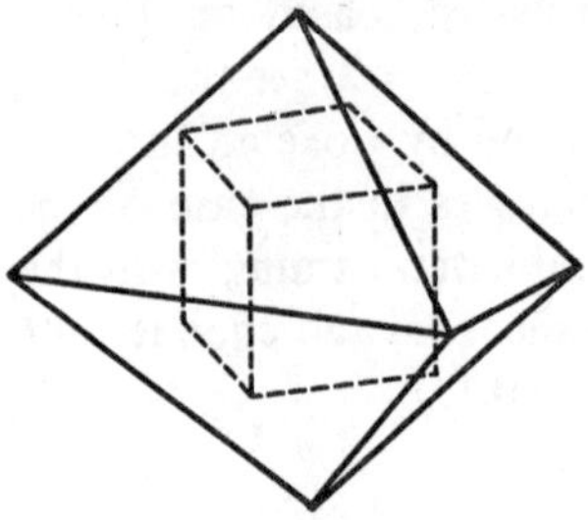

FIGURE 10.2

(33.4) THEOREM. *Let $\mathcal{G}$ be a finite group of rotations in R_3; then $\mathcal{G}$ is isomorphic with one of the groups in the following list:*

$$\begin{aligned} &\mathcal{C}_n, \quad n = 1, 2, \cdots, \\ &\mathcal{D}_n, \quad n = 1, 2, \ldots, \\ &\mathcal{T}, \mathcal{O}, \text{ or } \mathcal{I}. \end{aligned}$$

Proof. Let S be the unit sphere in R_3 consisting of all points† $x \in R_3$ such that $\|x\| = 1$. If $T \in \mathcal{G}$, then $Tx \in S$ for every $x \in S$, and T is completely determined by its action on the elements of the sphere S because S contains a basis for R_3. Every $T \in \mathcal{G}$ such that $T \neq 1$ leaves fixed two antipodal points on the sphere, by Corollary (33.3), and no others. These points are called the *poles* of T. Since $\mathcal{G}$ is finite, the set of all poles of all elements T of $\mathcal{G}$ such that $T \neq 1$ is a finite set of points on the sphere S; we proceed to examine this set in great detail.

Let p be a pole on S. The set of all elements T of $\mathcal{G}$ such that p is a pole of T, together with the identity element 1, is clearly a *subgroup* of $\mathcal{G}$, that is, a subset of $\mathcal{G}$ which itself forms a group under the operation defined on $\mathcal{G}$. The order of this subgroup is called the *order of the pole p* and is denoted by v_p.

Next we define two poles p and p' *equivalent* and we write $p \sim p'$ if and only if $p = Tp'$ for some $T \in \mathcal{G}$. The set of all poles equivalent to p is called the *equivalence class* of p. We prove that each pole belongs to one and only one equivalence class. Since $1 \in \mathcal{G}$, $p = 1p$, and so p belongs to the equivalence class of p. Now we have to show that if a pole p belongs to the equivalence classes of p' and p'' then these equivalence classes coincide. Let q belong to the equivalence class of p'; then $q \sim p'$, and $q = Tp'$ for some $T \in \mathcal{G}$. Since $p \sim p'$ and $p \sim p''$, we have $p = T'p'$ and $p = T''p''$ for T' and $T'' \in \mathcal{G}$, and hence $p' = (T')^{-1}T''p''$. Then $q = T(T')^{-1}T''p''$ so that $q \sim p''$. We have shown that the equivalence class of p' is contained in the equivalence class of p'', and

† We shall speak of the elements in R_3 as points in this discussion.

a similar argument establishes the reverse inclusion. This completes the proof that a pole belongs to one and only one equivalence class.

(33.5) LEMMA. *Equivalent poles have the same order.*

Proof. Let $p \sim p'$, let $\mathcal{H}$ be the subgroup of $\mathcal{G}$ consisting of the identity 1 together with the elements which have p as a pole, and let $\mathcal{H}'$ be the corresponding subgroup for p'. Let $p = Tp'$. Then an easy verification shows that the mapping $X \to TXT^{-1}$ is a one-to-one mapping of $\mathcal{H}'$ onto $\mathcal{H}$, and this establishes the lemma.

(33.6) LEMMA. *Let p be a pole of order v_p and let n_p be the number of poles equivalent to p; then $v_p n_p = N$ where N is the order of $\mathcal{G}$.*

Proof. Let $\mathcal{H}$ be the subgroup associated with p and, for $X \in G$, let $X\mathcal{H}$ denote the set of all elements $Y \in \mathcal{G}$ such that $Y = XT$ for some $T \in \mathcal{H}$. We observe first that, for all $X \in \mathcal{G}$, Xp is a pole and that poles Xp and $X'p$ are the same if and only if $X' \in X\mathcal{H}$. Thus the number of poles equivalent to p is equal to the number of distinct sets of the form $X\mathcal{H}$. The set $X\mathcal{H}$ is called the *left coset of $\mathcal{H}$ containing X*. We prove now that each element of $\mathcal{G}$ belongs to one and only one left coset and that the number of elements in each left coset is v_p. If $X \in \mathcal{G}$, then $X = X \cdot 1 \in X\mathcal{H}$ since $1 \in \mathcal{H}$. Now suppose $X \in X'\mathcal{H} \cap X''\mathcal{H}$; we show that $X'\mathcal{H} = X''\mathcal{H}$. We have, for $Y \in X'\mathcal{H}$, $Y = X'T$ for some $T \in \mathcal{H}$. Moreover, $X \in X'\mathcal{H} \cap X''\mathcal{H}$ implies that $X = X'T' = X''T''$ for T' and $T'' \in \mathcal{H}$. Then $Y = X'T = X(T')^{-1}T = X''T''(T')^{-1}T \in X''\mathcal{H}$. Thus, $X'\mathcal{H} \subset X''\mathcal{H}$ and, similarly, $X''\mathcal{H} \subset X'\mathcal{H}$. Therefore $X'\mathcal{H} = X''\mathcal{H}$, and we have proved that each element of $\mathcal{G}$ belongs to a unique left coset. Now let $X\mathcal{H}$ be a left coset. Then the mapping $T \to XT$ is a one-to-one mapping of $\mathcal{H}$ onto $X\mathcal{H}$, and each left coset contains v_p elements where v_p is the order of $\mathcal{H}$.

We have shown that the number of poles equivalent to p is equal to the number of left cosets of $\mathcal{H}$ in $\mathcal{G}$. From what has been shown, this number is equal to N/v_p, and the lemma is proved.

Now we can finish the proof of Theorem (33.4). Consider the set of all pairs (T, p) with $T \in \mathcal{G}$, $T \neq 1$, and p a pole of T. Counting the pairs in two different ways,† we obtain

$$2(N - 1) = \sum_p (v_p - 1)$$

† We use first the fact that $T \neq 1$ is associated with two poles and next that with each pole are associated $(v - 1)$ elements $T \in \mathcal{G}$ such that $T \neq 1$.

and collecting terms on the right-hand side according to the equivalence classes of poles, C, we have

$$2(N-1) = \sum n_C(v_C - 1)$$

where n_C is the number of poles in an equivalence class C, v_C is the order of a typical pole in C [see Lemma (33.5)], and the sum is taken over the different equivalence classes of poles. Applying (33.6), we obtain

$$2(N-1) = \sum_C (N - n_C) = \sum_C \left(N - \frac{N}{v_C}\right).$$

Dividing by N, we obtain

$$2 - \frac{2}{N} = \sum_C \left(1 - \frac{1}{v_C}\right). \tag{33.7}$$

The left side is >1 and <2; therefore there are at least two classes C and at most three classes. The rest of the argument is an arithmetical study of the equation (33.7).

Case 1. There are two classes of poles of order v_1, v_2. Then (33.7) yields

$$\frac{2}{N} = \frac{1}{v_1} + \frac{1}{v_2}, \qquad 2 = \frac{N}{v_1} + \frac{N}{v_2},$$

and we have $N/v_1 = N/v_2 = 1$. This case occurs if and only if $\mathcal{G}$ is cyclic.

Case 2. There are three classes of poles of orders v_1, v_2, v_3, where we may assume $v_1 \le v_2 \le v_3$. Then (33.7) implies that

$$2 - \frac{2}{N} = \left(1 - \frac{1}{v_1}\right) + \left(1 - \frac{1}{v_2}\right) + \left(1 - \frac{1}{v_3}\right)$$

or

$$1 + \frac{2}{N} = \frac{1}{v_1} + \frac{1}{v_2} + \frac{1}{v_3}.$$

Not all the v_i can be greater than 2; hence $v_1 = 2$ and we have

$$\frac{1}{2} + \frac{2}{N} = \frac{1}{v_2} + \frac{1}{v_3}.$$

Not both v_2 and v_3 can be ≥ 4; hence $v_2 = 2$ or 3.

Case 2a: $v_1 = 2, v_2 = 2$. Then $v_3 = N/2$, and in this case $\mathcal{G}$ is the dihedral group.

Case 2b: $v_1 = 2, v_2 = 3$. Then we have

$$\frac{1}{v_3} = \frac{1}{6} + \frac{2}{N},$$

and $v_3 = 3, 4$, or 5. For each possibility of v_3 we have:

$v_1 = 2, v_2 = 3, v_3 = 3$; then $N = 12$ and $\mathcal{G}$ is the tetrahedral group $\mathcal{T}$.
$v_1 = 2, v_2 = 3, v_3 = 4$; then $N = 24$ and $\mathcal{G}$ is the octahedral group $\mathcal{O}$.
$v_1 = 2, v_2 = 3, v_3 = 5$; then $N = 60$ and $\mathcal{G}$ is the icosahedral group $\mathcal{I}$.

EXERCISES

1. Let V be a real vector space with an inner product.

a. Prove that, if T_1 and T_2 are reflections with respect to the same hyperplane H, then $T_1 = T_2$.

b. Prove that, if T is an orthogonal transformation that leaves all elements of a hyperplane H fixed, then either $T = 1$ or T is the reflection with respect to H.

c. Let T be the reflection with respect to H and let $x_0 \in H^{\perp}$. Prove that T is given by the formula

$$T(x) = x - 2\frac{(x, x_0)}{(x_0, x_0)}x_0, \qquad x \in V.$$

2. Let $\mathcal{G}$ be a finite group of orthogonal transformations and let $\mathcal{H}$ be the rotations contained in $\mathcal{G}$. Prove that $\mathcal{H}$ is a subgroup of $\mathcal{G}$ and that if $\mathcal{H} \neq \mathcal{G}$ then, for any element $X \in \mathcal{G}$ and $X \notin \mathcal{H}$, we have

$$\mathcal{G} = \mathcal{H} \cup X\mathcal{H}, \qquad \mathcal{H} \cap X\mathcal{H} = \varnothing.$$

3. Let $\mathcal{G}$ be an arbitrary finite group of invertible linear transformations on R_3. Prove that there exists an inner product $((x, y))$ on R_3 such that the $T_i \in \mathcal{G}$ are orthogonal transformations with respect to the inner product $((x, y))$. [*Hint:* Let (x, y) be the usual inner product on R_3. Define

$$((x, y)) = \sum_{T_i \in \mathcal{G}} (T_i(x), T_i(y))$$

and prove that $((x, y))$ has the required properties. Note that the same argument can be applied to finite groups of invertible linear transformations in R_n for n arbitrary.]

4. Let $\mathcal{G}$ be the set of linear transformations on C_2 where C is the complex field whose matrices with respect to a basis are

$$\pm\begin{pmatrix} 1 & 0 \\ 0 & 1 \end{pmatrix}, \quad \pm\begin{pmatrix} 0 & 1 \\ -1 & 0 \end{pmatrix}, \quad \pm\begin{pmatrix} 0 & i \\ i & 0 \end{pmatrix}, \quad \pm\begin{pmatrix} i & 0 \\ 0 & -i \end{pmatrix}.$$

Prove that $\mathcal{G}$ forms a finite group and that there exists no basis of C_2 such that the matrices of the elements of $\mathcal{G}$ with respect to the new basis all have real coefficients. (*Hint:* If such a basis did exist, then $\mathcal{G}$ would be isomorphic either with a cyclic group or with a dihedral group, by Exercise 3 and the results of Section 14.)

34. APPLICATION TO DIFFERENTIAL EQUATIONS

We consider in this section a system of first-order linear differential equations with constant coefficients in the unknown functions $y_1(t), \ldots, y_n(t)$ where t is a real variable and the $y_i(t)$ are real-valued functions. All this means is that we are given differential equations

$$\begin{aligned} \frac{dy_1}{dt} &= \alpha_{11}y_1 + \cdots + \alpha_{1n}y_n \\ &\cdots\cdots\cdots\cdots\cdots \\ \frac{dy_n}{dt} &= \alpha_{n1}y_1 + \cdots + \alpha_{nn}y_n \end{aligned} \tag{34.1}$$

where $\mathbf{A} = (\alpha_{ij})$ is a fixed n-by-n matrix with real coefficients. We shall discuss the problem of finding a set of solutions $y_1(t), \ldots, y_n(t)$ which take on a prescribed set of initial conditions $y_1(0), \ldots, y_n(0)$. For example, we are going to show, if t is the time and if $y_1(t), \ldots, y_n(t)$ describe the motion of some mechanical system, how to solve the equations of motion with the requirement that the functions take on specified values at time $t = 0$.

The simplest case of such a system is the case of one equation

$$\frac{dy}{dt} = \alpha y,$$

and in this case we know from elementary calculus that the function

$$y(t) = y(0)e^{at}$$

solves the differential equation and takes on the initial value $y(0)$ when $t = 0$.

We shall show how matrix theory can be used to solve a general system (34.1) in an equally simple way. We should also point out that our discussion includes as a special case the problem of solving an nth-order linear differential equation with constant coefficients.

$$\alpha_0 \frac{d^n y}{dt^n} + \alpha_1 \frac{d^{n-1}y}{dt^{n-1}} + \cdots + \alpha_{n-1}\frac{dy}{dt} + \alpha_n y = 0, \quad \alpha_0 \neq 0, \tag{34.2}$$

where the α_i are real constants. This equation can be replaced by a system of the form (34.1) if we view $y(t)$ as the unknown function $y_1(t)$ and rename the derivatives as follows:

$$\frac{d^i y}{dt^i} = y_{i+1}(t), \qquad 1 \leq i \leq n-1.$$

Then the functions $y_i(t)$ satisfy the system

$$\begin{aligned}
&\frac{dy_1}{dt} = y_2 \\
&\frac{dy_2}{dt} = y_3 \\
(34.3)\qquad &\vdots \\
&\frac{dy_n}{dt} = -\frac{\alpha_n}{\alpha_0}y_1 - \frac{\alpha_{n-1}}{\alpha_0}y_2 - \cdots - \frac{\alpha_1}{\alpha_0}y_n
\end{aligned}$$

and, conversely, any set of solutions of the system (34.3) will also yield a solution $y_1(t) = y(t)$ of the original equation (34.2). The initial conditions in this case amount to specifying the values of $y(t)$ and its first $n - 1$ derivatives at $t = 0$.

Now let us proceed with the discussion. We may represent the functions $y_1(t), \ldots, y_n(t)$ in vector form (or as an n-by-1 matrix):

$$\mathbf{y}(t) = \begin{pmatrix} y_1(t) \\ \vdots \\ y_n(t) \end{pmatrix}.$$

We have here a function which, viewed abstractly, assigns to a real number t the vector $\mathbf{y}(t) \in R_n$. We may define limits of such functions as follows:

$$\lim_{t\to t_0} \mathbf{y}(t) = \begin{pmatrix} \lim_{t\to t_0} y_1(t) \\ \vdots \\ \lim_{t\to t_0} y_n(t) \end{pmatrix}$$

provided all the limits $\lim_{t\to t_0} y_i(t)$ exist.

It will be useful to generalize all this slightly. We may consider functions

$$t \to \mathbf{A}(t) = \begin{pmatrix} \alpha_{11}(t) & \cdots & a_{1r}(t) \\ \cdots & \cdots & \cdots \\ a_{s1}(t) & \cdots & a_{sr}(t) \end{pmatrix}$$

which assign to a real number t an s-by-r matrix $\mathbf{A}(t)$ whose coefficients are complex numbers $a_{ij}(t)$. For a 1-by-1 matrix function we may define

$$\lim_{t\to t_0} a(t) = u$$

for some complex number u, provided that for each $\epsilon > 0$ there exists a $\delta > 0$ such that $0 < |t - t_0| < \delta$ implies $|a(t) - u| < \epsilon$ where $|a(t) - u|$ denotes the distance between the points $a(t)$ and u in the complex plane. By using the fact (proved in Section 15) that

$$|u + v| \leq |u| + |v|$$

for complex numbers u and v, it is easy to show that the usual limit theorems of elementary calculus carry over to complex-valued functions. We may then define, as in the case of a vector function $\mathbf{y}(t)$,

$$\lim_{t\to t_0} \mathbf{A}(t),$$

$$\frac{d\mathbf{A}}{dt} = \lim_{h\to 0} \frac{\mathbf{A}(t+h) - \mathbf{A}(t)}{h} = \begin{pmatrix} \dfrac{da_{11}}{dt} & \cdots & \dfrac{da_{1r}}{dt} \\ \cdots & \cdots & \cdots \\ \dfrac{da_{s1}}{dt} & \cdots & \dfrac{da_{sr}}{dt} \end{pmatrix}$$

and

$$\lim_{n\to\infty} \mathbf{A}_n(t) = \begin{pmatrix} \lim\limits_{n\to\infty} a_{11}^{(n)}(t) & \cdots & \lim\limits_{n\to\infty} a_{1r}^{(n)}(t) \\ \cdots & \cdots & \cdots \\ \lim\limits_{n\to\infty} a_{s1}^{(n)}(t) & \cdots & \lim\limits_{n\to\infty} a_{sr}^{(n)}(t) \end{pmatrix}$$

where $\{\mathbf{A}_n(t)\}$ is a sequence of matrix-valued functions $\mathbf{A}_n(t) = (a_{ij}^{(n)}(t))$.

We can now express our original system of differential equations (34.1) in the more compact form

$$\text{(34.4)} \qquad \frac{d\mathbf{y}}{dt} = \mathbf{A}\mathbf{y}$$

where

$$\mathbf{y}(t) = \begin{pmatrix} y_1(t) \\ \vdots \\ y_n(t) \end{pmatrix}$$

is a vector-valued function and $\mathbf{Ay}$ denotes the product of the constant n-by-n matrix $\mathbf{A}$ by the n-by-1 matrix $\mathbf{y}(t)$.

The remarkable thing is that the equation (34.4) can be solved exactly as in the one-dimensional case. Let us begin with the following definition.

(34.5) DEFINITION. Let $\mathbf{B}$ be an n-by-n matrix with complex coefficients (β_{ij}). Define the sequences of matrices

$$\mathbf{E}^{(n)} = \mathbf{I} + \mathbf{B} + \frac{1}{2}\mathbf{B}^2 + \cdots + \frac{1}{n!}\mathbf{B}^n, \qquad n = 0, 1, 2, \ldots.$$

Define the *exponential matrix*

$$e^B = \lim_{n\to\infty} \mathbf{E}^{(n)} = \lim_{n\to\infty} \left(\mathbf{I} + \mathbf{B} + \frac{1}{2}\mathbf{B}^2 + \cdots + \frac{1}{n!}\mathbf{B}^n\right).$$

Of course, it must first be verified that e^B exists, that, in other words, the limit of the sequence $\{\mathbf{E}^{(n)}\}$ exists. Let ρ be some upper bound for the $\{|\beta_{ij}|\}$; then $|\beta_{ij}| \leq \rho$ for all (i,j). Let $\mathbf{B}^n = (\beta_{ij}^{(n)})$. Then the (i,j) entry of $\mathbf{E}^{(n)}$ is

$$\beta_{ij}^{(0)} + \beta_{ij}^{(1)} + \frac{1}{2}\beta_{ij}^{(2)} + \cdots + \frac{1}{n!}\beta_{ij}^{(n)}$$

and we have to show that this sequence tends to a limit. This means that the infinite series

$$\sum_{k=0}^{\infty} \frac{1}{k!}\beta_{ij}^{(k)}$$

converges, and this can be checked by making a simple comparison test as follows. By induction on k we show first that for $k = 1, 2, \ldots,$

$$|\beta_{ij}^{(k)}| \leq n^{k-1}\rho^k, \qquad 1 \leq i, \ j \leq n.$$

Then each term of the series

$$\sum_{k=0}^{\infty} \frac{1}{k!}\beta_{ij}^{(k)}$$

is dominated in absolute value by the corresponding term in the series of positive terms

$$\sum_{k=0}^{\infty} \frac{1}{k!} n^{k-1}\rho^k,$$

and this series converges for all ρ by the ratio test. This completes the proof that e^B exists for all matrices $\mathbf{B}$.

We now list some properties of the function $\mathbf{B} \to e^B$, whose proofs are left as exercises,

(34.6) $e^{A+B} = e^A e^B$ *provided that* $\mathbf{AB} = \mathbf{BA}$.

(34.7) $\dfrac{d}{dt} e^{tB} = \mathbf{B}\, e^{tB}$ *for all n-by-n matrices* $\mathbf{B}$.

(34.8) $\mathbf{S}^{-1}e^A\mathbf{S} = e^{S^{-1}AS}$ *for an arbitrary invertible matrix* $\mathbf{S}$.

The solution of (34.4) can now be given as follows.

(34.9) THEOREM. *The vector differential equation*

$$\frac{d\mathbf{y}}{dt} = \mathbf{A}\mathbf{y},$$

where $\mathbf{A}$ *is an arbitrary constant matrix with real coefficients, has the solution*

$$\mathbf{y}(t) = e^{tA} \cdot \mathbf{y}_0,$$

which takes on the initial value $\mathbf{y}_0$ *when* $t = 0$.

Proof. It is clear that $\mathbf{y}(0) = \mathbf{y}_0$, and it remains to check that $\mathbf{y}(t)$ is actually a solution of the differential equation. We note first that, if $\mathbf{A}(t)$ is a matrix function and $\mathbf{B}$ is a constant vector, then

$$\frac{d}{dt}[\mathbf{A}(t)\mathbf{B}] = \frac{d\mathbf{A}}{dt} \cdot \mathbf{B}.$$

Applying this to $\mathbf{y}(t) = e^{tA} \cdot \mathbf{y}_0$ and using (34.7), we have

$$\frac{d\mathbf{y}}{dt} = \left(\frac{d}{dt} e^{tA}\right) \cdot \mathbf{y}_0 = (\mathbf{A}\, e^{tA})\mathbf{y}_0 = \mathbf{A}\mathbf{y}(t).$$

This completes the proof of the theorem.

Theorem (34.9) solves the problem stated at the beginning of the section, but the solution is still not of much practical importance because of the difficulty of computing the matrix e^{tA}. We show now how Theorem (24.9) can be used to give a general method for calculating e^{tA}, provided that the complex roots of the minimal polynomial of the matrix $\mathbf{A}$ are known. Let

$$m(x) = (x - \alpha_1)^{e_1} \cdots (x - \alpha_s)^{e_s}, \qquad e_i > 0,$$

be the minimal polynomial of $\mathbf{A}$. Then there exists an invertible matrix $\mathbf{S}$, possibly with complex coefficients, such that, by Theorem (24.9),

$$\mathbf{S}^{-1}\mathbf{A}\mathbf{S} = \mathbf{D} + \mathbf{N}$$

where

$$\mathbf{D} = \left(\begin{array}{ccc|ccc|c} \alpha_1 & & 0 & & & & \\ & \ddots & & & 0 & & \\ 0 & & \alpha_1 & & & & \\ \hline & & & \alpha_2 & & 0 & \\ 0 & & & & \ddots & & \\ & & & 0 & & \alpha_2 & \\ \hline & & & & & & \end{array}\right)$$

is a diagonal matrix, $\mathbf{N}$ is nilpotent, and $\mathbf{DN} = \mathbf{ND}$. Moreover, $\mathbf{S}$, $\mathbf{D}$, and $\mathbf{N}$ can all be calculated by the methods of Sections 24. Then

$$\mathbf{A} = \mathbf{S}(\mathbf{D} + \mathbf{N})\mathbf{S}^{-1}$$

and, by (34.6) and (34.8), we have

$$\begin{aligned} e^{tA} &= e^{tS(D+N)S^{-1}} = \mathbf{S}(e^{tD+tN})\mathbf{S}^{-1} \\ &= \mathbf{S}\, e^{tD}\, e^{tN}\mathbf{S}^{-1}. \end{aligned}$$

The point of all this is that e^{tD} and e^{tN} are both easy to compute; e^{tD} is simply the diagonal matrix

$$\left(\begin{array}{c|c|c} \begin{matrix} e^{\alpha_1 t} & & 0 \\ & \ddots & \\ 0 & & e^{\alpha_1 t} \end{matrix} & & \\ \hline & \begin{matrix} e^{\alpha_2 t} & & 0 \\ & \ddots & \\ 0 & & e^{\alpha_2 t} \end{matrix} & \\ \hline & & \end{array}\right)$$

while $e^{tN} = \mathbf{I} + t\mathbf{N} + \dfrac{t^2\mathbf{N}^2}{2} + \cdots + \dfrac{t^{r-1}\mathbf{N}^{r-1}}{(r-1)!}$ if $\mathbf{N}^r = 0$. The solution vector $\mathbf{y}(t) = \mathbf{S}\, e^{tD}e^{tN}\mathbf{S}^{-1}\mathbf{y}_0$.

Example A. Consider the vector differential equation

$$\frac{dy}{dt} = \mathbf{A}\mathbf{y}, \qquad \text{with} \qquad \mathbf{A} = \begin{pmatrix} 2 & 2 \\ -2 & -3 \end{pmatrix}.$$

We shall calculate a solution $\mathbf{y}$ of the differential equation satisfying the initial condition $\mathbf{y}(0) = \mathbf{y}_0$. The minimal polynomial of A is $x^2 + x - 2$, and we see that A is diagonable (why?). The characteristic roots of A are -2 and 1, with corresponding characteristic vectors,

$$\begin{pmatrix} 1 \\ -2 \end{pmatrix} \quad \text{and} \quad \begin{pmatrix} 2 \\ -1 \end{pmatrix}.$$

Then $\mathbf{AS} = \mathbf{SD}$, where

$$\mathbf{S} = \begin{pmatrix} 1 & 2 \\ -2 & -1 \end{pmatrix}, \qquad \mathbf{D} = \begin{pmatrix} -2 & 0 \\ 0 & 1 \end{pmatrix}.$$

We have $\mathbf{A} = \mathbf{SDS}^{-1}$, and hence

$$e^{At} = \mathbf{S}e^{Dt}\mathbf{S}^{-1},$$

where

$$e^{Dt} = \begin{pmatrix} e^{-2t} & 0 \\ 0 & e^{t} \end{pmatrix}.$$

A solution $\mathbf{y}$ of the differential equation satisfying the initial condition $\mathbf{y}(0) = \mathbf{y}_0$ is given by

$$\mathbf{y}(t) = \mathbf{S}e^{Dt}\mathbf{S}^{-1}\mathbf{y}_0.$$

Example B. Compute e^{At}, in case A is the matrix

$$\begin{pmatrix} -1 & 1 \\ 0 & -1 \end{pmatrix}.$$

In this case, A is not diagonable (why?). The Jordan decomposition of A [from Theorem (24.9)] is given by

$$\mathbf{A} = \mathbf{D} + \mathbf{N},$$

with

$$\mathbf{D} = \begin{pmatrix} -1 & 0 \\ 0 & -1 \end{pmatrix}, \quad \mathbf{N} = \begin{pmatrix} 0 & 1 \\ 0 & 0 \end{pmatrix}.$$

Since ND = DN, we have

$$e^{At} = e^{Dt}e^{Nt},$$

where $e^{Dt} = \begin{pmatrix} e^{-t} & 0 \\ 0 & e^{-t} \end{pmatrix}$ and $e^{Nt} = \begin{pmatrix} 1 & t \\ 0 & 1 \end{pmatrix}$.

EXERCISES

1. Let

$$\mathbf{A} = \begin{pmatrix} 0 & 1 \\ 0 & 0 \end{pmatrix}, \quad \mathbf{B} = \begin{pmatrix} -1 & 0 \\ 0 & 0 \end{pmatrix}.$$

Show that $\mathbf{AB} \neq \mathbf{BA}$. Calculate e^A, e^B, e^{A+B}, and show that $e^{A+B} \neq e^Ae^B$. Thus (34.6) does not hold without some hypothesis like $\mathbf{AB} = \mathbf{BA}$.

2. Show that $D(e^A) = e^{\Sigma\alpha_i}$ where the $\{\alpha_i\}$ are the characteristic roots of A. [*Hint:* Use the triangular form theorem (24.1) together with (34.8).]
3. Show that e^A is always invertible and that $(e^A)^{-1} = e^{-A}$ for all $\mathbf{A} \in M_n(C)$.
4. Let $\mathbf{y}(t)$ be any solution of the differential equation $dy/dt = \mathbf{A}\mathbf{y}$ such that $\mathbf{y}(0) = \mathbf{y}_0$. Prove that

$$\frac{d}{dt}(e^{-At}\mathbf{y}) = 0$$

and hence that $\mathbf{y} = e^{At}\cdot\mathbf{y}_0$. Thus the solution of the differential equation $dy/dt = \mathbf{A}\mathbf{y}$ satisfying the given initial condition is uniquely determined.

5. Prove that ${}^t(e^A) = e^{{}^tA}$, where ${}^t\mathbf{A}$ is the transpose of A.
6. Define a matrix A to be skew symmetric if ${}^t\mathbf{A} = -\mathbf{A}$, for example,

$$\mathbf{A} = \begin{pmatrix} 0 & 1 \\ -1 & 0 \end{pmatrix},$$

Prove that if A is any real skew symmetric matrix then e^A is an orthogonal matrix of determinant $+1$.

7. Apply the methods of this section to prove that the differential equation (considered in Section 1)

$$\frac{d^2y}{dt^2} + m^2y = 0, \qquad m \text{ a positive constant},$$

has a unique solution of the form $A \sin mt + B \cos mt$ such that $y(0) = 0, y'(0) = 1$.

8. Solve the system

$$\begin{cases} \dfrac{dy_1}{dt} = -y_1 + y_2 \\ \dfrac{dy_2}{dt} = -y_1 - 3y_2 \end{cases} \qquad y_1(0) = 0, \quad y_2(0) = 1.$$

9. Compute e^{At}, in case A is any one of the following matrices:

$$\begin{pmatrix} 1 & 1 \\ -1 & -1 \end{pmatrix}, \quad \begin{pmatrix} 1 & -3 \\ 1 & -2 \end{pmatrix}, \quad \begin{pmatrix} 0 & 1 & 0 \\ 0 & 0 & 1 \\ 1 & 0 & 0 \end{pmatrix}.$$

10. Let A be the coefficient matrix of the vector differential equation equivalent to

$$\alpha_0 \frac{d^ny}{dt^n} + \alpha_1 \frac{d^{n-1}y}{dt^{n-1}} + \cdots + \alpha_n y = 0.$$

Prove that the characteristic polynomial of A is

$$\alpha_0 x^n + \alpha_1 x^{n-1} + \cdots + \alpha_n.$$

11. (*Optional.*) Verify the following derivation of a particular solution of the "nonhomogeneous" vector differential equation

$$\frac{d\mathbf{y}}{dt} = \mathbf{A}\mathbf{y} + \mathbf{f}(t)$$

where $\mathbf{f}(t)$ is a given continuous vector function of t. Attempt to find a solution of the form $\mathbf{y}(t) = e^{At}\mathbf{c}(t)$, where $\mathbf{c}(t)$ is a vector function to be determined. Differentiate to obtain

$$\frac{d\mathbf{y}}{dt} = \mathbf{A}\, e^{At}\mathbf{c}(t) + e^{At}\frac{d\mathbf{c}(t)}{dt} = \mathbf{A}\mathbf{y} + \mathbf{f}(t).$$

Since $\mathbf{A}e^{At}\mathbf{c}(t) = \mathbf{A}\mathbf{y}$, we obtain [since $(e^{At})^{-1} = e^{-At}$]

$$\frac{d\mathbf{c}}{dt} = e^{-At}\mathbf{f}(t).$$

Thus $\mathbf{c}(t)$ is an indefinite integral of $e^{-At}\mathbf{f}(t)$ and can be expressed as

$$\mathbf{c}(t) = \int_{t_0}^{t} e^{-As}\mathbf{f}(s)\, ds.$$

Then $\mathbf{y}(t) = e^{At}\int_{t_0}^{t} e^{-As}\mathbf{f}(s)\, ds = \int_{t_0}^{t} e^{A(t-s)}\mathbf{f}(s)\, ds$. Finally, verify that $\mathbf{y}(t)$ is a solution of the differential equation.

35. SUMS OF SQUARES AND HURWITZ'S THEOREM

In Section 21, we constructed the field of complex numbers by defining a multiplication in R_2. More precisely, given $z_1 = \langle\alpha, \beta\rangle$ and $z_2 = \langle\gamma, \delta\rangle$ in R_2, we defined

$$z_1 z_2 = \langle\alpha\gamma - \beta\delta, \alpha\delta + \beta\gamma\rangle,$$

and proved that this operation, together with vector addition, satisfied the axioms for a field. It was also shown that the product operation has the following interesting connection with the length of vectors in R_2:

(35.1) $$|z_1 z_2| = |z_1|\,|z_2|,$$

for all z_1 and z_2 in R_2, where $|z_1| = (\alpha^2 + \beta^2)^{1/2}$, etc.

Complex numbers are so useful in many parts of mathematics (including linear algebra!) that it is natural to ask whether there are other kinds of "complex numbers," obtained by multiplying vectors in $R_3, R_4, \ldots$, which might be of equal value to their cousins in R_2. We shall consider in this section one of the milestones in the long search for a satisfactory answer to this basic question.

Let us begin by examining more closely some properties of the multiplication in R_2 defined above. First of all, the multiplication is *bilinear*, in the sense that the mapping

$$(z_1, z_2) \to z_1 z_2$$

is a bilinear mapping of $R_2 \times R_2 \to R_2$. The proof of this fact is immediate from the definition and will be left to the reader. Second, there is the mysterious law (35.1). Of course there are other properties, but these two are already surprisingly powerful. For example, using *only* the bilinearity of the product and (35.1), we shall prove that if $z \neq 0$, then the equation $zx = z'$ has a solution for every z' in R_2. This is equivalent to the assertion that the left multiplication L_z, defined by $L_z(u) = zu$, for $u \in R_2$, maps R_2 onto itself. The bilinearity of the multiplication implies that L_z is a linear transformation on R_2. The identity (35.1) implies that the null space of L_z is $\{0\}$, if $z \neq 0$. It then follows that the range of L_z is all of R_2, as required.

The problem we shall consider in this section is the following one. We use the notation $|x|$ for the length $(x_1^2 + \cdots + x_n^2)^{1/2}$ of an arbitrary vector $x = \langle x_1, \ldots, x_n\rangle$ in R_n.

Problem I. For which values of n does there exist a bilinear mapping $(x, y) \to x * y$ of $R_n \times R_n \to R_n$, such that

$$|x * y| = |x|\,|y|,$$

for all x and y in R_n?

The argument given above, in the case of R_2, shows that if Problem I has a solution for some integer n, then the equations $z * x = z'$ and $x * z = z'$, for $z \neq 0$, have solutions for all z' in R_n. Thus R_n, under vector addition and the product operation $*$, would satisfy all properties of a field, except possibly the commutative and associative laws and the existence of an identity element for multiplication.

Our first task is to translate Problem I into a more manageable form. Suppose Problem I has a solution for some n. Let $e_1, \ldots, e_n$ be an orthonormal basis for R_n. Then there exist scalars $\{\gamma_{pq}^{(i)}\}$, uniquely determined by the multiplication $*$, such that

$$e_p * e_q = \sum_{i=1}^{n} \gamma_{pq}^{(i)} e_i,$$

for $1 \leq p, q \leq n$. Products of arbitrary vectors can now be calculated in terms of the scalars $\{\gamma_{pq}^{(i)}\}$ as follows. Let $x = \sum x_p e_p$ and $y = \sum y_q e_q$ be vectors in R_n, expressed as linear combinations of the basis vectors. Then the bilinearity of the mapping implies that

$$\begin{aligned} x * y = (\sum x_p e_p) * (\sum y_q e_q) &= \sum_{p=1}^{n} \sum_{q=1}^{n} x_p y_q (e_p * e_q) \\ &= \sum_{p=1}^{n} \sum_{q=1}^{n} \sum_{i=1}^{n} x_p y_q \gamma_{pq}^{(i)} e_i \\ &= \sum z_i e_i, \end{aligned}$$

where

$$z_i = \sum_{p,q=1}^{n} \gamma_{pq}^{(i)} x_p y_q .$$

Since $|x|$, $|y|$ and $|x * y|$ are the square roots of the sums of squares of their components, Problem I can now be reformulated as follows.

Problem II. For which values of n does there exist an identity

$$(x_1^2 + \cdots + x_n^2)(y_1^2 + \cdots + y_n^2) = (z_1^2 + \cdots + z_n^2),$$

holding for all choices of real numbers $\{x_i\}$ and $\{y_j\}$, where

$$z_i = \sum_{p,q=1}^{n} \gamma_{pq}^{(i)} x_p y_q ,$$

and the matrices $\mathbf{C}^{(i)} = (\gamma_{pq}^{(i)})$, $1 \leq i \leq n$, are constant matrices (independent of the x's and y's), with real entries?

Example A. In the identity for sums of two squares we have

$$(x_1^2 + x_2^2)(y_1^2 + y_2^2) = (x_1 y_1 - x_2 y_2)^2 + (x_1 y_2 + x_2 y_1)^2,$$

so that $$z_1 = 1x_1y_1 + 0x_1y_2 + 0x_2y_1 - 1x_2y_2$$

and $$z_2 = 0x_1y_1 + 1x_1y_2 + 1x_2y_1 + 0x_2y_2 .$$

The matrices C_1 and C_2 in this case are given by

$$C_1 = \begin{pmatrix} 1 & 0 \\ 0 & -1 \end{pmatrix}, \quad C_2 = \begin{pmatrix} 0 & 1 \\ 1 & 0 \end{pmatrix}.$$

The first progress on Problem II was achieved by Euler. He proved that the problem has a solution when $n = 4$, namely,

$$(35.2) \quad (x_1^2 + x_2^2 + x_3^2 + x_4^2)(y_1^2 + y_2^2 + y_3^2 + y_4^2) = z_1^2 + z_2^2 + z_3^2 + z_4^2,$$

where

$$\begin{aligned} z_1 &= x_1y_1 - x_2y_2 - x_3y_3 - x_4y_4, \\ z_2 &= x_1y_2 + x_2y_1 + x_3y_4 - x_4y_3, \\ z_3 &= x_1y_3 - x_2y_4 + x_3y_1 + x_4y_2, \\ z_4 &= x_1y_4 + x_2y_3 - x_3y_2 + x_4y_1. \end{aligned}$$

The matrices in this case are

$$C_1 = \begin{pmatrix} 1 & 0 & 0 & 0 \\ 0 & -1 & 0 & 0 \\ 0 & 0 & -1 & 0 \\ 0 & 0 & 0 & -1 \end{pmatrix}, \quad C_2 = \begin{pmatrix} 0 & 1 & 0 & 0 \\ 1 & 0 & 0 & 0 \\ 0 & 0 & 0 & 1 \\ 0 & 0 & -1 & 0 \end{pmatrix},$$

$$C_3 = \begin{pmatrix} 0 & 0 & 1 & 0 \\ 0 & 0 & 0 & -1 \\ 1 & 0 & 0 & 0 \\ 0 & 1 & 0 & 0 \end{pmatrix}, \quad C_4 = \begin{pmatrix} 0 & 0 & 0 & 1 \\ 0 & 0 & 1 & 0 \\ 0 & -1 & 0 & 0 \\ 1 & 0 & 0 & 0 \end{pmatrix}.$$

The skeptical reader may wish to check (35.2) before proceeding.

The identity (35.2) was used by Euler to prove the theorem of Lagrange which states that every positive integer is a sum of four squares of integers. The identity (35.2) shows that it is sufficient to prove that every prime number is a sum of four squares, since every positive integer is a product of prime numbers.

The identity (35.2) was rediscovered in the nineteenth-century by Hamilton as a by-product of his discovery of quaternions. A solution to the problem was found for $n = 8$ by Arthur Cayley in 1845. There the matter stood until 1898, when Hurwitz proved the following theorem.

(35.3) THEOREM (HURWITZ, 1898). *The equivalent problems I and II on composition of sums of squares can have a solution only when* $n = 1, 2, 4,$ *or* 8.

The proof we shall give is closely modeled after Hurwitz's original proof, with some modifications by Dickson. A fuller historical discussion, and a different proof, is given in the article by Curtis in the first book listed in the Bibliography.

Before giving the proof of Hurwitz's theorem, we need some preliminary facts about symmetric and skew symmetric matrices.

(35.4) DEFINITION. An n-by-n matrix $\mathbf{A}$ with real entries is called *symmetric* if ${}^t\mathbf{A} = \mathbf{A}$, and *skew symmetric* if ${}^t\mathbf{A} = -\mathbf{A}$, where ${}^t\mathbf{A}$ denotes the transpose of $\mathbf{A}$.

(35.5) LEMMA.

a. *Every n-by-n matrix* $\mathbf{A}$ *can be expressed*

$$\mathbf{A} = \tfrac{1}{2}(\mathbf{A} + {}^t\mathbf{A}) + \tfrac{1}{2}(\mathbf{A} - {}^t\mathbf{A})$$

as a sum of a symmetric matrix $\tfrac{1}{2}(\mathbf{A} + {}^t\mathbf{A})$ *and a skew symmetric matrix* $\tfrac{1}{2}(\mathbf{A} - {}^t\mathbf{A})$. *Moreover, if* $\mathbf{A} = \mathbf{S}_1 + \mathbf{S}_2$, *with* $\mathbf{S}_1$ *symmetric and* $\mathbf{S}_2$ *skew symmetric, then* $\mathbf{S}_1 = \tfrac{1}{2}(\mathbf{A} + {}^t\mathbf{A})$, *and* $\mathbf{S}_2 = \tfrac{1}{2}(\mathbf{A} - {}^t\mathbf{A})$.

b. *Suppose* $\mathbf{S}$ *is an invertible n-by-n skew symmetric matrix. Then n is even,* $n = 2m$, *for some m.*

Proof. We first check that $\tfrac{1}{2}(\mathbf{A} + {}^t\mathbf{A})$ and $\tfrac{1}{2}(\mathbf{A} - {}^t\mathbf{A})$ are symmetric and skew symmetric, respectively. We have

$${}^t[\tfrac{1}{2}(\mathbf{A} + {}^t\mathbf{A})] = \tfrac{1}{2}({}^t\mathbf{A} + {}^{tt}\mathbf{A}) = \tfrac{1}{2}(\mathbf{A} + {}^t\mathbf{A}),$$

and

$${}^t[\tfrac{1}{2}(\mathbf{A} - {}^t\mathbf{A})] = \tfrac{1}{2}({}^t\mathbf{A} - {}^{tt}\mathbf{A}) = -\tfrac{1}{2}(\mathbf{A} - {}^t\mathbf{A}),$$

as required. The fact that $\mathbf{A}$ is the sum of the two matrices in question is clear. Now suppose

$$\mathbf{A} = \mathbf{S}_1 + \mathbf{S}_2 = \mathbf{T}_1 + \mathbf{T}_2,$$

where $\mathbf{S}_1$ and $\mathbf{T}_1$ are symmetric, and $\mathbf{S}_2$ and $\mathbf{T}_2$ are skew symmetric. Then

$$\mathbf{S}_1 - \mathbf{T}_1 = \mathbf{T}_2 - \mathbf{S}_2$$

is both symmetric and skew symmetric, and the only such matrix is the zero matrix. This completes the proof of part (a).

To prove (b), we have, from Chapter 5,

$$\mathbf{D}(\mathbf{S}) = \mathbf{D}({}^t\mathbf{S}) = \mathbf{D}(-\mathbf{S}) = (-1)^n\mathbf{D}(\mathbf{S}).$$

Since $\mathbf{D}(\mathbf{S}) \neq 0$ by assumption, we have

$$(-1)^n = 1,$$

and n is even.

Proof of Theorem (35.3). We derive consequences of the assumption that an identity as in Problem II holds for all x's and y's. We may assume $n > 1$.

We first put, for a given set of x's,

$$\alpha_{ij} = \sum_{p=1}^{n} \gamma_{pj}^{(i)} x_p, \qquad 1 \le i,\ j \le n.$$

Then the equation $(\sum x_i^2)(\sum y_i^2) = \sum z_i^2$, where

$$z_i = \sum_{p,q=1}^{n} \gamma_{pq}^{(i)} x_p y_q ,$$

becomes

$$(35.6)\quad (x_1^2 + \cdots + x_n^2)(y_1^2 + \cdots + y_n^2) = \left(\sum_{j=1}^{n} \alpha_{1j} y_j\right)^2 + \cdots + \left(\sum_{j=1}^{n} \alpha_{nj} y_j\right)^2.$$

For a fixed set of x's, this identity holds for all y's. Therefore we can set each y_j in turn equal to one, and the rest equal to zero, to obtain the equations

$$x_1^2 + \cdots + x_n^2 = \alpha_{11}^2 + \alpha_{21}^2 + \cdots + \alpha_{n1}^2$$
$$\cdots\cdots\cdots\cdots\cdots\cdots\cdots\cdots$$
$$x_1^2 + \cdots + x_n^2 = \alpha_{1n}^2 + \alpha_{2n}^2 + \cdots + \alpha_{nn}^2 .$$

Cancelling the terms involving y_i^2 from both sides of the equation (35.6), we have

$$0 = 2 \sum_{i=1}^{n} \sum_{j=1}^{n} (\alpha_{1i}\alpha_{1j} + \alpha_{2i}\alpha_{2j} + \cdots + \alpha_{ni}\alpha_{nj}) y_i y_j .$$

Now set $y_i = y_j = 1$ and the other y's equal to zero to obtain

$$0 = 2(\alpha_{1i}\alpha_{1j} + \cdots + \alpha_{ni}\alpha_{nj}),$$

for $1 \le i, j \le n$. The equations we have derived can be put in the economical form:

$$(35.7)\qquad {}^t\mathbf{A}(\mathbf{A}) = \left(\sum_{i=1}^{n} x_i^2\right)\mathbf{I},$$

where $\mathbf{A}$ is the n-by-n matrix (α_{ij}).

We now use the fact that (35.7) is an identity in the x's. From the definition of the matrix coefficients $\{\alpha_{ij}\}$, we can write

$$\mathbf{A} = x_1\mathbf{A}_1 + x_2\mathbf{A}_2 + \cdots + x_n\mathbf{A}_n ,$$

where $\mathbf{A}_1, \mathbf{A}_2, \ldots, \mathbf{A}_n$ are constant matrices, independent of the x's and the y's.

As an example of how this process works, we have when $n = 2$,

$$\begin{pmatrix} 3x_1 + 2x_2 & x_1 \\ 0 & -x_1 + x_2 \end{pmatrix} = x_1 \begin{pmatrix} 3 & 1 \\ 0 & -1 \end{pmatrix} + x_2 \begin{pmatrix} 2 & 0 \\ 0 & 1 \end{pmatrix}.$$

Substituting the expression for $\mathbf{A}$ in the formula (35.7), we have

$$(35.8)\quad (x_1{}^t\mathbf{A}_1 + \cdots + x_n{}^t\mathbf{A}_n)(x_1\mathbf{A}_1 + \cdots x_n\mathbf{A}_n) = \left(\sum_{i=1}^{n} x_i^2\right)\mathbf{I}.$$

Comparing coefficients of x_i^2 on both sides (as we did before with the y's), we obtain

$$^t\mathbf{A}_i\mathbf{A}_i = \mathbf{I}, \qquad 1 \le i \le n,$$

and hence $(\mathbf{A}_i)(^t\mathbf{A}_i) = \mathbf{I}$, $1 \le i \le n$.

Now define $\mathbf{B}_i = {}^t\mathbf{A}_n\mathbf{A}_1$, $i = 1, 2, \ldots, n-1$. Then (35.8) becomes

$$(x_1{}^t\mathbf{A}_1 + \cdots + x_n{}^t\mathbf{A}_n)(\mathbf{A}_n{}^t\mathbf{A}_n)(x_1\mathbf{A}_1 + \cdots + x_n\mathbf{A}_n) = \left(\sum_{i=1}^{n} x_i^2\right)\mathbf{I},$$

and hence

(35.9) $$(x_1{}^t\mathbf{B}_1 + \cdots + x_{n-1}{}^t\mathbf{B}_{n-1} + x_n\mathbf{I})(x_1\mathbf{B}_1 + \cdots + x_{n-1}\mathbf{B}_{n-1} + x_n\mathbf{I}) = \left(\sum_{i=1}^{n} x_i^2\right)\mathbf{I},$$

using the fact that $^t\mathbf{A}_i\mathbf{A}_n = {}^t\mathbf{B}_i$.

We can now compare coefficients of x_i^2 on both sides of (35.9) to obtain

$$^t\mathbf{B}_i\mathbf{B}_i = \mathbf{I}, \qquad 1 \le i \le n-1.$$

Cancelling these terms, and setting $x_i = x_j = 1$ for $i \neq j$, and the other x's equal to zero, we have

$$^t\mathbf{B}_i + \mathbf{B}_i = 0, \qquad 1 \le i \le n-1,$$

and $$^t\mathbf{B}_i\mathbf{B}_j + {}^t\mathbf{B}_j\mathbf{B}_i = 0, \qquad 1 \le i,\ j \le n-1, \quad i \neq j.$$

We have shown at this point that if the composition of sums of squares problem has a solution for a given n, then there exist $n-1$ skew symmetric matrices $\mathbf{B}_1, \ldots, \mathbf{B}_{n-1}$ satisfying the equations

(35.10) $$\mathbf{B}_i^2 = -\mathbf{I}, \qquad \mathbf{B}_i\mathbf{B}_j = -\mathbf{B}_j\mathbf{B}_i,$$

for all i and j, with $i \neq j$.

We shall now prove that such a set of matrices can exist only if $n = 2$, 4, or 8.

First of all, the equation (35.10) shows that the matrices $\mathbf{B}_i$ are invertible skew symmetric matrices and, therefore, by Lemma (35.5) n is even. Note that we have now shown that the problem has no solution in the first unsolved case, $n = 3$.

We now assume n is even, and consider the set of all products

$$\{\mathbf{B}_{i_1}\mathbf{B}_{i_2} \cdots \mathbf{B}_{i_r}, \quad 1 \le i_1, \ldots, i_r \le n-1, \quad r \ge 1\}$$

which can be formed from the $\mathbf{B}_i$'s. Notice that in such a product, if $\mathbf{B}_1$ occurs it can be moved past the other $\mathbf{B}_i$'s to the first position, by (35.10), leaving the product otherwise unchanged except for a $\pm$ sign in front. Continuing with other occurrences of $\mathbf{B}_1$, and then with $\mathbf{B}_2$, etc., we can write an arbitrary product of the $\mathbf{B}$'s in the form

$$\pm \mathbf{B}_1^{p_1} \cdots \mathbf{B}_{n-1}^{p_{n-1}}, \qquad p_j \geq 0.$$

Finally, since $\mathbf{B}_i^2 = -\mathbf{I}$, $1 \leq i \leq n$, such a product can be written in the form

$$\pm \mathbf{B}_1^{e_1} \cdots \mathbf{B}_{n-1}^{e_{n-1}}, \qquad e_i = 0 \text{ or } 1.$$

There are 2^{n-1} possible products. We shall now investigate the linear dependence of these products.

(35.11) LEMMA. *At least half* (2^{n-2}) *of the matrices* $\{\mathbf{B}_1^{e_1} \cdots \mathbf{B}_{n-1}^{e_{n-1}} \mid e_i =$ 0 *or* $1\}$ *are linearly independent.*

Proof. We first have to decide which of the matrices are symmetric and which are skew symmetric. Let

(35.12) $$\mathbf{M} = \mathbf{B}_{i_1}\mathbf{B}_{i_2} \cdots \mathbf{B}_{i_r}, \qquad r \leq n-1, \quad i_1 < i_2 < \cdots < i_r.$$

Then, using (35.10), we have

$$\begin{aligned}
{}^t\mathbf{M} &= {}^t(\mathbf{B}_{i_1} \cdots \mathbf{B}_{i_r}) = {}^t\mathbf{B}_{i_r} \cdots {}^t\mathbf{B}_{i_1} \\
&= (-1)^r \mathbf{B}_{i_r} \cdots \mathbf{B}_{i_1} \\
&= (-1)^{r+(r-1)} \mathbf{B}_{i_1}\mathbf{B}_{i_r} \cdots \mathbf{B}_{i_2} \\
&= (-1)^{r+(r-1)+(r-2)} \mathbf{B}_{i_1}\mathbf{B}_{i_2}\mathbf{B}_{i_r} \cdots \mathbf{B}_{i_3} \\
&= (-1)^{r+(r-1)+\cdots+1} \mathbf{B}_{i_1}\mathbf{B}_{i_2} \cdots \mathbf{B}_{i_r} \\
&= (-1)^{r(r+1)/2} \mathbf{B}_{i_1} \cdots \mathbf{B}_{i_r} = (-1)^{r(r+1)/2}\mathbf{M}.
\end{aligned}$$

Therefore $\mathbf{M}$ is symmetric if and only if $\frac{1}{2}r(r+1)$ is even, or $r(r+1)$ is divisible by 4. This case occurs when either r or $r+1$ is divisible by 4.

Suppose we have a relation of linear dependence

(35.13) $$\alpha_1\mathbf{M}_1 + \cdots + \alpha_k\mathbf{M}_k = 0$$

among matrices of the form (35.12), where we may assume that all $\alpha_i \neq 0$. Such a relation is called an *irreducible relation of linear dependence* if all proper subsets of $\{\mathbf{M}_1, \ldots, \mathbf{M}_k\}$ are linearly independent. Because of the first part of Lemma (35.5), it follows that all the matrices in an irreducible relation of linear dependence are either symmetric, or all are skew symmetric.

Now let (35.13) be an irreducible relation of linear dependence. Multiplying by $\alpha_1^{-1}\mathbf{M}_1^{-1}$, we may assume the relation has the form

(35.14) $$\mathbf{I} = \beta_1\mathbf{M}_1 + \cdots + \beta_{k-1}\mathbf{M}_{k-1},$$

where all the matrices $\mathbf{M}_i$ are symmetric. Suppose $\mathbf{M}_1$ involves the smallest number r of factors, and that $r < n-1$. If r is divisible by 4, and

(35.15) $$\mathbf{M}_1 = \mathbf{B}_i \cdots \mathbf{B}_{i_r},$$

then we can choose $j \neq i_1, \ldots, i_r$, and multiply by $\mathbf{B}_j$ to obtain

$$\mathbf{B}_j = \beta_1 \mathbf{M}_1 \mathbf{B}_j + \cdots + \beta_{k-1} \mathbf{M}_{k-1} \mathbf{B}_j$$

where $\mathbf{B}_j$ is skew symmetric and $\mathbf{M}_1\mathbf{B}_j$ is symmetric, contradicting the assumption that we started from an irreducible relation. On the other hand, if $r + 1$ is divisible by 4, then we can multiply both sides by $\mathbf{B}_{i_1}$, so that $\mathbf{B}_{i_1}\mathbf{M}_1$ is again symmetric while $\mathbf{B}_{i_1}$ is skew symmetric, and again we contradict the assumption that we had an irreducible relation of linear dependence. Therefore the only possible irreducible relation of the form (35.14) is

$$\mathbf{I} = a\mathbf{B}_1 \cdots \mathbf{B}_{n-1},$$

and the above argument shows that n must be divisible by 4.

Therefore all the matrices $\mathbf{M}$ in (35.12) are linearly independent if n is even but not divisible by 4. Now suppose n is divisible by 4. We assert that the set of all products of at most $\frac{1}{2}(n-2)$ factors is linearly independent. The number of such matrices is, by the binomial theorem,

$$1 + \binom{n-1}{1} + \binom{n-1}{2} + \cdots + \binom{n-1}{\frac{1}{2}(n-2)} = \tfrac{1}{2}2^{n-1} = 2^{n-2}.$$

The reason for the linear independence is that if an irreducible relation of linear dependence among them should exist, then by multiplying by a product of at most $\frac{1}{2}(n-2)$ factors, we can make one term equal to $\mathbf{I}$, but it is impossible to obtain $\mathbf{B}_1 \cdots \mathbf{B}_{n-1}$ for the other term. This completes the proof of the lemma.

Returning to the proof of the theorem, we have constructed a set of 2^{n-2} linearly independent n-by-n matrices, and have therefore the inequality

$$2^{n-2} \leq n^2,$$

which is false for $n = 10$ (and true for $n = 2, 4, 6, 8$). Since $2^{n-2} > n^2$ implies

$$2^{n+1-2} = 2 \cdot 2^{n-2} > 2n^2 > (n+1)^2$$

if $n > 3$, the only possible values for n are 2, 4, 6, or 8.

Finally we take up the special case $n = 6$. In this case all $2^5 = 32$ matrices $\mathbf{M}$ in (35.12) are linearly independent. Among these there are

$$\binom{5}{1} + \binom{5}{2} + 1 = 16$$

skew symmetric ones. But we can prove directly (see the exercise) that there are exactly

$$1 + 2 + 3 + 4 + 5 = 15$$

linearly independent skew symmetric matrices among all 6-by-6 matrices. Therefore the case $n = 6$ is not possible. This completes the proof of Hurwitz's theorem.

Exercise. Prove that the set of symmetric matrices in $M_n(R)$ forms a subspace of dimension

$$1 + 2 + \cdots + n = \frac{n(n + 1)}{2},$$

and the skew symmetric matrices form a subspace of dimension

$$1 + 2 + \cdots + n - 1 = \frac{n(n - 1)}{2}.$$

Bibliography

Albert, A. A. (ed.), *Studies in Mathematics*, Vol. II: *Studies in Modern Algebra* (Buffalo: Mathematical Association of America, 1963).

Artin, E., *Geometric Algebra* (New York: Interscience, 1957).

Benson, C. T., and L. C. Grove, *Finite Reflection Groups* (Tarrytown-on Hudson, N.Y.: Bogden and Quigley, 1971).

Birkhoff, G., and S. MacLane, *Survey of Modern Algebra*, rev. ed. (New York: Macmillan, 1953).

Bourbaki, N., *Algèbre*, Chapitre 2, "Algèbre linéaire," 3rd ed. (Paris: Hermann, Actualités et Industrielles, no. 1144).

Boyce, W. E., and R. C. DiPrima, *Elementary Differential Equations and Boundary Value Problems* (New York: John Wiley, 1969).

Courant, R., and H. Robbins, *What is Mathematics*? (New York: Oxford University Press, 1941).

Gruenberg, K., and A. Weir, *Linear Geometry* (Princeton: Van Nostrand, 1967).

Halmos, P. R., *Finite Dimensional Vector Spaces*, 2nd ed. (Princeton: Van Nostrand, 1958).

Jacobson, N., *Lectures in Abstract Algebra*, Vol. II: *Linear Algebra* (Princeton: Van Nostrand, 1953).

Kaplansky, I., *Linear Algebra and Geometry* (Boston: Allyn and Bacon, 1969).

MacLane, S., and G. Birkhoff, *Algebra* (New York: Macmillan, 1967).

Noble, B., *Applied Linear Algebra* (Englewood Cliffs, N.J.: Prentice-Hall, 1969).

Noble, B., *Applications of Undergraduate Mathematics in Engineering* (New York: The Mathematical Association of America and The Macmillan Company, 1967).

Polya, G., *How to Solve It* (Princeton: Princeton University Press, 1945).

Schreier, O., and E. Sperner, *Modern Algebra and Matrix Theory*, English translation (New York: Chelsea, 1952).

Smith, K. T., *Primer of Modern Analysis* (Tarrytown-on Hudson, N.Y.: Bogden & Quigley, 1971).

Synge, J. L., and B. A. Griffith, *Principles of Mechanics* (New York: McGraw-Hill, 1949).

Van der Waerden, B. L., *Modern Algebra*, Vols. I and II, English translation (New York: Ungar, 1949 and 1950).

Weyl, H., *Symmetry* (Princeton: Princeton University Press, 1952).

NOTES ON THE BIBLIOGRAPHY

Some books which influenced the writing of the present one and which develop the subject of linear algebra more deeply in some respects, are those of Bourbaki, Halmos, Jacobson, Kaplansky, Noble (*Applied Linear Algebra*) and Schreier and Sperner. The books of Birkhoff and MacLane, MacLane and Birkhoff, and Van der Waerden all contain thorough investigations of the axiomatic systems of modern algebra (groups, rings, fields, modules, etc.), which have all appeared naturally, and in a concrete way, in this book.

The real purpose of these notes, however, is to show the reader where he can study the connections between linear algebra and other parts of mathematics. The books of Artin, Benson and Grove, Gruenberg and Weir, and Kaplansky explore many aspects of the fascinating interplay between geometry and linear algebra. The two books by Noble will provide the reader with an orientation towards the computational aspects of linear algebra, and to the many, and sometimes unexpected, applications of linear algebra to the sciences and engineering. The usefulness of linear algebra in analysis is illustrated from several points of view in the books of Boyce and DePrima, Halmos, and Smith.

Solutions of Selected Exercises

In the case of numerical problems where a simple check is available, no answers are supplied. On theoretical problems, comments or hints on one method of solution are given, but not all the details. Of course there are often many valid ways to do a particular problem, and a correct solution may not always agree with the one given below.

SECTION 2

1. (a) The statement holds for $k = 1$. Assume it holds for $k \geq 1$. Then

$$1 + 3 + 5 + \cdots + (2k - 1) + [2(k+1) - 1]$$
$$= k^2 + 2(k+1) - 1 = (k+1)^2.$$

2. (a) $u = \alpha/\beta$ and $v = \gamma/\delta$ are solutions of the equations $\beta u = \alpha$ and $\delta v = \gamma$, respectively. Multiply the equations by δ and β and add, obtaining $\beta\delta(u + v) = \alpha\delta + \beta\gamma$.

3. Suppose for some n,

$$(\alpha + \beta)^n = \binom{n}{0}\alpha^n + \binom{n}{1}\alpha^{n-1}\beta + \cdots + \binom{n}{n}\beta^n.$$

Then

$$(\alpha + \beta)^{n+1} = (\alpha + \beta)^n(\alpha + \beta) = \binom{n}{0}\alpha^{n+1} + \binom{n}{1}\alpha^n\beta + \cdots$$
$$+ \binom{n}{n}\alpha\beta^n + \binom{n}{0}\alpha^n\beta + \cdots + \binom{n}{n}\beta^{n+1}.$$

Collect terms and use the definition of $\binom{n+1}{k}$ to complete the proof.

SECTION 3

1. (a) $\langle 1, 4, -2\rangle$.
   (b) $\langle -2, 4, 2\rangle + \langle -2, -1, 3\rangle + \langle 0, 1, 0\rangle = \langle -4, 4, 5\rangle$.
   (d) $\langle -\alpha + 2\beta, 2\alpha + \beta + \gamma, \alpha - 3\beta\rangle$.
2. Check your answers by substitution in the equations.
6. Since $\overrightarrow{AB} = B - A$, and $\overrightarrow{CD} = D - C$, $\overrightarrow{AB} = \overrightarrow{CD}$ implies that $B - A = D - C$. The vector $X = C - A = D - B$ behaves as required. Conversely, if $C = A + X$, $D = B + X$, then $C - D = (A + X) - (B + X) = A - B$, from which $\overrightarrow{AB} = \overrightarrow{CD}$ follows by multiplying both sides by -1.
7. (a) $\langle 1, \frac{1}{2}\rangle$. (b) $\langle 2, \frac{1}{2}\rangle$.
8. (a) and (c) are vertices of parallelograms; (b) is not.

SECTION 4

1. (a) Not a subspace; (b) subspace; (c) subspace; (d) not a subspace; (e) subspace; (f) this set is a subspace if and only if $B = 0$; (g) not a subspace.
3. (a) Subspace; (b) not a subspace; (c) subspace; (d) not a subspace; (e) subspace; (f) subspace; (g) subspace; (h) this set is a subspace if and only if g is the zero function: $g(x) = 0$ for all x.
4. (a) Linearly independent; (b) linearly dependent, $-3\langle 1, 1\rangle + \langle 2, 1\rangle + \langle 1, 2\rangle = 0$; (c) linearly independent; (d) linearly dependent, $\beta\langle 0, 1\rangle + \alpha\langle 1, 0\rangle - \langle \alpha, \beta\rangle = 0$; (e) linearly dependent, $\langle 1, 1, 2\rangle - \langle 3, 1, 2\rangle - 2\langle -1, 0, 0\rangle = 0$; (f) linearly independent; (g) linearly dependent.
5. $\{\langle 1, 1, 0\rangle, \langle 0, 1, 1\rangle\}$ is one solution of the problem.
6. Let $f_1, \ldots, f_n$ be a finite set of polynomial functions. Let x^n be the highest power of x appearing with a nonzero coefficient in any of the polynomials $\{f_i\}$. Then every linear combination of $f_1, \ldots, f_n$ has the form $\alpha_0 + \alpha_1 x + \ldots + \alpha_n x^n$. But there certainly exist polynomials, such as x^{n+1}, which cannot be expressed in this form. To be certain on this point, we observe that upon differentiating $n + 1$ times, all linear combinations of $f_1, \ldots, f_n$ become zero, while there are polynomials whose $(n + 1)$st derivative is different from zero.
7. Yes. Let S and T be subspaces. Let $a, b \in S \cap T$. Then $a + b \in S$ and $a + b \in T$ so $a + b \in S \cap T$. Similarly, if $a \in S \cap T$ and $\alpha \in F$, $\alpha a \in S \cap T$.
8. No. For example, in R_2, let $S = S(\langle 1, 0\rangle)$, $T = S(\langle 0, 1\rangle)$. Then $\langle 1, 1\rangle = \langle 1, 0\rangle + \langle 0, 1\rangle \notin S \cup T$.

SECTION 5

1. Suppose $b_1 = 0$, for example. Then $1 \cdot b_1 + 0 \cdot b_2 + \cdots + 0 \cdot b_r = 0$, and the vectors $\{b_1, \ldots, b_r\}$ are linearly dependent.

4. If df/dt is identically zero, then f is a constant, and $f = cf_0$, where f_0 is the function everywhere equal to 1. The dimension of the subspace consisting of all f whose second derivative is zero is two.

5. Suppose $\alpha_1 a_1 + \cdots + \alpha_m a_m = \alpha_1' a_1 + \cdots + \alpha_m' a_m$. Then $(\alpha_1 - \alpha_1')a_1 + \cdots + (\alpha_m - \alpha_m')a_m = 0$. By the linear independence of $a_1, \ldots, a_m$, we have $\alpha_1 = \alpha_1', \ldots, \alpha_m = \alpha_m'$.

SECTION 6

2. and 3. (a) Linearly dependent (check your relation of linear dependence in this and subsequent problems), basis: $\langle -1, 1\rangle, \langle 0, 3\rangle$; (b) linearly dependent, basis $\langle 2, 1\rangle, \langle 0, 2\rangle$; (c) linearly dependent, basis $\langle 1, 4, 3\rangle, \langle 0, 3, 2\rangle$; (d) linearly dependent, basis $\langle 1, 0, 0\rangle$, $\langle 0, 1, 0\rangle, \langle 0, 0, 1\rangle$; (e) linearly dependent; (f) linearly independent.

5. (a) Row equivalent; (b) not row equivalent.

SECTION 7

1. It does not. The subspace spanned by $\langle 1, 3, 4\rangle, \langle 4, 0, 1\rangle$, and $\langle 3, 1, 2\rangle$ has a basis $\langle 1, 3, 4\rangle, \langle 0, 4, 5\rangle$. An arbitrary linear combination of these vectors has the form $\langle \alpha, 3\alpha + 4\beta, 4\alpha + 5\beta\rangle$. If $\langle 1, 1, 1\rangle$ is one of these linear combinations, then $\alpha = 1$, $3 + 4\beta = 1$, $\beta = -\frac{1}{2}$, and $4\alpha + 5\beta = 4 - \frac{5}{2} \neq 1$.

2. It does. A basis for the subspace, in echelon form, is $\langle 1, -1, 1, 0\rangle$, $\langle 0, 2, 1, -1\rangle$, $\langle 0, 0, -\frac{7}{2}, -\frac{1}{2}\rangle$. A typical element in the subspace is $\langle \alpha, -\alpha + 2\beta, \alpha + \beta - \frac{7}{2}\gamma, -\beta - \frac{1}{2}\gamma\rangle$. Comparing with $\langle 2, 0, -4, -2\rangle$, we obtain $\alpha = 2$, $\beta = 1$, $\gamma = 2$.

3. Let $\{t_1, \ldots, t_m\}$ be a basis for T, and let $S \subset T$. Then every set of $m + 1$ vectors in S is linearly dependent, by Theorem (5.1). Let $\{s_1, \ldots, s_k\}$ be a set of linearly independent vectors in S such that every set of $k + 1$ vectors is linearly dependent. By Lemma (7.1), every vector in S is a linear combination of $\{s_1, \ldots, s_k\}$. Therefore $\{s_1, \ldots, s_k\}$ is a basis for S, and $\dim S \leq \dim T$. Finally, let $\dim S = \dim T$ and $S \subset T$. Let $\{s_1, \ldots, s_m\}$ be a basis for S. Then by Theorem (5.1), $\{s_1, \ldots, s_m, t\}$ is linearly dependent for all $t \in T$, since $\dim T = m$. By Lemma (7.1) again, $t \in S$ and $S = T$.

4. $\dim(S + T) \leq 3$. Therefore $\dim(S \cap T) = \dim S + \dim T - \dim(S + T) \geq 1$, by Theorem (7.5).

5. $\dim S = \dim T = 3$, $\dim(S + T) = 4$, $\dim(S \cap T) = 2$.

6. There are 4 vectors in V, 3 one-dimensional subspaces, and 3 different bases.

SECTION 8

1. (a) Solvable; (b) solvable; (c) solvable; (d) solvable; (e) solvable; (f) not solvable; (g) not solvable.
2. There is a solution if and only if $\alpha \neq 1$.
3. It is sufficient to prove that the column vectors of an m-by-n matrix, with $n > m$, are linearly dependent. The column vectors belong to R_m, and since there are n of them, they are linearly dependent by Theorem (5.1).
4. This result is also a consequence of Theorem (5.1).

SECTION 9

1. The dimension of the solution space is 2. Let c_1, c_2, c_3, c_4 be the column vectors. Then $c_1 + 3c_2 - 4c_3 = 0$ and $c_1 + c_2 - 2c_4 = 0$. Therefore a basis for the solution space is $\langle 1, 3, -4, 0\rangle$ and $\langle 1, 1, 0, -2\rangle$.
2. The dimensions of the solution spaces are as follows. The actual solutions should be checked:

(a) zero	(b) one	(c) two	(d) one
(e) two	(f) one	(g) zero	

3. By Theorem (8.9), every solution has the form $x_0 + x$ where x_0 is the solution of the nonhomogeneous system and x is a solution of the homogeneous system. The dimension of the solution space of the homogeneous system is two, and the actual solutions found may be checked by substitution.
5. A, B, and C must satisfy the equations

$$\begin{aligned} 3A + B + C &= 0 \\ -A \qquad + C &= 0 \end{aligned}$$

or

$$A\begin{pmatrix} 3 \\ -1 \end{pmatrix} + B\begin{pmatrix} 1 \\ 0 \end{pmatrix} + C\begin{pmatrix} 1 \\ 1 \end{pmatrix} = 0.$$

The solution space of the system has dimension one, so that any two nontrivial solutions are multiples of each other.
6. We have to find a nontrivial solution to the system

$$A\begin{pmatrix} \alpha \\ \gamma \end{pmatrix} + B\begin{pmatrix} \beta \\ \delta \end{pmatrix} + C\begin{pmatrix} 1 \\ 1 \end{pmatrix} = 0.$$

The rank of the matrix is at most two, so there certainly exists a nonzero solution. To show that two such solutions are proportional, we have to show that the rank is two. If the rank is one, then $\alpha = \gamma$ and $\beta = \delta$, and the points are not distinct, contrary to assumption.

SECTION 10

3. The result is immediate from Theorem (10.6).
4. Since L is one-dimensional, $L = p + V$ where V is the directing space, and $p \in L$. By (10.2) $q - p \in V$, and since $\dim V = 1$, V consists of all scalar multiples of $q - p$.

5. This result follows from the definition of hyperplane in Exercise 3, and Theorem (10.6).

6. In the case of the first problem, for example, the solution involves finding two distinct solutions of the system $x_1 + 2x_2 - x_3 = -1$, $2x_1 + x_2 + 4x_3 = 2$. This is done by the methods of the previous sections.

7. The line through p and q consists of all vectors $p + \lambda(q - p)$, by Exercise 4. By (10.2), $q - p$ belongs to the directing space of V, and hence $p + \lambda(q - p) \in V$ for all λ.

9. $\dim(S_1 + S_2) = 4$, $\dim(S_1 \cap S_2) = \dim S_1 + \dim S_2 - \dim(S_1 + S_2) = 2$.

10. By Exercise 4, a typical point on the line has the form $x = p + \lambda(q - p)$, where $p = \langle 1, -1, 0\rangle$ and $q = \langle -2, 1, 1\rangle$. Substitute the coordinates of x in the equation of the plane and solve for λ.

SECTION 11

1. The mappings in (c), (d), (e) are linear transformations; the others are not.

3.
$$\begin{aligned} 2T: y_1 &= 6x_1 + 2x_2 \\ y_2 &= 2x_1 - 2x_2 \\ T - U: y_1 &= -4x_1 \\ y_2 &= -x_2 \\ T^2: y_1 &= 10x_1 - 4x_2 \\ y_2 &= -4x_1 + 2x_2 \end{aligned}$$

To find the system for TU, we let $U: \langle x_1, x_2\rangle \to \langle y_1, y_2\rangle$ where $y_1 = x_1 + x_2$, $y_2 = x_1$; $T: \langle y_1, y_2\rangle \to \langle z_1, z_2\rangle$ where $z_1 = -3y_1 + y_2$, $z_2 = y_1 - y_2$. Then TU: $\langle x_1, x_2\rangle \to \langle z_1, z_2\rangle$, where $z_1 = -2x_1 - 3x_2$, $z_2 = x_2$. $TU \neq UT$.

4. $(DM)f(x) = D[xf(x)] = xf'(x) + f(x)$.
$MDf(x) = (Mf')(x) = xf'(x)$. $DM \neq MD$

5. From $0 + 0 = 0$, we obtain $T(0) + T(0) = T(0)$, hence $T(0) = 0$. From $v + (-v) = 0$, we obtain $T(v) + T(-v) = T(0) = 0$, hence $T(-v) = -T(v)$.

6. The linear transformations defined in (a) and (c) are one to one; the others are not.

8. The linear transformations defined in (a) and (b) are onto; the others are not.

9. Suppose T is one-to-one. Then the homogeneous system of equations defined by setting all y_i's equal to zero, has only the trivial solution. Therefore the rank of the coefficient matrix is n, and it follows from Section 8 that T is onto. Conversely, if T is onto, the rank of the coefficient matrix is n, and from Section 9, the homogeneous system has only the trivial solution. Therefore T is one-to-one. These remarks prove the equivalence of parts (a), (b) and (c). The equivalence with (d) is proved in Theorem (11.13).

10. D maps the constant polynomials into zero, so that D is not one-to-one. I maps no polynomial onto a constant polynomial $\neq 0$, so that I is not onto. The equation $DI = 1$ implies that D is onto and that I is one-to-one.

SECTION 12

1. $$\begin{pmatrix} -1 & 2 \\ -1 & 0 \end{pmatrix} \begin{pmatrix} 1 & 1 \\ 0 & 1 \end{pmatrix} = \begin{pmatrix} -1 & 1 \\ -1 & -1 \end{pmatrix}$$

$$\begin{pmatrix} -1 & 2 & 3 \\ 1 & 1 & 1 \end{pmatrix} \begin{pmatrix} 1 & 1 \\ 1 & 0 \\ 2 & -1 \end{pmatrix} = \begin{pmatrix} 7 & -4 \\ 4 & 0 \end{pmatrix}, (1, 1, 2) \begin{pmatrix} -1 \\ 0 \\ 1 \end{pmatrix} = (1),$$

$$\begin{pmatrix} 0 & 1 \\ 0 & 0 \end{pmatrix} \begin{pmatrix} 0 & 1 \\ 0 & 0 \end{pmatrix} = \begin{pmatrix} 0 & 0 \\ 0 & 0 \end{pmatrix},$$

$$\begin{pmatrix} 1 & 0 & 0 \\ 0 & -1 & 0 \\ 0 & 0 & 2 \end{pmatrix} \begin{pmatrix} 1 & 1 & 0 \\ -1 & 2 & 1 \\ 1 & 1 & 3 \end{pmatrix} = \begin{pmatrix} 1 & 1 & 0 \\ 1 & -2 & -1 \\ 2 & 2 & 6 \end{pmatrix}$$

4. Multiplying **DA** is equivalent to multiplying the ith row of **A** by δ_i, while multiplying on the right multiplies the ith column by δ_i.

5. If $\mathbf{A} = (\alpha_{ij})$ commutes with all the diagonal matrices **D**, then by Problem 5, we have $\delta_i \alpha_{ij} = \alpha_{ij} \delta_j$ for all δ_i and δ_j in F. If $i \neq j$, it follows that $\alpha_{ij} = 0$.

7. The matrices in (b), (c), (d) and (e) are invertible; (a) is not. Check the formulas for the inverses by matrix multiplication.

8. Every linear transformation on the space of column vectors has the form $\mathbf{x} \to \mathbf{B} \cdot \mathbf{x}$ for some n-by-n matrix **B**. The linear transformation $\mathbf{x} \to \mathbf{A} \cdot \mathbf{x}$ is invertible if and only if there exists a matrix **B** such that $\mathbf{B}(\mathbf{A} \cdot \mathbf{x}) = \mathbf{A}(\mathbf{B} \cdot \mathbf{x}) = \mathbf{x}$ for all **x**. By the associative law (Exercise 3), these equations are equivalent to $(\mathbf{BA})\mathbf{x} = (\mathbf{AB})\mathbf{x} = \mathbf{x}$ for all **x**. Then show that for an n-by-n matrix **C**, $\mathbf{Cx} = \mathbf{x}$ for all **x** is equivalent to $\mathbf{C} = \mathbf{I}$. Thus the equations simply mean that the matrix **A** is invertible. If **A** is invertible, then $\mathbf{x} = \mathbf{A}^{-1}\mathbf{b}$ is a solution of the equation $\mathbf{Ax} = \mathbf{b}$ because $\mathbf{A}(\mathbf{A}^{-1}\mathbf{b}) = (\mathbf{AA}^{-1})\mathbf{b} = \mathbf{Ib} = \mathbf{b}$ by the associative law. If $\mathbf{x}'$ is another solution, then $\mathbf{Ax} = \mathbf{Ax}'$, and multiplying by $\mathbf{A}^{-1}$, we obtain $\mathbf{A}^{-1}(\mathbf{Ax}) = \mathbf{A}^{-1}(\mathbf{Ax})$. It follows that $\mathbf{x} = \mathbf{x}'$.

SECTION 13

1. The matrices with respect to the basis $\{u_1, u_2\}$ are

$$\mathbf{S} = \begin{pmatrix} 1 & 1 \\ -1 & 0 \end{pmatrix}, \quad \mathbf{T} = \begin{pmatrix} 0 & 1 \\ 1 & 0 \end{pmatrix}, \quad \mathbf{U} = \begin{pmatrix} 2 & 0 \\ 0 & -2 \end{pmatrix}.$$

With respect to the new basis $\{w_1, w_2\}$ the matrix of S is

$$\mathbf{S}' = \begin{pmatrix} \frac{5}{4} & \frac{3}{4} \\ -\frac{7}{4} & -\frac{1}{4} \end{pmatrix} = \mathbf{X}^{-1}\mathbf{S}\mathbf{X}, \qquad \text{where } \mathbf{X} = \begin{pmatrix} 3 & 1 \\ -1 & 1 \end{pmatrix}.$$

2.

	a	b	c	d
S	rank 1, nullity 1	not invertible	$u_1 + u_2$	$u_1 + u_2$
T	rank 2, nullity 0	invertible	—	—
U	rank 3, nullity 0	invertible	—	—

3.

$$\mathbf{D} = \begin{pmatrix} 0 & 1 & 0 & & 0 \\ 0 & 0 & 2 & & \\ \vdots & & & \ddots & \\ & & & & k-1 \\ 0 & & & & 0 \end{pmatrix}, \qquad \begin{aligned} \text{rank} &= k - 1, \\ \text{nullity} &= 1. \end{aligned}$$

4. The rank of T is the dimension of $T(V)$. A basis for $T(V)$ can be selected from among the vectors $T(v_i)$, $i = 1, \ldots, n$. Since $T(v_i) = \sum_j \alpha_{ji} w_j$, a subset of the vectors $\{T(v_i)\}$ forms a basis for $T(V)$ if and only if the corresponding columns of the matrix $\mathbf{A}$ of $\mathbf{T}$ form a basis for the column space of $\mathbf{A}$. Therefore $\dim T(V) = \text{rank}(\mathbf{A})$.

5. Since $\dim R_1 = 1$, the rank of T is one. Therefore $n(f) = n - 1$, by Theorem (13.9).

6. Let $V_1 = n(f)$; then $\dim V_1 = n - 1$, by Exercise 5. Let v_0 be a fixed solution of the equation $f(v) = \alpha$. (Why does one exist?) Then the set of all solutions is $v_0 + V_1$, and is a linear manifold of dimension $n - 1$.

9. If $TS = 0$ for $S \neq 0$, then $v_1 = S(v) \neq 0$ for some $v \in V$, and $Tv_1 = 0$. Conversely, let $T(v_1) = 0$, for some nonzero vector v_1. There exists a basis $\{v_1, v_2, \ldots, v_n\}$ of V starting with v_1. Define S on this basis by setting $S(v_1) = v_1$, $S(v_2) = \cdots = S(v_n) = 0$. Then $S \neq 0$ and $TS = 0$.

10. Let $ST = 1$. Let $\{v_1, \ldots, v_n\}$ be a basis for V. Since $ST = 1$, it follows that $\{Tv_1, \ldots, Tv_n\}$ is also a basis of V. On each of these basis elements, $TS(Tv_i) = Tv_i$, so that TS agrees with the identity transformation on a basis. Therefore $TS = 1$.

SECTION 14

1. (e) Two vectors are perpendicular if and only if the cosine of the angle between them is zero. From the law of cosines, we have

$$\|b - a\|^2 = \|a\|^2 + \|b\|^2 - 2\|a\|\,\|b\| \cos\theta.$$

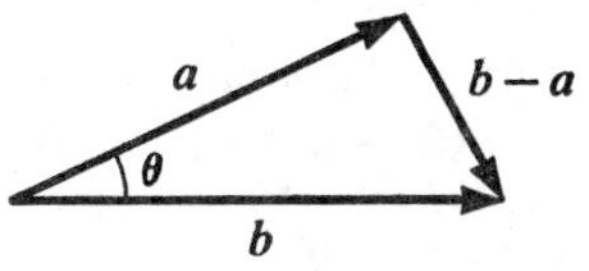

Since

$$\|b - a\|^2 = (b - a, b - a) = \|a\|^2 + \|b\|^2 - 2(a, b)$$

we have

$$\cos\theta = \frac{(a, b)}{\|a\|\,\|b\|}.$$

Therefore $a \perp b$ if and only if $(a, b) = 0$.
(f) $\|a + b\| = \|a - b\|$ if and only if $(a + b, a + b) = (a - b, a - b)$. This statement holds if and only if $(a, b) = 0$.

2. (a) If x belongs to both $p + S$ and $q + S$, then $x = p + s_1 = q + s_2$, with s_1 and $s_2 \in S$. It follows that $p \in q + S$ and $q \in p + S$, and hence that $p + S = q + S$.
(b) The set $L = p + S$ of all vectors of the form $\{p + \lambda(q - p)\}$ is easily shown to be a line containing p and q. Let $L' = p + S'$ be any line containing p and q. Then $q - p \in S'$ and since S and S' are one-dimensional $S = S'$. Since the lines L and L' intersect and have the same one-dimensional subspace, $L = L'$ by (a).
(c) p, q, r are collinear if they belong to the same line $L = p + S$. In that case $q - p \in S$ and $q - r \in S$ and since S is one-dimensional, $S(q - p) = S(q - r)$. Conversely, if $S(q - p) = S(q - r)$, the lines determined by the points p and q, and q and r have the same one-dimensional spaces. Hence they coincide by (a).
(d) Let $L = p + S$ and $L' = p' + S'$ be the given lines. From parts (a) and (b) we may assume $S \neq S'$. Since $\dim S = \dim S' = 1$, $S \cap S' = 0$. If x_1 and x_2 belong to $L \cap L'$, then $x_1 - x_2 \in S \cap S'$, hence $x_1 = x_2$. In order to show that $L \cap L'$ is not empty, let $S = S(s)$, $S' = S(s')$; then we have to find λ and λ' such that $p + \lambda s = p' + \lambda' s'$, and this is true since any 3 vectors in R_2 are linearly dependent.

3. $\lambda = \dfrac{-(p - r, q - p)}{\|q - p\|^2}$. The perpendicular distance is

$$\|u - r\| = \left\| p - r - (p - r, q - p)\frac{q - p}{\|q - p\|^2} \right\|.$$

SECTION 15

1. (a) $\dfrac{1}{\sqrt{3}}\langle 1, 1, 1, 0\rangle, \dfrac{1}{\sqrt{51}}\langle 5, -1, -4, -3\rangle$.

2. $1, \sqrt{12}\,(x - \frac{1}{2}), a(x^2 - x + \frac{1}{6})$,
where a is chosen to make $\|f\| = 1$.

3. The distance is given by $\|v - (v, u_1)u_1\|$, where v and u_1 are as follows:
(a) $v = \langle -1, -1\rangle, u_1 = \langle 2, -1\rangle/\sqrt{5}$;
(b) $v = \langle 1, 0\rangle, u_1 = \langle 1, 1\rangle/\sqrt{2}$;
(c) $v = \langle 2, 2\rangle, u_1 = \langle 1, -1\rangle/\sqrt{2}$.

5. (a) $(v, w) = \sum_{i,j} \xi_i\eta_j(u_i, u_j) = \sum \xi_i\eta_i$.
(b) Let $v = \sum_{i=1}^{n} \xi_i u_i$. Then $(v, u_k) = \sum_{i=1}^{n} \xi_i(u_i, u_k) = \xi_k$

6. By the Gram-Schmidt theorem, there exists an orthonormal basis $\{u_1, \ldots, u_n\}$ of R_n starting with u_1. The matrix with rows $u_1, \ldots, u_n$ has the required properties.

9. Let $\{v_1, \ldots, v_n\}$ be an orthonormal basis of V, and let

$$\begin{aligned} w_1 &= \alpha_{11}v_1 + \cdots + \alpha_{1n}v_n \\ &\cdots\cdots\cdots\cdots\cdots \\ w_d &= \alpha_{d1}v_1 + \cdots + \alpha_{dn}v_n \end{aligned}$$

be a basis for W. A vector $x = x_1v_1 + \cdots + x_nv_n$ belongs to $W^\perp$ if and only if $(x, w_1) = \cdots = (x, w_d) = 0$. Thus $x_1, \ldots, x_n$ have to be solutions of the homogeneous system

$$\begin{aligned} \alpha_{11}x_1 + \cdots + \alpha_{1n}x_n &= 0 \\ \cdots\cdots\cdots\cdots\cdots& \\ \alpha_{d1}x_1 + \cdots + \alpha_{dn}x_n &= 0. \end{aligned}$$

The result is now immediate from Corollary (9.4).

10. It can be shown that there exist orthonormal bases $\{v_i\}$ and $\{w_i\}$ of V such that $\{v_1, \ldots, v_d\}$ is a basis for W_1, and $\{w_1, \ldots, w_d\}$ is a basis for W_2. By Theorem 15.11, there exists an orthogonal transformation T such that $Tv_i = w_i$ for each i. Then $T(W_1) = W_2$.

11. (a) By Exercise 7, dim $S(n)^\perp = 2$. It is sufficient to prove that the set P of all p such that $(p, n) = \alpha$ is the set of solutions of a linear equation. Let $\{v_1, v_2, v_3\}$ be an orthonormal basis for R_3, and let $n = \alpha_1v_1 + \alpha_2v_2 + \alpha_3v_3$. Then $x = x_1v_1 + x_2v_2 + x_3v_3$ satisfies $(p, n) = \alpha$ if and only if $\alpha_1x_1 + \alpha_2x_2 + \alpha_3x_3 = \alpha$. This shows that the set of all p such that $(p, n) = \alpha$ is a plane. Using the results of Section 10, we can assert that $P = p_0 + S(n)^\perp$, where p_0 is a fixed solution of $(p, n) = \alpha$.
(b) Normal vector: $\langle 3, -1, 1\rangle$.
(c) $x_1 - x_2 + x_3 = -2$, or $(n, p) = -2$.
(d) The plane with normal vector n, passing through p is the set of all vectors x such that $(x, n) = (p, n)$, or $(x - p, n) = 0$.
(f) The normal vector n must be perpendicular to $\langle 2, 0, -1\rangle - \langle 1, 1, 1\rangle$ and to $\langle 2, 0, -1\rangle - \langle 0, 0, 1\rangle$ by (d). Then use the method of part (e).
(g) Following the hint, we write the second equation in the form $p_0 - p = \lambda n + (u - p)$. Taking the inner product with n yields $0 = \lambda(n, n) + (u - p, n)$.
(h) We have $u = \langle 1, 1, 2\rangle$, $p = \langle 0, 0, 1\rangle$, $n = \langle 1, 1, -1\rangle$. Then $3\lambda + (u - p, n) = 0$, $\lambda = -\frac{1}{3}$. Then the distance is $\|p_0 - u\| = \frac{1}{3}\|n\|$.

SECTION 16

1. (a) 2; (b) 0; (c) 0; (d) the determinants are: (a) 0, (b) 1, (c) 13, (d) -2, (e) 3.

2. Using the properties of the determinant function, we have $D(\mathbf{A}) = \alpha_1 \cdots \alpha_n D(\mathbf{A}')$, where

$$\mathbf{A}' = \begin{pmatrix} 1 & * & \cdots & * \\ 0 & \ddots & & \vdots \\ \vdots & & \ddots & * \\ 0 & \cdots & 0 & 1 \end{pmatrix}.$$

Then show that, by applying elementary row operations to $\mathbf{A}'$, we obtain

$$D(\mathbf{A}') = \begin{vmatrix} 1 & 0 & * & & * \\ & 1 & * & & * \\ \vdots & & \ddots & & \\ 0 & & \cdots & 0 & 1 \end{vmatrix} = \cdots = \begin{vmatrix} 1 & 0 & & \cdots & 0 \\ 0 & 1 & 0 & & 0 \\ & & \ddots & & \\ 0 & & \cdots & & 1 \end{vmatrix} = 1$$

3. Applying elementary row operations to the rows involving $\mathbf{A}_1$, we obtain

$$D(\mathbf{A}) = \left|\begin{array}{ccc|c|c|c} \alpha_{11} * & & * & & & \\ \vdots & & & * & & \\ 0 \cdots & & \overset{*}{\alpha}_{1d_1} & & & \\ \hline & \mathbf{O} & & \mathbf{A}_2 & & \\ \hline & \mathbf{O} & & \mathbf{O} & \ddots & \mathbf{A}_r \end{array}\right|$$

for some scalars $\alpha_{11}, \ldots, \alpha_{1d_1}$ (where $\mathbf{A}_1$ is a d_1-by-d_1 matrix) and $D(\mathbf{A}_1) = \alpha_{11} \cdots \alpha_{1d_1}$. Similarly we can apply elementary row operations to the rows involving $\mathbf{A}_2$, and obtain

$$D(\mathbf{A}) = \left|\begin{array}{ccc|ccc|c} \alpha_{11} * & & * & & & & \\ \vdots & & & & * & & \\ 0 \cdots & & \overset{*}{\alpha}_{1d_1} & & & & \\ \hline & & & \alpha_{21} * & & * & \\ & \mathbf{O} & & \vdots & & & \\ & & & 0 \cdots & & \overset{*}{\alpha}_{2d_2} & \\ \hline & \mathbf{O} & & & \mathbf{O} & & \ddots \end{array}\right|$$

where $\alpha_{21} \cdots \alpha_{2d_2} = D(\mathbf{A}_2)$. Continuing in this way, $\mathbf{A}$ is reduced to triangular form. Applying Exercise 2, we obtain the final result.

4. All that has to be proved is that

$$D^*(a_1, \ldots, a_n) = D^*(a_1, \ldots, a_i + a_j, \ldots, a_n)$$

and this is immediate from assumptions (c) and (d) about D^*.

SECTION 17

1. Using the definition and Theorem 16.6, we have $D(\langle\xi, \eta\rangle, \langle\lambda, \mu\rangle) = D(\langle\xi, 0\rangle, \langle\lambda, 0\rangle) + D(\langle\xi, 0\rangle, \langle 0, \mu\rangle) + D(\langle 0, \eta\rangle, \langle\lambda, 0\rangle) + D(\langle 0, \eta\rangle, \langle 0, \mu\rangle) = \xi\mu - \eta\lambda$. $D(\langle\xi, 0\rangle, \langle\lambda, 0\rangle) = 0$ because the vectors involved are linearly dependent. $D(\langle\xi, 0\rangle, \langle 0, \mu\rangle) = \xi\mu\, D(e_1, e_2)$, while $D(\langle 0, \eta\rangle, \langle\lambda, 0\rangle) = -D(\langle\lambda, 0\rangle, \langle 0, \eta\rangle) = -\lambda\eta$.

SECTION 18

2. $\mathbf{AA}^{-1} = \mathbf{I}$. Apply Theorem (18.3) to this equation.
3. $D(\mathbf{P}_{ij}) = -1$; $D(\mathbf{B}_{ij}(\lambda)) = 1$; $D(\mathbf{D}_i(\mu)) = \mu$.

   The fact that $\mathbf{A}$ is a product of elementary matrices was shown in Section 12.
5. Let T be an orthogonal transformation. If $\mathbf{A}$ is the matrix of T with respect to an orthonormal basis, then $\mathbf{A}^t\mathbf{A} = \mathbf{I}$, and since $D(\mathbf{A}) = D({}^t\mathbf{A})$ we have $D(\mathbf{A})^2 = 1$.

SECTION 19

4. Expanding the determinant along the first row, we see that it has the form $Ax_1 + Bx_2 + C$. Substituting (α, β) or (γ, δ) for (x_1, x_2) makes two rows of the determinant equal, and hence the equation $Ax_1 + Bx_2 + C = 0$ is satisfied by the points (α, β) and (γ, δ).
6. The image of the square is the parallelogram $\{\lambda T(e_1) + \mu T(e_2) : 0 \le \lambda, \mu \le 1\}$. The area of the parallelogram is $|D(T(e_1), T(e_2))|$, where $T(e_1)$ and $T(e_2)$ are the columns of the matrix of T with respect to the basis $\{e_1, e_2\}$ of R_2.
7. Since

$$\begin{vmatrix} \alpha_1 & \alpha_2 & 1 \\ \beta_1 & \beta_2 & 1 \\ \gamma_1 & \gamma_2 & 1 \end{vmatrix} = \begin{vmatrix} \alpha_1 & \alpha_2 & 1 \\ \beta_1 - \alpha_1 & \beta_2 - \alpha_2 & 0 \\ \gamma_1 - \alpha_1 & \gamma_2 - \alpha_2 & 0 \end{vmatrix},$$

   the determinant is

$$\begin{vmatrix} \beta_1 - \alpha_1 & \beta_2 - \alpha_2 \\ \gamma_1 - \alpha_1 & \gamma_2 - \alpha_2 \end{vmatrix},$$

   which is the area of the parallelogram with edges $\langle\beta_1, \beta_2\rangle - \langle\alpha_1, \alpha_2\rangle$, $\langle\gamma_1, \gamma_2\rangle - \langle\alpha_1, \alpha_2\rangle$. The area of the triangle is one-half the area of this parallelogram.
11. There exists an orthogonal transformation T such that $T(a_1/\|a_1\|) = e_1, \ldots, T(a_n/\|a_n\|) = e_n$, where $\{e_1, \ldots, e_n\}$ are the unit vectors. Then $D(a_1, \ldots, a_n) = |D(T)|\,D(e_1, \ldots, e_n)\|a_1\| \ldots \|a_n\| = \|a_1\| \ldots \|a_n\|$.

SECTION 20

1. $Q = \frac{2}{3}x - \frac{1}{9}$, $R = -\frac{1}{9}x^2 - x - \frac{2}{3}$.
2. Let $\alpha_1, \ldots, \alpha_k$ be distinct zeros of f. Then $f = (x - \alpha_1)g_1$. Since $\alpha_2 \neq \alpha_1$, $x - \alpha_2$ is a prime polynomial which divides f but not $x - \alpha_1$. Therefore $x - \alpha_2$ divides g_1 and $f = (x - \alpha_1)(x - \alpha_2)g_2$. Continuing in this way, $f = (x - \alpha_1) \cdots (x - \alpha_k)g$, and hence $\deg f \geq k$.
4. If the degree of f is two or three, any nontrivial factorization will involve a linear factor, and hence a zero of f. The result is false if $\deg f > 3$. For example, $f = (x^2 + 1)^2$ has no zeros in R, but is not a prime in $R[x]$.
5. Suppose m/n satisfies the equation. Multiplying the resulting equation by n^r we obtain

$$a_0m^r + a_1m^{r-1}n + \cdots + a_rn^r = 0.$$

Then n divides a_0m^r, and since n does not divide m^r, $n|a_0$. Similarly $m|a_r$.

7. In $Q[x]$, the prime factors are: (a) $(2x + 1)(x^2 - x + 1)$; (c) $(x^2 + 1)(x^4 - x^2 + 1)$.
   In $R[x]$, the prime factors are: (a) $(2x + 1)(x^2 - x + 1)$; (c) $(x^2 + 1)(x^2 + \sqrt{3}\,x + 1)(x^2 - \sqrt{3}\,x + 1)$.
8. The process must terminate, otherwise we have an infinite decreasing sequence of nonnegative integers, contrary to the principle of well-ordering. Now suppose $r_{i_0} \neq 0$ and $r_{i_0+1} = 0$. From the way the r_i's are defined, $r_{i_0}|r_{i_0}$ and r_{i_0-1}. From the preceding equation we see that $r_{i_0}|r_{i_0-2}$. Continuing in this way we obtain $r_{i_0}|a$ and $r_{i_0}|b$. On the other hand, starting from the top, if $d|a$ and $d|b$, then $d|r_0$. From the next equation we get $d|r_1$. Continuing we obtain eventually $d|r_{i_0}$. Thus $r_{i_0} = (a, b)$.
9. (a) $2x + 1$.

SECTION 21

1. $-10 + 10i$, $\frac{1}{13}(3 - 2i)$, $\frac{1}{5}(3 + 4i)$.
2. $\cos 3\theta = (\cos\theta)^3 - 3\cos\theta(\sin\theta)^2$, obtained by taking the real part of both sides in the formula $(\cos\theta + i\sin\theta)^3 = \cos 3\theta + i\sin 3\theta$.
3. $\lambda(\cos\theta_k + i\sin\theta_k)$ where λ is a real fifth root of 2 and $\theta_k = 2\pi k/5$, $k = 0, 1, 2, 3, 4$.
5. The one-to-one mapping that produces the isomorphism is

$$\alpha + i\beta \to \begin{pmatrix} \alpha & -\beta \\ \beta & \alpha \end{pmatrix}.$$

9. The equation $x^2 = -1$ has a solution in the field of complex numbers, but cannot have a solution in an ordered field.

SECTION 22

2. Let $\{v_1, \ldots, v_m\}$ be a basis for V and $\{w_1, \ldots, w_n\}$ a basis for W. For each pair (i, j), let E_{ij} be the linear transformation such that $E_{ij}v_j = w_i$, and $E_{ij}v_k = 0$ if $k \neq j$. Then the mn linear transformations $\{E_{ij}\}$ form a basis for $L(V, W)$.

3. The minimal polynomials are $(x - 2)(x + 1)$, $x^3 - 1$, $x^2 + x - 1$, and x^3, respectively.

4. (a) For each v_i, $f(T)v_i = (T - \xi_1) \cdots (T - \xi_n)v_i = 0$ since the factors $T - \xi_k$ commute, and $(T - \xi_i)v_i = 0$.
(b) Let $m(x) = \prod(x - \xi_j)$, where the ξ_j are the distinct characteristic roots of T. By the argument of part (a), $m(T) = 0$. Then the minimal polynomial of T divides $m(x)$. It is enough to show that if $m'(x) = \prod_{\xi_j \neq \xi_k} (x - \xi_j)$, then $m'(T) \neq 0$. We have

$$m'(T)x_k = \prod_{\xi_j \neq \xi_n} (\xi_k - \xi_j)x_k \neq 0,$$

and the result is proved.

5. Suppose T is invertible, and let $m(x) = x^r + \cdots + \alpha_1 x + \alpha_0$ be the minimal polynomial. Suppose $\alpha_0 = 0$. Then $m(x) = xm_1(x)$. Moreover, $m_1(T) \neq 0$ since $m(x)$ is the minimal polynomial. Then $m(T) = Tm_1(T) = 0$, contradicting the fact that T is invertible.
Conversely, suppose $\alpha_0 \neq 0$. Then $m(x) = m_1(x) \cdot x + \alpha_0$ for some polynomial $m_1(x)$, and $m_1(T)T = -\alpha_0 1$. Then $-\alpha_0^{-1}m_1(T) = T^{-1}$.

6. (a) Suppose α is a characteristic root of T. Then for some $v \neq 0$, $Tv = \alpha v$. Then $\alpha \neq 0$ (why?), and $T^{-1}Tv = \alpha T^{-1}v$. Then $T^{-1}v = \alpha^{-1}v$. The proof of the converse is the same.
(b) Suppose $Tv = \alpha v$. Then $f(T)v = f(\alpha)v$.

8. This result follows from the fact that similar matrices can be viewed as matrices of a single linear transformation with respect to different bases, and that their characteristic roots are the characteristic roots of the linear transformation. A direct proof can be made as follows. Let $\mathbf{A} = \mathbf{S}^{-1}\mathbf{B}\mathbf{S}$ for some invertible $\mathbf{S}$. If $\mathbf{x} \neq 0$ and $\mathbf{A}\mathbf{x} = \alpha\mathbf{x}$, then $\mathbf{B}(\mathbf{S}\mathbf{x}) = \alpha(\mathbf{S}\mathbf{x})$ and $\mathbf{S}\mathbf{x} \neq 0$.

9. The characteristic roots are as follows: (a) 3, -1; (b) -1, -1; (c) 1, -2.
Characteristic vectors are found by solving the equations

$$\mathbf{A}\begin{pmatrix} x_1 \\ x_2 \end{pmatrix} = \alpha\begin{pmatrix} x_1 \\ x_2 \end{pmatrix},$$

where $\mathbf{A}$ is the given matrix, and α a characteristic root of $\mathbf{A}$. Be sure to check your results.

10. Suppose $Tv = \alpha v$, for $\alpha \neq 0$. Then $0 = T^m = \alpha^m v$, and hence $\alpha = 0$.

11. and 12. are both easy consequences of Theorem (22.8).

SECTION 23

2. The minimal polynomial of $\begin{pmatrix} 1 & -2 \\ 1 & -1 \end{pmatrix}$ is $x^2 + 1$, which can be factored into distinct linear factors in $C[x]$ but not in $R[x]$.
3. The minimal polynomial is $x^3 - 1 = (x - 1)(x^2 + x + 1)$. The reasoning from this point on is the same as in Exercise 2.
4. The minimal polynomial of D is x^{n+1}.
8. Since $d(x)|f(x)$, $n[d(T)] \subset n[f(T)]$. Conversely, let $f(T)v = 0$. Since $d(x) = a(x)m(x) + b(x)f(x)$ for some polynomials $a(x)$ and $b(x)$, we have

$$d(T)v = a(T)m(T)v + b(T)f(T)v = 0,$$

since $f(T)v = 0$ and since $m(x)$ is the minimal polynomial. Thus $n[f(T)] \subset n[d(T)]$ and the result is established.

SECTION 24

1. (a) $(x + 2)^2$.
(b) $(x + 2)^2$.
(c) -2 (appearing twice in the characteristic and minimal polynomials).
(d) No. Because the minimal polynomial is not a product of distinct linear factors.
(e) Let $v = x_1v_1 + x_2v_2$ be a vector with unknown coefficients such that $(T + 2)v = 0$. Using the definition of T, this leads to a system of homogeneous equations with the nontrivial solution, $(1, -1)$. Then $v_1 - v_2$ is a characteristic vector for T.
(f) A basis which puts the matrix of T in triangular form is $w_1 = v_1 - v_2$, $w_2 = v_2$. Then $\mathbf{SB} = \mathbf{AS}$ where

$$\mathbf{B} = \begin{pmatrix} -2 & 1 \\ 0 & -2 \end{pmatrix}, \quad \mathbf{S} = \begin{pmatrix} 1 & 0 \\ -1 & 1 \end{pmatrix}.$$

(g) The Jordan decomposition of T is $T = D + N$, where D and N are the linear transformations whose matrices are $-2\mathbf{I}$ and $\begin{pmatrix} 0 & 1 \\ 0 & 0 \end{pmatrix}$ with respect to the basis $\{w_1, w_2\}$.
3. The minimal polynomial of T is $x^2 + \alpha\beta$, where $\alpha\beta > 0$. The minimal polynomial is not a product of distinct linear factors in $R[x]$, and the answer to the question is no.
4. Use the triangular form theorem.
5. The minimal polynomial of T divides $x^2 - x$, and therefore has distinct linear factors in $C[x]$. There does exist a basis of V consisting of characteristic vectors.

6. The minimal polynomial divides $x^r - 1$ which has distinct linear factors in $C[x]$, namely $x - (\cos\theta_k + i\sin\theta_k)$, $k = 0, \ldots, r-1$, where $\theta_k = 2\pi k/r$.
7. V has a basis consisting of characteristic vectors of T if and only if $\lambda \neq 0$.

SECTION 25

1. (a)
$$\left(\begin{array}{cc|cc} 1 & 1 & & \\ 0 & 1 & \multicolumn{2}{c}{\mathbf{O}} \\ \hline & & 0 & -1 \\ \multicolumn{2}{c|}{\mathbf{O}} & 1 & 1 \end{array}\right).$$

(b)
$$\left(\begin{array}{c|cc|cc} -1 & 0 & 0 & 0 & 0 \\ \hline 0 & 0 & -1 & 0 & 1 \\ 0 & 1 & -1 & 0 & 0 \\ \hline \mathbf{O} & \multicolumn{2}{c|}{\mathbf{O}} & 0 & -1 \\ & & & 1 & -1 \end{array}\right).$$

2. The rational canonical forms are
$$\begin{pmatrix} 0 & 1 \\ 1 & 1 \end{pmatrix}, \quad \begin{pmatrix} 1 & 0 & 0 \\ 0 & 0 & -1 \\ 0 & 1 & -1 \end{pmatrix},$$
respectively.
3. The rational canonical forms over R are
$$\begin{pmatrix} \frac{1}{2} + \frac{\sqrt{5}}{2} & 0 \\ 0 & \frac{1}{2} - \frac{\sqrt{5}}{2} \end{pmatrix} \quad \text{and} \quad \begin{pmatrix} 1 & 0 & 0 \\ 0 & 0 & -1 \\ 0 & 1 & -1 \end{pmatrix},$$
respectively.
4. The canonical forms over C are
$$\begin{pmatrix} \frac{1}{2} + \frac{\sqrt{5}}{2} & 0 \\ 0 & \frac{1}{2} - \frac{\sqrt{5}}{2} \end{pmatrix} \quad \text{and} \quad \begin{pmatrix} 1 & 0 & 0 \\ 0 & -\frac{1}{2} + \frac{\sqrt{-3}}{2} & 0 \\ 0 & 0 & -\frac{1}{2} - \frac{\sqrt{-3}}{2} \end{pmatrix}$$
respectively.
5. Let $\{v, Tv, \ldots, T^{d-1}v\}$ be a basis for V, and suppose $T^d v = \alpha_0 v + \alpha_1 Tv + \cdots + \alpha_{d-1}T^{d-1}v$. The matrix of T with respect to this basis is
$$\mathbf{A} = \begin{pmatrix} 0 & & \cdots & \alpha_0 \\ 1 & 0 & \cdots & \alpha_1 \\ \cdots & \cdots & \cdots & \cdots \\ 0 & & \cdots & \alpha_{d-1} \end{pmatrix}.$$
Then $D(\mathbf{A} - x\mathbf{I}) = \pm(x^d - \alpha_{d-1}x^{d-1} - \cdots - \alpha_0)$. On the other hand, $x^d - \alpha_{d-1}x^{d-1} - \cdots - \alpha_0$ is the minimal polynomial of T by Lemma (25.6).
6. The matrices are similar over C in all three cases.

Symbols
(Including Greek Letters)

Lower-case Greek Letters Used in This Book

α	alpha
β	beta
γ	gamma
δ	delta
ϵ	epsilon
ζ	zeta
η	eta
θ	theta
λ	lambda
μ	mu
ν	nu
ξ	xi (ksi)
π	pi
ρ	rho
σ	sigma
τ	tau
φ	phi
ψ	psi
ω	omega

Partial List of Symbols Used

R	field of real numbers
$a \in A$	set membership
$A \subset B$	set inclusion
$f: A \to B$	function (or mapping) from A into B
$\langle \alpha_1, \ldots, \alpha_n \rangle$	vector with components $\{\alpha_1, \ldots, \alpha_n\}$.
$\sum x_i$	summation
$\prod u_i$	product
$S(v_1, \ldots, v_n)$	vector space generated (or spanned) by $\{v_1, \ldots, v_n\}$.
$\mathscr{F}(R)$	vector space of all real-valued functions on R.
$C(R)$	continuous real valued functions on R
$P(R)$	polynomial functions on R
$L(V, W)$	linear transformations from V into W
$\mathbf{A}, \mathbf{a}$	matrices (write by hand $\underset{\sim}{A}, \underset{\sim}{a}$)
$\mathbf{A} \sim \mathbf{B}$	row equivalence of matrices
${}^t\mathbf{A}, {}^tT$	transpose of a matrix $\mathbf{A}$, of a linear transformation T
$\lvert\alpha\rvert$	absolute value
(u, v)	inner product
$\lVert u \rVert$	length of a vector u
$D(u_1, \ldots, u_n)$, $D(\mathbf{A})$, $\det \mathbf{A}$, $D(T)$	determinant of a set of vectors, of a matrix, of a linear transformation
$\operatorname{Tr}(\mathbf{A})$, $\operatorname{Tr}(T)$	trace of a matrix $\mathbf{A}$, or of a linear transformation T.
$n(T)$, $T(V)$	null space, range of a linear transformation $T: V \to W$.
$F[x]$	polynomials with coefficients in a field F
$\bar{z}$	conjugate of a complex number z
$V_1 \oplus V_2$	direct sum of vector spaces
$f(T)$	polynomial in a linear transformation T
$S^\perp$	set of vectors orthogonal to the vectors in S
e^A	exponential of a matrix $\mathbf{A}$
V^*	dual space of V
$V \times W$	cartesian product of V and W
$V \otimes W$, $T \otimes U$, $\mathbf{A} \dot{\times} \mathbf{B}$	tensor product of vector spaces, linear transformations, matrices.
$\bigwedge^k V$, $v \wedge w$	wedge product of vector spaces, vectors.
T'	adjoint of a linear transformation

Index

N

O

P

Q

R

S

T

U

V

W